Arms Industry Limited

Stockholm International Peace Research Institute

SIPRI is an independent international institute for research into problems of peace and conflict, especially those of arms control and disarmament. It was established in 1966 to commemorate Sweden's 150 years of unbroken peace.

The Institute is financed mainly by the Swedish Parliament. The staff, the Governing Board and the Scientific Council are international.

The Governing Board and the Scientific Council are not responsible for the views expressed in the publications of the Institute.

Stockholm International Peace Research Institute
Pipers väg 28 S-170 73 Solna Sweden
Cable SIPRI
Telephone 46 8/655 97 00 *Telefax* 46 8/655 97 33

Arms Industry Limited

Edited by
Herbert Wulf

sipri

OXFORD UNIVERSITY PRESS
1993

Oxford University Press, Walton Street, Oxford OX2 6DP
Oxford New York Toronto
Delhi Bombay Calcutta Madras Karachi
Kuala Lumpur Singapore Hong Kong Tokyo
Nairobi Dar es Salaam Cape Town
Melbourne Auckland Madrid
and associated companies in
Berlin Ibadan

Oxford is a trade mark of Oxford University Press

Published in the United States
by Oxford University Press Inc., New York

British Library Cataloguing in Publication Data
Data available

Library of Congress Cataloging in Publication Data
Data available

ISBN 0–19–829164–7

Typeset and originated by Stockholm International Peace Research Institute
Printed in Great Britain
on acid-free paper by
Biddles Ltd, Guildford and King's Lynn

Contents

Part IV. Europe

Part V. Other countries

Annexes

Preface

The changes that have taken place the world over, in particular the end of the cold war and the collapse of the bipolar system, have not left arms industries unaffected. Diminishing arms production and restructuring the arms industry, however, have hardly brought about the expected 'peace dividend'. In this book the reader will find at least a partial answer to the intriguing question: Why has this happened?

This compendium of case studies is intended to give an overview of the present state of the world arms industry. It is an empirical account of the size of the industry and particularly its present problems, as signalled in the title of the book.

The origins of this publication go back to many discussions in the SIPRI research project on the arms trade in 1989, when a new research project on arms production was initiated. Momentous political changes have occurred since then in the international system that have affected arms production world-wide. Governments are in the process of restructuring national armed forces and revising military equipment procurement plans. The changed political climate in Europe—with the end of the cold war, the Treaty on Conventional Armed Forces in Europe, financial constraints in many countries, reduced possibilities for arms exports—has contributed to fundamentally changed conditions for arms production. Thus, this book is a timely account of the arms industry at a time when it is being transformed in many parts of the world.

The idea for the book was twofold. First, SIPRI has traditionally been reporting in great detail about the size and flow of the arms trade, arms procurement and production, and military expenditures. The focus of these topics was mainly government activities in buying or selling, procuring or controlling arms. In this book the activities of another important actor are described and analysed: the industry that produces the weapons. Second, as early as in 1989, when the first concept of such a publication was discussed, it was clear that the arms industry was experiencing difficulties. Although at that time the cold war was not yet over, the end of the boom period of the arms industry, nurtured for many decades by the cold war, was already visible.

The interest of SIPRI in this issue stems from the fact that all too often economic pressures have contributed to arms races or prevented a dynamic disarmament process. The question raised in this book is whether the arms industry is dependent on arms production and arms exports or if there are alternatives. Arms production capacities have begun to be cut, but the gradual reductions in military manpower, military expenditures and the rate of arms production are probably only the beginning of a long-term development. Employment in arms production will decrease substantially.

This will create pressures to continue to produce weapons without proper military rationales and to 'liberalize' arms export controls at a time when the control of arms transfers are being seriously discussed for the first time among the five major arms exporters and in the United Nations. How the dilemma of a strong political motivation for reduced arms production and economic pressures to continue the production of weapons on a high level is going to be solved is still an open question.

This book could not have come into existence without the commitment, persistence and organizational talents of Herbert Wulf, who in 1989–92 led the SIPRI research project on arms trade and arms production.

This book, with case studies of the important centres of arms production in the world, is a companion volume to several previous SIPRI studies: Michael Brzoska and Thomas Ohlson (eds), *Arms Production in the Third World* (1986); Ian Anthony, Agnès Courades Allebeck and Herbert Wulf, *West European Arms Production: Structural Changes in the New Political Environment*, a SIPRI Research Report (1990); and Michael Brzoska and Peter Lock (eds), *Restructuring of Arms Production in Western Europe* (1991).

Adam Daniel Rotfeld
Director of SIPRI
November 1992

Acknowledgements

I would like to acknowledge the efforts of several people and institutions in the preparation of this book. Plainly, the book would not have been possible without the efforts of the contributing authors. However, the contributions of the members of the SIPRI research project on arms trade and arms production—Ian Anthony, Agnès Courades Allebeck and Elizabeth Sköns—went far beyond the preparation of the chapters which bear their names. They discussed and shaped the concept of the book and, more important, spent a fair amount of their time in building up the SIPRI arms production data base, which is the empirical backbone of this book.

Other researchers and research assistants at SIPRI who are not contributing authors of this book have equally contributed to the SIPRI arms production data base, particularly Gerd Hagmeyer-Gaverus, Paolo Miggiano and Espen Gullikstad (who also contributed the bibliography to this book). Cynthia Loo, the secretary for this research group, typed some of the manuscripts and assisted in keeping the archives up to date, which was an invaluable precondition for quick and easy reference.

Connie Wall shouldered a tremendous responsibility for this book. She edited the manuscripts and set the book in camera-ready copy but, more than that, she checked manuscripts for consistency and transformed them into a coherent and readable book.

When I joined the SIPRI arms trade team in 1989 it was not difficult to motivate the research group to go beyond the research focus on arms transfers and initiate a major data collection effort and a long-term research project on the arms industry. This book, like almost all the other products of the research group, is a true team effort.

Both Adam Daniel Rotfeld, the present Director of SIPRI, and Walther Stützle, the previous Director, have supported and encouraged this work.

Without generous financial assistance from the John D. and Catherine T. MacArthur foundation to operate the SIPRI arms production data bank, the book would not have been possible in its present form.

Herbert Wulf
November 1992

Abbreviations, acronyms and conventions

AAM	Air-to-air missile
ACM	Advanced cruise missile
ADATS	Air defense anti-tank system
AIP	Australian Industry Participation (programme)
ASM	Air-to-surface missile
AWACS	Airborne warning and control system
BA	Budget Authority
CBO	(US) Congressional Budget Office
CFE	Conventional Armed Forces in Europe (Treaty)
CFSP	Common Foreign and Security Policy
CIS	Commonwealth of Independent States
CMC	Central Military Commission
CMEA	Council for Mutual Economic Assistance
CNAD	(NATO) Conference of National Armaments Directors
COCOM	Co-ordinating Committee on Multilateral Export Controls
DARPA	Defense Advanced Projects Research Agency
DoD	(US) Department of Defense
EC	European Community
ECU	European Currency Unit
EFA	European Fighter Aircraft
EFTA	European Free Trade Association
EP	European Parliament
EPC	European Political Co-operation
EUCLID	European Co-operative Long-term Initiative for Defence
FDS	Fixed distribution system
FSU	Former Soviet Union
FY	Fiscal year
GAO	(US) General Accounting Office
GATT	General Agreement on Tariffs and Trade
GDP	Gross domestic product
GERD	Gross expenditure on R&D
GNP	Gross national product
ICBM	Intercontinental ballistic missile
IEPG	Independent European Programme Group
IRBM	Intermediate-range ballistic missile
KGB	*Komitet Gosudarstvennoy Bezopasnosti,* Committee for State Security
LDDI	Less-developed defence industries
LOSAT	Line-of-sight anti-tank system
MBT	Main battle tank

MRLS	Multiple rocket launcher system
MTCR	Missile Technology Control Regime
NASA	(US) National Aeronautical and Space Administration
NATO	North Atlantic Treaty Organization
O&M	Operations and maintenance
OECD	Organization for Economic Co-operation and Development
PLA	People's Liberation Army
PRC	People's Republic of China
R&D	Research and development
RDT&E	Research, development, testing and evaluation
SAM	Surface-to-air missile
SICBM	Small intercontinental ballistic missile
TOA	Total Obligational Authority
WEU	Western European Union
WTO	(former) Warsaw Treaty Organization

Conventions used in the tables

$	US $, unless otherwise indicated
A$	Australian $
b.	Billion (thousand million)
m.	Million
. .	Data not available or not applicable
–	Nil or a negligible figure
0	Less than 0.5

Part I
Introduction

1. Arms industry limited: the turning-point in the 1990s

Herbert Wulf

I. The changed environment

'So long as the communist threat was alive and well, we could expect eventually to be pulled out of the doldrums.'[1] This statement by William A. Anders, Chairman and Chief Executive Officer of General Dynamics, the second largest arms-producing company in the world, illustrates that arms producers have 'got the message': the vast arms industry in East and West, for decades nurtured by the cold war, is now faced with unprecedented contractions in arms procurement orders. Arms industries world-wide must in some way adapt to a new and decidedly different environment.

Arms-producing companies, especially those in the United States, have experienced previous declines in arms procurement: after World War II, after the Korean War and after the Viet Nam War. However, these declines were short-term and were reversed by the next cycle of massive arms buildups. 'Our industry', said William A. Anders, 'would be sick for a while, and then the continued threat of communism would allow it to regain its strength on the next cycle.'[2] Today, neither strong supporters of the arms industry and staunch cold war warriors nor disarmament proponents predict a repeat of the cyclical pattern of ever-increasing military expenditures. The industry has rapidly—within a five-year period—reached a situation in which fundamental changes are required. There is no alternative to 'down-sizing' production capacity: the arms industry has reached its limits under the current political and economic conditions.

A gradual decline in arms procurement is a global trend that began in 1988 and continued uninterrupted until 1992; it is projected to continue at the same rate, at least for the medium term. Reduced arms production and large production over-capacities are a consequence of military budget cuts. The extent to which global military expenditures have declined, after the peak of nearly $1000 billion in 1987, is illustrated in figure 1.1. In both the industrialized and the less-developed countries, military budgets are being cut by 3–4 per cent per year.

[1] Anders, W., 'Rationalising America's defense industry', Keynote Address, *Defense Week*, 12th Annual Conference, 30 Oct. 1991, p. 4.

[2] See note 1.

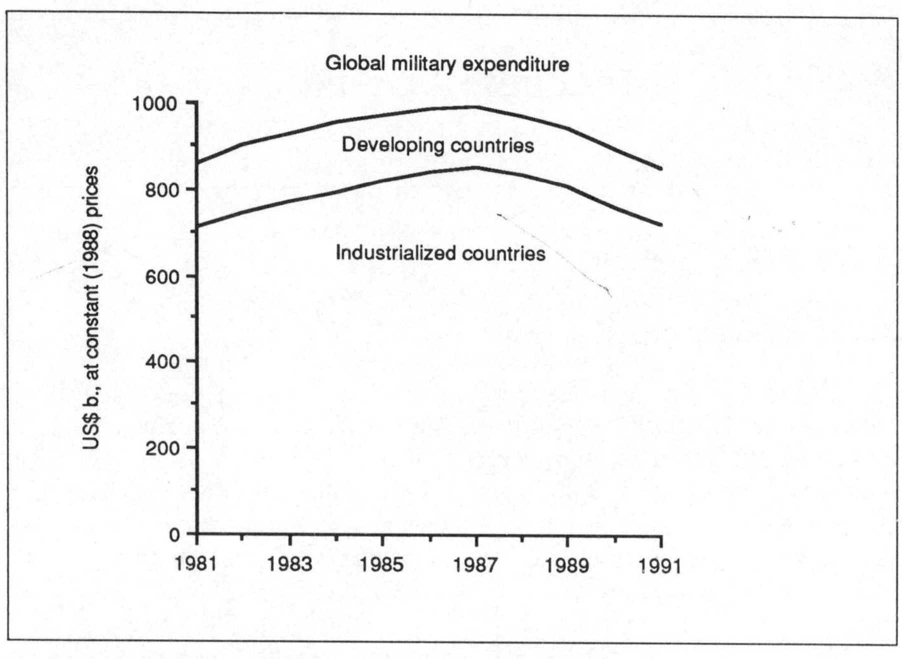

Figure 1.1. Development of military expenditures, 1981–91

Sources: Author's estimates, based on SIPRI military expenditure data; *SIPRI Yearbooks 1991* and *1992: World Armaments and Disarmament* (Oxford University Press: Oxford, 1991, 1992).

Several factors have contributed to the squeeze the arms industry is experiencing and to the change in the long established trend that lasted for over four decades.

1. *The changed political climate in Europe, culminating in the dismantlement of the Warsaw Treaty Organization (WTO) and the collapse of the Soviet Union.* As a result of political reforms in the Soviet Union since the mid-1980s and the disappearance of the former superpower in 1991, European and US threat perceptions have changed drastically. Compared to these fundamental changes, however—unprecedented since the end of colonialism—disarmament and reductions in armed forces have, so far, been modest. There is every indication that gradual reductions in military manpower, spending and production rates are the beginning of a long-term trend.

2. *Financial constraints*. In this more benign security environment, military budgets have fallen in real terms (after adjustment for inflation) since other economic priorities have competed more successfully for allocations. Huge arms buildups such as were common in the two antagonistic blocs are no longer credible. Planned weapon programmes have been stopped, deferred or reduced in numbers. This process has for several domestic reasons been opposed in many countries. Traditional attitudes to disarmament were in many cases reversed after the end of the cold war: special interests often override pro-

disarmament positions, such as when economic and social interests are threatened by the cancellation of a particular weapon system that affects employment and income in a region, city or community.

On the basis of the past record of arms procurement growth rates, arms-producing companies have been over-enthusiastic in their capacity planning. As of 1987 the reality has changed. The actual development of procurement budgets and five-year procurement projections has varied greatly. Each projection has had to be corrected downwards the following year. Growth rates were last projected in 1987 for the period 1988–92; they were never realized, as the actual trend in military expenditure shows.

The arms industry was stimulated in many countries by these optimistic budget projections and by procurement plans. The experience during the last few years of the cold war had led companies into initiating plans to expand existing production facilities. This ill-fated policy has exacerbated the present strains on industry. The arms industry in both East and West was unprepared for reversed growth rates. In a belated effort to repair economic weakness, the arms industry is trying to reduce over-capacities: the process of down-sizing production capacities has begun, but there is no question that the arms industry has to rationalize existing capacities further.

3. *Reduced possibilities for arms exports*. Expanding arms exports is not a viable alternative for the arms industry. Changing international arms market conditions are making it more difficult to export the same volume of weapons. The value of exports of major conventional weapon systems[3] declined in 1991 for the fourth consecutive year, to $22 billion, compared to the peak of nearly $46 billion in 1987. Scarce hard currency reserves in many countries, increased expansion of indigenous arms production facilities, and restraint in the supply of weapons to several previously important recipients involved in armed conflicts (such as Iraq, Angola and Ethiopia, in rank order of weapon deliveries in 1987–91) have contributed to the downward trend in arms exports (see figure 1.2).[4]

Furthermore, as a consequence of arms control agreements and disarmament, particularly in Europe, some governments are trying to 'unload' surplus arms, thus saturating the arms market for arms-producing companies which often directly compete with these weapons for their sales. Only in exceptional cases can arms exports ease the economic situation of the military industry. In the

[3] SIPRI arms trade data cover five categories of major conventional weapons or systems: aircraft, armour and artillery, guidance and radar systems, missiles and warships. See *SIPRI Yearbook 1992: World Armaments and Disarmament* (Oxford University Press: Oxford, 1992), appendix 8D, pp. 353–54, for further methodological information.

[4] Figure 1.2 shows only the export of major conventional weapon systems as defined by SIPRI. The decline in these exports is not paralleled by a similar trend in arms production and dual-use technology. On the contrary, as major recipient countries increasingly insist on indigenous arms production, the export of production technology and the supply of components and sub-systems seem to be increasing, while exports of finished weapons are decreasing. Furthermore, greater emphasis is placed on upgrading and modernizing existing weapon platforms. These trends have been described in the arms trade chapters of the *SIPRI Yearbooks 1991: World Armaments and Disarmament* (Oxford University Press: Oxford, 1992) and *1992* (note 3), chapters 7 and 8, respectively.

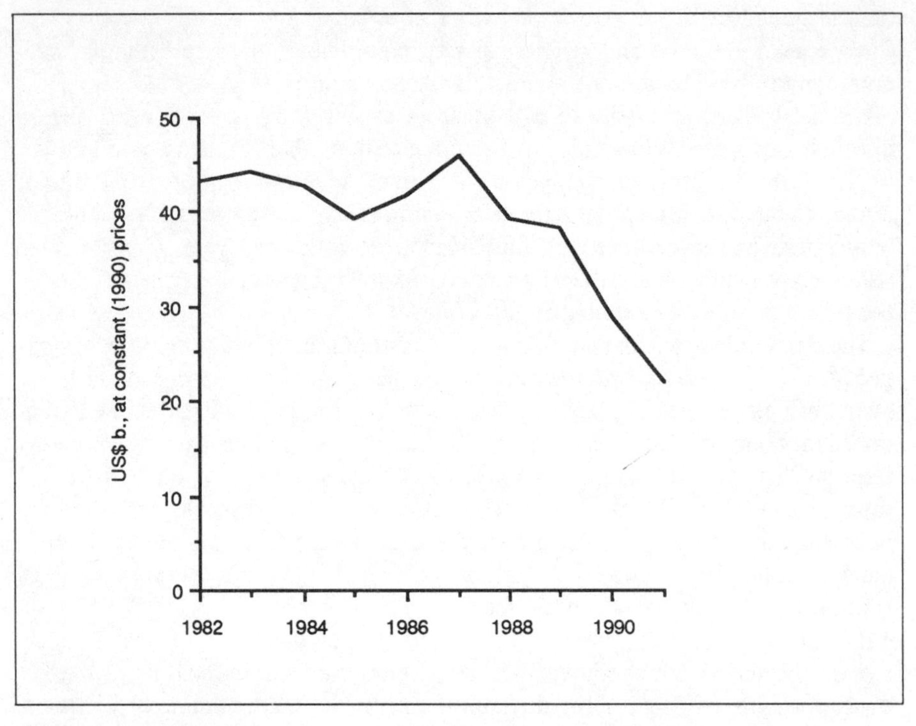

Figure 1.2. Global exports of major conventional weapons, 1982–91
Source: SIPRI data base.

past, arms producers often compensated for reduced domestic demand by aggressively marketing abroad and by lobbying at home to increase domestic demand when arms exports declined. Although governments still promote exports, not least to improve the balance of trade, this is no longer possible on a global scale. Today the arms industry is faced with reductions in both the domestic and the international demand for arms.

In summary, political and economic factors have mutually contributed to a drastically changed political and economic environment, with reduced business perspectives for the arms industry. These limits are felt in several areas—shrinking arms sales, loss of jobs and reduced profit opportunities—which are discussed below.

II. Diverse regional developments

The effects on arms production and arms-producing companies have varied greatly. While severe cuts in arms procurement and production were made in some countries, a gradual process has just begun in others. However, there are also developments in the opposite direction. Governments in some countries have made long-term commitments and have decided to retain or even expand their arms industrial bases at a time when over-capacities are an economic and a political burden in global terms.[5] Compared to other regions of the world, the most far-reaching reductions have been experienced in the two largest centres of arms production: in the former Soviet Union—with the bulk of the arms industry located in Russia—and in the United States.

Of all the countries in the world, the process in the successor republics of the *Soviet Union* is probably the most fundamental. The changes there have not only entailed quantitative reductions, as in many other countries, but have also taken place while the entire economy is in turmoil and during the difficult period in which the Union was dissolved into independent republics.[6] The severe economic problems throughout the old system led to the initiation of economic reforms, and the process of reduced military spending was begun even before the dissolution of the USSR: the beginning of this process can be traced to the address of President Mikhail Gorbachev at the United Nations General Assembly on 7 December 1988, in which he announced the conversion to civilian production of the Soviet arms industry.[7] Conversion of the arms industry was caught in the economic chaos of changing from a planned to a market economy. The political collapse of the former system and the ensuing economic and political turmoil prevented successful conversion of the arms industry.

With the independence of the former Soviet republics and the initiation of economic reforms, procurement expenditure was slashed even more severely. According to Russian sources, arms production is being cut back on a huge scale—by 50–60 per cent, and for particular weapon categories by even 90 per cent. It has been stated that in 1991, in the former Soviet Union, tank production was reduced by 60 per cent, the production of strategic missiles by one-third and military aircraft by 30 per cent.[8] The Government of the Russian Federation has placed an order for only 20 tanks in 1992 for its armed forces, while in 1988, according to the First Deputy Minister of Defence, Andrei Kokoshin, more than 3200 tanks were produced, of which 2800 were for the Soviet armed forces and 400 were for export.[9] Exports of major conventional weapon systems have reverted to a fraction of the deliveries in 1987.[10] This

[5] The goal of this book is to present a comprehensive overview of the world arms industry. All the countries with an appreciable arms industry are covered except for North Korea, Iran and Iraq (because of a lack of information it was impossible adequately to cover these countries; see chapter 17 in this volume) and except for Canada (it was beyond the control of the editor of this volume that the commissioned chapter on Canada could not be delivered before the book went to print). A brief summary of arms production in the former Yugoslavia is presented in annexe A in this volume.

[6] The consequences for the arms industry in the different republics is described in Cooper, J., 'The Soviet Union and the successor republics: defence industries coming to terms with disunion', chapter 5 in this volume.

[7] For the analysis of the conversion process see Izyumov, A., 'The Soviet Union: arms control and conversion—plan and reality', chapter 6 in this volume.

[8] Foreign Broadcast Information Service, *Daily Report–Soviet Union (FBIS-Sov)*, FBIS-SOV-92-141, 23 July 1992, p. 26.

[9] First Deputy Defense Minister Andrey Kokoshin, according to *Krasnaya Zvezda*, quoted in FBIS-SOV-92-142, 23 July 1992, p. 2; and in FBIS-SOV-92-145, 28 July 1992, p. 36.

[10] For details, see Wulf, H., 'The Soviet Union and the successor republics: arms exports and the struggle with the heritage of the military–industrial complex', chapter 7 in this volume.

reduction—although creating difficulties for arms factories—is a necessary condition for the economic changes required in the Commonwealth of Independent States (CIS), since arms production of the past is considered to be one of the major causes of the current economic crisis. Every tank that is not produced might create difficulties for the arms-producing factories, but it is on the other hand a gain for the economy.

With limited possibilities for non-military production, many arms-producing factories in the former Soviet Union are in great difficulties. This is the price for past omissions. Many jobs are endangered, but large-scale layoffs of millions of employees have so far been avoided, as can be seen in table 1.1 below. This is due to a number of social and economic compensatory measures, including production for stocks, unpaid leave, part-time work, and so on.

In the *United States,* the arms industry has been faced with cuts in procurement and research and development (R&D) expenditures since 1986. Procurement and R&D expenditure combined, which had been declining moderately (in real terms), began to take a sharper downward turn in 1990. From a peak of over $168 billion in 1985, procurement and R&D spending was reduced to less than $100 billion in 1992.[11] US exports of major conventional weapons have not declined much during the past five years, and the United States is likely to dominate the world arms market. The economic importance to US companies of international sales, while limited, is likely to grow since domestic procurement is predicted to continue to be cut.[12] However, arms exports are not the remedy for a reduction in the large US arms industry. The US domestic demand was far more important than its exports—although in 1991 the United States was the largest exporter of arms.[13]

At the government policy level, there is an inherent conflict between plans to cut procurement and R&D expenditure and plans to maintain a viable and competitive defence industrial base.[14] Companies are facing a shrinking market at home and a rupture in the close link between weapon development and arms production. A smaller percentage of weapon systems than was historically the case will go from the development stage into production.

Companies are of course reacting to the new environment.[15] The fact that the arms sales of major arms-producing companies have not shrunk drastically is due to a number of factors. First, the full extent of cuts (or announced reductions) will be experienced in industry after a time lag, since projects often extend over a period of five years or more. Companies are working through

[11] The empirical evidence of the long-term trends in procurement and R&D expenditures is documented in Adams, G. and Kosiak, S., 'The United States: trends in defence procurement and research and development programmes', chapter 2 in this volume.

[12] Anthony, I., 'The United States: arms exports and implications for arms production', chapter 4 in this volume.

[13] *SIPRI Yearbook 1992* (note 3), chapter 9, pp. 275–79.

[14] Congress of the United States, Office of Technology Assessment, *Redesigning Defense. Planning the Transition to the Future U.S. Defense Industrial Base* (US Government Printing Office: Washington, DC, 1991).

[15] See Reppy, J., 'The United States: unmanaged change in the defence industry', chapter 3 in this volume.

their backlog of orders. Second, the larger contractors are teaming together to share the remaining contracts, often at the expense of the smaller arms producers.[16] Third, other contractors are buying up competitors, thus further reducing competition. Fourth, some arms producers have specialized in certain niches of the budget that were not affected by cuts.

The effects, however, are being felt in arms industry employment. From the peak of 3.36 million jobs in 1987 until 1992, over 600 000 jobs were lost, and it is expected that perhaps a total of 1.4 million jobs will be lost by 1995.[17]

Comparing arms industry developments in the former Soviet Union and the United States, generalizations are perhaps not valid. The major difference is probably that rapid restructuring (with plant closures, rearrangements among the top producers and massive layoffs) is taking place in the United States. The contractions in Russia were more drastic than in the United States but have not led to the same restructuring of industry—not the least because of a lack of alternatives. Thus, production facilities in Russia are basically kept intact—although not fully utilized—and the work-force is largely retained—although not fully occupied.

The experience in *China* has been different from that in the Soviet Union or in the West. The Chinese Government is sending mixed signals to the international community about its arms production.[18] Military spending has been substantially increased since 1988, after a period of reductions in the early 1980s and moderate growth in the mid-1980s. The government announced the success of the process of conversion from military to civil production during the 1980s. However, for China, conversion is not only intended to reduce the quantity of military production or—as is often proclaimed by the leadership—to contribute to world peace; it is also a policy to modernize the Chinese arms industrial base through imports of modern technology. Conversion in China means 'guns *and* butter'. The modernization drive during the 1980s was a result of the general weakness of industry. The arms industry—which could not be debated until the late 1970s—was reformed. Economic factors were given priority over political ideology.

Arms industry responses in *Western Europe* are occurring in a more benign economic environment than in Russia, the Central and East European states or even the United States. The West European arms industrial base is smaller than in the United States or Russia, in both absolute terms (size of arms production) and relative terms (procurement expenditure per capita, or the share of arms industry employment in the total industrial work-force). Economic readjustment problems are therefore less severe. However, one point should be stressed: despite a number of co-operative arrangements, arms production and arms industrial policy in Western Europe is still largely decided at the national

[16] The process of concentration is not restricted to the USA alone but is taking place in Western Europe as well; see Sköns, E., 'Western Europe: internationalization of the arms industry', chapter 9 in this volume.

[17] Note 15.

[18] Frankenstein, J., 'The People's Republic of China: arms production, industrial strategy and problems of history', chapter 14 in this volume.

level. None of the existing multinational institutions (NATO, the European Community or the Western European Union) has the sovereignty to enforce a unified arms procurement or industrial policy.[19]

Both the larger European arms producers (the United Kingdom, France and Germany, in rank order of size of industry) and the medium-size producers that are capable of developing and producing a number of modern major conventional weapon systems (Italy, Sweden, the Netherlands and Spain) are faced with an inherent policy dilemma. Procurement budgets and arms exports are too small to maintain the present arms industrial base.[20] The answer could be a co-operative and streamlined European industrial base with international companies.[21] This issue has been on the agenda for more than 30 years, and policy recommendations for a co-ordinated West European arms industry have frequently been made. However, European governments have not managed to 'down-size' their large over-capacities. These policy recommendations have never been put into practice on a large scale because of national idiosyncrasies and powerful industrial interests. With the increasing economic pressure on industry, severe budget constraints and the disappearance of the Soviet threat, the answer to retaining the European arms industrial base might be different during the remainder of the 1990s. If a more co-ordinated procurement policy and a co-operative industrial structure are implemented, the problem of over-capacities becomes even more pressing. The number of producers of tanks, fighter aircraft, ships and missiles is too large to be maintained with current military budgets. Difficult decisions will have to be made. The process of shrinkage has already begun—with a time lag in comparison to the United States; however, this is less the result of conscious government planning than a consequence of industrial responses to economic pressures.

The recent experience of the arms industry in *Poland* and *Czechoslovakia*[22] is closely connected to the dissolution of the WTO and similar to the fate of the industry in Russia—although on a totally different quantitative level. Both the Polish and the Czechoslovakian industry worked in a highly developed division of labour within the WTO. At the end of the 1980s their industries were faced simultaneously with a breakdown of this division of labour, a substantial reduction in domestic demand and a shrinking export market. The results are drastic cuts in employment, desperate attempts to convert facilities to non-military production and a return to previous levels of arms exports. This is an economically painful process that, within the next few years, will further reduce the arms industrial base of these two countries.

[19] Courades Allebeck, A., 'The European Community: from the EC to the European Union', chapter 10 in this volume, describes the potential role and the limitations of the EC.

[20] See Wulf, H., 'Western Europe: facing over-capacities', chapter 8 in this volume.

[21] See Sköns (note 16).

[22] See Perczynski, M. and Wieczorek, P., 'Poland: declining industry in a period of difficult economic transformation', chapter 11 in this volume; and Cechak, O., Selesovsky, J. and Stembera, M., 'Czechoslovakia: reductions in arms production in a time of economic and political transformation', chapter 12 in this volume. The experience in Yugoslavia was different, since the arms industry in that country was not linked to the WTO. Annexe A documents the structure of the industry and its locations in the former Yugoslavia and explains why the UN arms embargo could not possibly have immediate effects.

Although *Japan* is often cited as an example of a country with a restrained and small arms industry, limited by the Constitution and restrictive export regulations, its industry has become important in international comparison. Fostered by developing commercial technologies and generous government R&D funding, many of the large Japanese industrial conglomerates are engaged in arms production projects—partly in co-operation with US companies. Japan's domestic arms market has been steadily growing parallel to its GNP (gross national product) growth rate, and in contrast to most other countries military spending is still planned to grow—although at a reduced rate.[23]

By international comparison, the arms industry in *Australia* is very small. However, in contrast to most other countries, its arms industry has grown during the past few years as a consequence of the implementation of the government policy of 'defence self-reliance'. On the basis of advanced technologies obtained as part of licensing or co-production agreements, Australian companies are building a number of major weapon systems.[24] The success of the arms industry depends on Australia being able to fund and maintain a reasonably sophisticated arms industrial base to serve the 'self-reliance' policy, which depends in turn on arms exports. As a logical result of these policies, the industry is bound to undergo a process of down-sizing after completion of current ongoing projects, if export plans do not materialize.

The experiences in the over a dozen *'third tier' countries* that have an arms industrial base are rather mixed.[25] Five types of experience can be distinguished.[26]

1. Countries that have developed ambitious arms development and production facilities and have encountered serious difficulties both in keeping abreast of modern technology and fully utilizing existing capacities (Argentina, Brazil, and to some extent South Africa).
2. Countries that have developed substantial arms development and production facilities at a very high cost and are prepared for political and security reasons to continue to pay the price for these programmes (India, Israel and Pakistan).
3. Countries that have started late to develop arms industrial bases which are still in the process of expansion despite the negative experience of other countries (Egypt, South Korea, Taiwan and to a lesser degree Indonesia).
4. Countries that have specialized in certain sectors of arms production and try to co-operate internationally, with some commercial success (Singapore and to some extent also Israel). Other countries are also moving towards

[23] Ikegami-Anderson, M., 'Japan: a latent but large supplier of dual-use technology', chapter 15 in this volume.

[24] Cheseman, G., 'Australia: an emerging arms supplier?', chapter 16 in this volume.

[25] For a detailed analysis of Turkey's arms industry, see Gülük-Senesen, G., 'Turkey: the arms industry modernization programme', chapter 13 in this volume.

[26] Ian Anthony analyses the experience of these countries in 'The "third tier" countries: production of major weapons', chapter 17 in this volume.

privatization and commercialization of arms production, particularly South Africa, South Korea, Singapore and Taiwan.

5. Countries that are known to have or are claimed to have important arms-production capacities, but for which knowledge about their facilities is limited and information about it highly speculative (North Korea, Iran and Iraq).

III. Employment in the arms industry

An area in which the constraints on the arms industry are already felt is employment. Two approaches are chosen here to give an indication of the order of magnitude of employment effects today and possibly in the future: global statistics on a country-by-country basis; and a comparison of employment in the largest arms-producing companies for different time periods.[27]

Global employment effects

Information on the military sector has greatly improved during the past few years, but this is not true for industry employment data. Important caveats remain. In addition to a general reluctance in many countries to publish information on the military sector, a further problem with employment data is of a methodological nature. Since the arms industry is not a clearly defined industrial branch and since there is no clear-cut borderline between military and civil industry, comparative global data are not available. Furthermore, data collection in the various countries reported below is based on different methodologies and often on rough estimates. To evaluate and compare the number of jobs created by military expenditure in the economy requires estimation of both the direct and indirect effects. Sometimes it is not even known whether the data published in open sources refer to only direct or to both direct and indirect employment effects.

The total number of arms industry jobs is not the only statistic of interest. As was pointed out in a comparative study: 'Employment consequences need not simply mean the number of jobs created; the composition of the change in employment is also of importance. The sex mix of the jobs, the number of full–time and part–time, the number of skilled and unskilled, the geographical distribution, even the duration, will all have differing economic implications.'[28] The fact remains, however, that this important information is available for only a limited number of countries.

[27] The most systematic overview of employment effects in the arms industry is documented in Pauker, L. and Richard, P. (eds), *Defence Expenditure, Industrial Conversion and Local Employment* (International Labour Office: Geneva, 1991). See also Renner, M., *Economic Adjustments after the Cold War: Strategies for Conversion* (United Nations Institute for Disarmament Research [UNIDIR]: Dartmouth Publishing, Bookfield, Vermont, 1992).

[28] On the different methodologies, see Dunne, J. P., 'Quantifying the relation of defence expenditure to employment', in Pauker and Richard (note 27).

Given these limitations, the statistics can be no more than an estimate of the order of magnitude of employment in the arms industry at different times. From the data presented in table 1.1, four conclusions can be drawn:

1. Arms industry employment world-wide fluctuated during the 1980s at around 15 million employees.
2. The number of arms industry employees is lower than the number of soldiers in the armed forces and para-military forces, which were estimated to be approximately 32 million in 1990, not taking into account reserves.[29]
3. The peak in arms industry employment was reached in the mid-1980s, at around 16 million. The figure dropped below 15 million at the beginning of the 1990s.
4. In terms of employment figures, the largest arms production centres are in the successor republics of the former Soviet Union (primarily Russia, Ukraine and Belarus, in rank order of size), China, the United States and the European Community countries (with a high concentration in the UK, France and Germany, in rank order of size of employment). These four centres, or eight countries, account for nearly 90 per cent of all employment in the industrial manufacture of arms. The rest of the countries of the world (around 40 additional countries that have arms-producing facilities) account for the remaining 10–11 per cent (around 1.5 million jobs) of total employment in the arms industry.

Employment in the largest arms-producing companies

A second indicator for assessing recent effects on employment in the arms industry is a comparison of employment figures in the largest arms-producing companies over time. The impact of the procurement of fewer weapons and of falling arms exports is now clearly visible in arms industry employment. The list of companies that have announced layoffs in 1991 for the next few years is long. SIPRI has reported employment cuts in its *Yearbooks*.[30] How has employment in the largest arms-producing companies developed since 1988?

Most companies are not exclusively engaged in arms production; changes in employment might thus not be the result of changes in arms sales. Many large automobile manufacturers, for example, are also large arms producers. These companies' employment cuts, however, are usually only to a very limited extent due to reduced arms sales. To arrive at a conclusion on employment effects in arms business, only those companies are chosen here that depended heavily on arms production. Among the 100 largest arms-producing companies in 1988 and their 25 largest subsidiaries, there were 43 companies or company

[29] US Arms Control and Disarmament Agency, *World Military Expenditures and Arms Trade 1990* (US Government Printing Office: Washington, DC, 1991); see also International Institute for Strategic Studies (IISS), *Military Balance 1991–92* (Brassey's: London, 1991); SIPRI arms production archive.

[30] For the most recent developments, see Miggiano, P., Sköns, E. and Wulf, H., 'Arms production', *SIPRI Yearbook 1992* (note 3), pp. 365–69.

Table 1.1. Global arms industry employment, early 1980s to 1992[a]

Countries are listed in order of their employment figures for 1990–92, column 4.

Country	Early 1980s	Mid-1980s	1990–92
Former Soviet Union	*5 800 000*	*6 000 000*	*5 900 000*
Russia	..	..	4 500 000
Ukraine	..	..	800 000
Belarus	..	..	150 000
Kazakhstan	..	..	75 000
Baltic Republics	..	..	100 000
All other former Soviet republics	..	..	275 000
China	4 000 000	..	3 000 000–5 000 000
USA	2 085 000	3 100 000	2 750 000
UK	560 000	470 000	(400 000)
France	340 000	290 000	255 000
Germany	*268 000*	*347 000*	*241 000*
Former West	268 000	307 000	200 000
Former East	..	40 000	41 000
India	235 000	240 000	250 000
Poland	..	260 000	180 000
Spain	(40 000)	66 000	100 000
Romania	..	..	90 000
Italy	78 000	86 000	80 000
South Africa	100 000	100 000	(80 000)
Czechoslovakia	..	145 000	75 000
Egypt	75 000	100 000	75 000
Israel	90 000	90 000	60 000
North Korea	55 000	..	(55 000)
Canada	46 000	(50 000)	50 000
Pakistan	40 000	50 000	(50 000)
Yugoslavia	..	..	(50 000)
Japan	33 000	39 000	45 000
Iran	..	..	45 000
Brazil	75 000	75 000	(40 000)
Taiwan	50 000	..	(40 000)
South Korea	30 000	..	(40 000)
Sweden	29 000	35 000	30 000
Hungary	..	..	30 000
Belgium	20 000–30 000	35 000	25 000
Switzerland	..	30 000	25 000
Turkey	..	..	(25 000)
Indonesia	26 000	..	(25 000)
Netherlands	13 000	18 000	(20 000)
Singapore	11 000	..	20 000
Argentina	60 000	(60 000)	20 000
Greece	..	15 000	14 000
Australia	..	..	12 000
Norway	..	15 000	10 000
Portugal	..	..	10 000
Finland	..	..	10 000

Country	Early 1980s	Mid-1980s	1990–92
Chile	3 000	. .	(10 000)
Denmark	. .	. .	7 000
Peru	5 000	. .	(5 000)
Philippines	5 000	. .	(5 000)
Thailand	5 000	. .	(5 000)
Malaysia	3 000	. .	(5 000)
New Zealand	. .	. .	500
Iraq	. .	. .	. .
Syria	. .	. .	. .
Saudi Arabia	. .	. .	. .

[a] More detailed information on the development of employment in a number of the countries mentioned in table 1.1 can be found in the various country studies in this book. Owing to uncertainty of data, many figures are rough estimates only. Figures in parentheses are the author's estimates.

Sources: Data are based on four sources: (*a*) the country studies published in this volume; (*b*) previous SIPRI studies, particularly Brzoska, M. and Ohlson, T. (eds), SIPRI, *Arms Production in the Third World* (Taylor & Francis: London, 1986); Anthony, I., Courades Allebeck, A. and Wulf, H., *Western European Arms Production*, A SIPRI Research Report (SIPRI: Stockholm, 1990); and Brzoska, M. and Lock, P. (eds), SIPRI, *Restructuring of Arms Production in Western Europe* (Oxford University Press: Oxford, 1992); (*c*) the SIPRI company data base; and (*d*) the SIPRI arms production archive (in which employment figures have been collected from open sources). In addition to these sources, in which actual figures, often contradictory, are given, estimates have been made on general descriptions of the size of the arms industry in a given country. On the different methodologies, see Dunne, J. P., 'Quantifying the relation of defence expenditure to employment', in eds L. Pauker and P. Richard, *Defence Expenditure, Industrial Conversion and Local Employment* (International Labour Office: Geneva, 1991).

subsidiaries whose share of arms sales was more than half of their total sales. Information on employment changes up to 1990 or 1991 is available for 35 of these companies.[31] The employment changes for 1988–91 are given in table 1.2.[32]

Of the 35 companies, 28 companies (80 per cent) reported redundancies of up to one-quarter of their 1988 work-force. This list of 35 companies contains 11 US arms manufacturers which also recorded the largest number of job losses. This trend, clearly illustrated in table 1.2, is the result of the fact that US companies were more severely affected by budget cuts and because it is much more

[31] Five companies or subsidiaries no longer exist as independent units; they have been bought or restructured, so comparisons are not possible. For 3 of the 43 companies, no recent employment figures are available.

[32] This analysis is based on the SIPRI list of the 100 largest arms-producing companies. Data have been collected systematically on over 400 arms-producing companies in OECD and developing countries since 1988. For the most recent list, see *SIPRI Yearbook 1992* (note 3), chapter 9, pp. 391–97.

Table 1.2. Employment losses and gains in the largest arms-producing companies, 1988–91

Figures in column 3 are in US$ m.; in column 4 in percentages, and in the last column in percentages.

Company (parent company)	Country	Arms sales 1988	Share of arms sales in total sales, 1988	Total employ-ment, 1988	% employ-ment change 1988–91
Loss in employment					
Grumman	USA	3 000	*82*	32 000	*26*
General Dynamics	USA	8 000	*84*	102 800	*22*
Armscor	S. Africa	1 870	*90*	26 000	*20*[a]
Lockheed	USA	8 400	*79*	86 800	*17*
Dassault Electronique	France	510	*75*	4 100	*17*
VSEL	UK	830	*100*	15 520	*16*
HSA	Netherlands	410	*90*	5 300	*15*[a]
Northrop	USA	5 200	*90*	42 000	*14*
EN Bazan (INI)	Spain	380	*82*	10 908	*12*[a]
Hindustan Aeronautics	India	480	*97*	48 833	*12*[a]
McDonnell Douglas	USA	8 680	*60*	121 000	*10*
Martin Marietta	USA	4 300	*75*	67 500	*10*
Loral	USA	1 050	*88*	14 000	*9*
Thiokol	USA	580	*54*	12 600	*9*
CASA (INI)	Spain	500	*72*	10 372	*8*[a]
Hughes Electronics	USA	6 860	*61*	100 000	*7*
British Aerospace	UK	5 470	*54*	131 300	*6*
Raytheon	USA	5 500	*67*	75 000	*5*
Litton Industries	USA	2 920	*60*	55 000	*5*
Eidgenössische Rüstungsbetriebe	Switzerland	550	*92*	4.900	*5*[a]
Oto Melara	Italy	530	*98*	2 329	*4*[a]
Dornier	FRG	570	*52*	9 800	*3*
FFV	Sweden	500	*72*	10 037	*3*[a]
FIAT Aviazione	Italy	660	*82*	4 749	*2*[a]
Krauss-Maffei	FRG	380	*53*	5 100	*2*
MTU (DASA)	FRG	970	*52*	17 200	*1*
Israel Military Industries	Israel	470	*98*	12 150	*1*[a]
Westland Group	UK	450	*71*	9 163	*1*
Gain in employment					
Hunting	UK	440	*62*	6 834	*1*
GIAT Industries	France	1 150	*100*	14 740	*2*[a]
Israel Aircraft Industries	Israel	800	*75*	16 500	*4*
Dassault	France	2 080	*70*	13 818	*5*
Saab Aircraft	Sweden	380	*57*	6 490	*6*
DCN	France	3 030	*100*	28 000	*7*
Thomson-CSF	France	4 320	*77*	41 400	*11*

[a] Figures are for the change up to 1990.

Source: SIPRI data base.

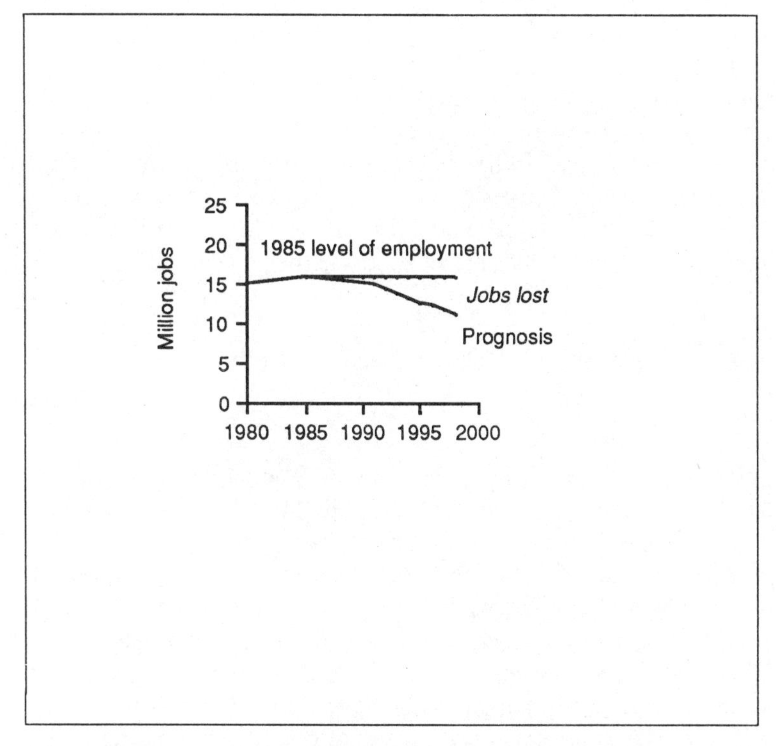

Figure 1.3. Prognosis for global employment in the arms industry, 1980–98

Source: Author's projection on the basis of points *a*, *b* and *c* on page 18.

common immediately to apply a 'hire and fire' policy in the United States. However, companies from other countries are among the list of those that reported job losses.

Only seven companies increased their work-force during the period 1988–91. The employment increases were on average much smaller than the job losses. While only one French company laid off employees, four French companies are among the seven companies that increased employment. This trend is partly due to company restructuring but not due to hiring new people in arms production. Companies such as Thomson and GIAT have been purchasing smaller companies abroad in recent years, thus increasing their arms production base and employment. In addition, French procurement expenditure has began to stagnate, with a time-lag of two years behind other major producing countries. The French state-owned and -protected arms industry has only recently begun to react to shrinking sales.

Only a few companies from other countries are listed among those with a high dependency on arms sales. It is a typical feature of US and French arms-producing companies to rely heavily on arms production. In most other countries, arms producers are more diversified and therefore less dependent on arms manufacturing. Contractions in employment in arms production in these cases do not necessarily mean redundancies. It is easier for diversified companies with a limited interest in arms production to relocate workers in other, non-military, production divisions of the company.

Based on (*a*) the development in employment during the period from the mid-1980s to the beginning of the 1990s, (*b*) an extrapolation of the trend in procurement expenditure and arms transfers and (*c*) likely productivity increases in the arms industry, it is estimated that, of the current 15 million jobs in the arms industry, 3–4 million jobs might be lost between 1993 and 1998.

IV. Effects on companies

Profits

It seems that the massive production over-capacity built up on the basis of over-optimistic planning is bearing down on the financially troubled arms-producing companies. Table 1.3 lists 97 major arms-producing companies and subsidiaries that have reported profits in 1988 and 1991.[33] The results of the development of profits within this group of companies are mixed. The period of more or less permanently improved profits is over. Over half of the companies and subsidiaries (50 of 97) could not improve their profits in absolute terms, and many of them ran into serious difficulties. However, this was not in all cases due to reduced arms sales alone but also the result of financial difficulties in the non-military sectors of the companies. Since only a few companies report their profits in both business sectors (military and non-military) separately, these profit figures are not a clear indicator of the state of the arms business.

Table 1.3 clearly illustrates that within the group of the 10 companies that reduced profits most heavily or incurred losses, there is not a single company which is highly dependent on arms sales. Included in this group are automobile manufacturers, business conglomerates, and electronics and computer companies that not only lost in their arms business. Of the 97 companies, however, 47 companies improved their profits.

How did profitability (profit per dollar of sales) of the industry develop? In 1991 the profitability of the 10 largest US arms-producing companies was only about one-third of the 1988 profit ratio. Profit after tax per dollar of sales was 4.5 per cent in 1988 and 1.2 per cent in 1991. In comparison, the 10 largest non-US arms-producing companies (mainly from Western Europe) were able to keep their profitability position at 2.6 per cent during the same period.[34]

[33] Not all of the 100 largest arms-producing companies in the 'SIPRI 100' list continued to exist in the form they had in 1988. Only those companies and subsidiaries are included here that can be compared over this 3-year period.

[34] All calculations are based on the SIPRI arms production data base as recorded annually in the *SIPRI Yearbook*.

Table 1.3. Changes in profits of the largest arms-producing companies, 1988–91

Figures in column 3 are percentages; columns 4 and 5 are in US$ m. Companies are listed in order of profit changes, column 5.

Company (parent company)	Country	Share of arms sales dependence 1991	Profits 1991	Profit change 1988–91
General Motors	USA	*6*	– 4 453	– 9 309
IBM	USA	*2*	– 2 827	– 8 673
Ford Motor	USA	*1*	– 363	– 5 663
Unisys	USA	*20*	– 1 393	– 2 074
Westinghouse Electric	USA	*18*	– 1 086	– 1 909
United Technologies	USA	*19*	– 1 021	– 1 680
FIAT	Italy	*17*	897	– 1 595
Tenneco	USA	*16*	– 732	– 1 554
IRI	Italy	*36*	– 254	– 984
Nobel Industries	Sweden	*21*	– 835	– 934
INI	Spain	*9*	– 574	– 843
Allied Signal	USA	*19*	– 873	– 736
Texas Instruments	USA	*29*	– 249	– 615
British Aerospace	UK	*40*	– 269	– 546
General Dynamics	USA	*80*	505	– 452
Teledyne	USA	*16*	– 25	– 417
TRW	USA	*35*	35	– 401
Lockheed	USA	*75*	308	– 316
Hughes Electronics (General Motors)	USA	*57*	559	– 243
Rockwell International	USA	*33*	601	– 211
Lucas Industries	UK	*15*	92	– 209
Ferranti–International Signal[a]	UK	*54*	– 175	– 176
Dowty Group[a]	UK	*33*	1	– 140
NEC	Japan	*1*	385	– 118
Oerlikon-Bührle	Switzerland	*44*	– 130	– 106
Olin	USA	*13*	– 13	– 105
Litton Industries	USA	*58*	64	– 103
Saab-Scania	Sweden	*11*	231	– 91
Gencorp[a]	USA	*45*	63	– 85
Harris	USA	*25*	20	– 81
Toshiba	Japan	*1*	855	– 77
Sequa	USA	*32*	– 6	– 75
Thorn EMI[a]	UK	*7*	462	– 67
Bofors (Nobel Industries)[a]	Sweden	*65*	– 14	– 58
VSEL Consortium	UK	*100*	85	– 57
Agusta (EFIM)[a]	Italy	*60*	– 33	– 56
Hawker Siddeley[a]	UK	*12*	145	– 53
Smiths Industries[a]	UK	*41*	126	– 50
Dassault Aviation	France	*74*	18	– 48

Table 1.3 *contd*

Company (parent company)	Country	Share of arms sales dependence 1991	Profits 1991	Profit change 1988–91
Martin Marietta	USA	*75*	313	– 46
Rheinmetall	FRG	*41*	20	– 27
Racal Electronics	UK	*13*	224	– 27
Dornier (DASA)	FRG	*24*	– 2	– 26
Hercules	USA	*20*	95	– 25
MBB (DASA)[a]	FRG	*50*	37	– 20
Matra Groupe	France	*26*	45	– 12
Oto Melara (EFIM)[a]	Italy	*100*	4	– 10
MTU (DASA)	FRG	*55*	14	– 4
CEA Industrie	France	*25*	233	– 2
ITT	USA	*6*	817	0
Hindustan Aeronautics[a]	India	*90*	90	2
SAGEM (SAGEM Groupe)	France	*43*	30	5
CASA (INI)[a]	Spain	*80*	– 45	7
Motorola	USA	*5*	454	9
Krauss-Maffei (Mannesmann)	FRG	*47*	11	10
E-Systems[a]	USA	*75*	86	11
Grumman	USA	*72*	99	12
Computer Sciences[a]	USA	*32*	65	13
Westland Group	UK	*64*	34	13
Kawasaki Heavy Industries	Japan	*14*	108	17
Thiokol	USA	*55*	53	20
EFIM[a]	Italy	*79*	0	20
EN Bazan (INI)[a]	Spain	*87*	29	20
Mannesmann	FRG	*3*	158	23
FMC	USA	*28*	164	35
Hunting[a]	UK	*31*	69	36
Ishikawajima-Harima	Japan	*14*	43	39
Bombardier	Canada	*1*	94	40
Israel Aircraft Industries	Israel	*67*	32	43
SAGEM Groupe	France	*28*	75	44
Aérospatiale	France	*40*	38	49
Newport News (Tenneco)	USA	*100*	225	50
SNECMA (SNECMA Groupe)	France	*33*	14	57
Textron	USA	*25*	300	66
McDonnell Douglas	USA	*55*	423	73
FIAT Aviazione (FIAT)[a]	Italy	*55*	21	74
Thomson-CSF (Thomson S.A.)	France	*77*	416	79
Northrop	USA	*86*	201	97
Thyssen	FRG	*3*	313	101
Emerson Electric	USA	*8*	632	103

Company (parent company)	Country	Share of arms sales dependence 1991	Profits 1991	Profit change 1988–91
Bremer Vulkan	FRG	*39*	45	108
Raytheon	USA	*57*	592	112
GEC	UK	*25*	918	115
Mitsubishi Electric	Japan	*3*	565	150
Morrison Knudsen	USA	*16*	35	163
Rolls-Royce	UK	*27*	90	168
Sundstrand	USA	*22*	109	186
Mitsubishi Heavy Industries	Japan	*17*	684	200
Daimler Benz	FRG	*7*	1 170	201
Thomson S.A.	France	*38*	479	278
Siemens	FRG	2	1 080	289
GTE[a]	USA	7	1 541	316
General Electric	USA	*10*	2 636	750
Honeywell	USA	*6*	331	760
Boeing	USA	*18*	1 567	953
AT&T[a]	USA	2	2 700	982
LTV	USA	*30*	74	3 228

[a] Figures in column 3 for this company are for 1990 and those in column 4 are for 1988–90.
Source: SIPRI data base.

Company managers know that growth is no longer inevitable. They are choosing different strategies to improve their profit situation. Some companies engage in what is euphemistically called 'divestment': they sell part of their assets to improve liquidity—to survive and to improve profits for the benefit of their stockholders. Part of this strategy is meant to reduce dependence on arms sales and strengthen the non-military part of company activities by selling arms-production facilities and attempting to diversify the range of products.[35]

In contrast, the opposite strategy has been chosen by companies that have decided to increase the performance of their arms market operations. Some firms with a mix of military and civil products have even sold their civil production segments in order to raise capital in times of financial difficulty. The most prominent example is General Dynamics, the second largest arms-producing company. Others—for example, Finmeccanica, a subsidiary of IRI in

[35] The strategies of US companies are analysed in Reppy (note 15). For West European companies, see Huffschmid, J. and Voß, W., *Militärische Beschaffungen–Waffenhandel–Rüstungskonversion in der EG, Eine Studie im Auftrag des Europäischen Parlamentes,* PIW-Studien no. 7 (PIW: Bremen, 1991); and Anthony, I. and Wulf, H., 'The economics of the West European arms industry', eds M. Brzoska and P. Lock, SIPRI, *Restructuring of Arms Production in Western Europe* (Oxford University Press: Oxford, 1992), pp. 17–35.

Italy—follow a similar strategy.[36] The idea is to sell so-called 'non-strategic' operations that do not 'fit' the expectation to generate capital. The hope of such companies is, as one company executive described it, to 'focus on what we know best: our core defense competencies'[37] and that other companies will have to leave the military market, thus reducing the number of competitors in the shrinking market.

Particularly the large producers hope to increase their share in the shrinking market at the expense of small and medium-sized producers. Referring to the effects of arms control in January 1990, a spokesman for the French firm Matra was convinced that there was still growth:

> The necessary reconversion of the arms industries is perceived, in a confused manner, by public opinion as an ineluctable reduction in size, while rather it is necessary to prepare for a change of state ['un changement de nature']: the arms limitation treaties contain clauses for quantitative limitation but cannot eliminate qualitative improvement. Every treaty signature has therefore as an immediate consequence a significant growth in the credits for research and development . . .[38]

Developments in various industrial sectors

Are there industrial sectors within the arms industry that have been less affected by the changed environment than others? The largest arms-producing companies are divided here into six groups, according to what they produce: (*a*) land systems and infantry weapons, (*b*) aerospace, (*c*) electronics, (*d*) diversified companies, (*e*) ships and (*f*) others.[39]

As table 1.4 indicates, only the small shipbuilding sector, with five companies, has been able to increase its arms sales during the past four years (1988–91). Within the other sectors the result is mixed. Of the 10 land system and infantry weapon producers, six increased their sales. Of the 27 aerospace companies, 12 increased and 15 decreased their arms sales. The comparable figures for the electronics sector are: 13 companies increased, 14 companies decreased and 2 companies had stable arms sales. Finally, within the group of diversified companies that are operating in several of the sectors, 14 companies increased their arms sales between 1988 and 1991, while six companies lost business during this period.[40]

[36] Airaghi, A., 'Opportunities and problems of West–East and North–South cooperation in conversion of aerospace industry', unpublished paper presented at the United Nations Conference on Technology Assessment in Conversion for Development, Moscow, 12–16 Oct. 1992.

[37] Anders (note 1), p. 14.

[38] Quoted by Gummett, P. and Walker, W., 'Science, technology and the peace dividend,' in Brzoska and Lock (note 35), p. 77.

[39] The division between these groups is not always clear: some of the companies in the electronics sector do produce missiles and vice versa. They have been grouped according to their quantitatively important sector of production.

[40] To prevent misinterpretations of these figures it should be pointed out that some of the increases and decreases that companies report are the consequence of mergers and acquisitions and are not necessarily the result of changed business performance.

Table 1.4. Changes in arms sales 1988–91, by arms industry sector

Figures for arms sales are in US$ m.; arms sales dependencies are expressed as percentages.

Production sector/company	Country	Industry	Arms sales 1991	Arms sales 1988	Change	Arms sales dependency
Land systems, infantry weapons						
GIAT Industries	France	A MV SA/O	1 430[a]	1 150	+ 280	*97[a]*
Ordnance Factories	India	A SA/O Oth	1 430[a]	1 590	– 160	*97[a]*
FMC	USA	MV Sh Oth	1 060[a]	950	+ 110	*28[a]*
TRW	USA	MV Oth	2 800	2 900	– 100	*35*
Diehl	FRG	A MV El SA/O	860[a]	610	+ 250	*48[a]*
Rheinmetall	FRG	A SA/O	860	650	+ 210	*41*
Israel Military Industries	Israel	A MV SA/O	640[a]	470	+ 170	*98[a]*
FFV	Sweden	A El SA/O Oth	500[a]	500	0	*47[a]*
Hunting	UK	SA/O	420[a]	440	– 20	*31[a]*
Morrison Knudsen	USA	MV Oth	320	300	+ 20	*16*
Aerospace						
McDonnell Douglas	USA	Ac El Mi	10 200	8 680	+ 1 520	*55*
General Dynamics	USA	Ac MV El Mi Sh	7 620	8 000	– 380	*80*
British Aerospace	UK	Ac A El Mi SA/O	7 550	5 470	+ 2 080	*40*
Lockheed	USA	Ac	7 500[a]	8 400	– 900	*75[a]*
General Electric	USA	Ac Eng	6 120	6 250	– 130	*10*
Boeing	USA	Ac El Mi	5 100[a]	4 500	+ 600	*18[a]*
Northrop	USA	Ac	4 700[a]	5 200	– 500	*86[a]*
Martin Marietta	USA	Mi	4 600[a]	4 300	+ 300	*75[a]*
Rockwell International	USA	Ac El Mi	4 100[a]	5 000	– 900	*33[a]*
Aérospatiale	France	Ac Mi	3 450	2 300	+ 1 150	*40*
Grumman	USA	Ac El	2 900	3 000	– 100	*72*
Dassault Aviation	France	Ac	1 870	2 080	– 210	*74*
Rolls-Royce	UK	Eng	1 680	1 410	+ 270	*27*
SNECMA Groupe	France	Eng Oth	1 320	1 270	+ 50	*31*
FIAT	Italy	Eng	1 180[a]	1 530	– 350	*17[a]*
Allied Signal	USA	Ac El Oth	1 100	1 500	– 400	*9*
Kawasaki Heavy Industries	Japan	Ac Eng Sh	1 010[a]	1 170	– 160	*14[a]*
Israel Aircraft Industries	Israel	Ac El Mi	1 080	800	+ 280	*67*
Lucas Industries	UK	Ac	630[a]	530	+ 100	*15[a]*
Sequa	USA	Eng El Oth	610	700	– 90	*32*
Hercules	USA	Ac Mi SA/O Oth	600	890	– 290	*20*
Westland Group	UK	Ac	530	450	+ 80	*64*
Saab–Scania	Sweden	Ac Eng El Mi	520	570	– 50	*11*
Teledyne	USA	Eng El Mi	500	600	– 100	*16*
Hindustan Aeronautics	India	Ac Mi	500[a]	480	+ 20	*97[a]*
Sundstrand	USA	Ac	370	470	– 100	*22*
Bombardier	Canada	Ac	440	80	+ 360	*16*

Table 1.4 *contd*

Production sector/company	Country	Industry	Arms sales 1991	Arms sales 1988	Change	Arms sales dependency
Electronics						
Raytheon	USA	El Mi	5 500[a]	5 500	0	*57[a]*
Thomson S.A.	France	El Mi	4 800	4 470	+ 330	*38*
GEC	UK	El	4 280[a]	2 970	+ 1 310	*25[a]*
Westinghouse Electric	USA	El	2 300	2 600	– 300	*18*
Unisys	USA	El	2 000	2 500	– 500	*20[a]*
Texas Instruments	USA	El Mi Oth	1 950	2 150	– 200	*29*
Loral	USA	El	1 920[a]	1 050	+ 870	*90[a]*
IBM	USA	El Oth	1 300	2 100	– 800	*2*
GTE	USA	El	1 250[a]	1 100	+ 150	*7[a]*
ITT	USA	El	1 200	1 390	– 190	*6*
Matra Groupe	France	El Mi Oth	1 180[a]	1 040	+ 140	*26[a]*
Siemens	FRG	El	900	800	+ 100	*2*
Harris	USA	El	760	1 000	– 240	*25*
AT&T	USA	El	700[a]	650	+ 50	*2[a]*
Mitsubishi Electric	Japan	El Mi	690[a]	790	– 100	*3[a]*
Motorola	USA	El	600	600	0	*5*
SAGEM Groupe	France	El	590	350	+ 240	*28*
Computer Sciences	USA	El	560[a]	510	+ 50	*32[a]*
Dassault Electronique	France	El	490	510	– 20	*72*
Smiths Industries	UK	El	490[a]	530	– 40	*41[a]*
Emerson Electric	USA	El	480[a]	480	0	*13[a]*
Hawker Siddeley	UK	El	480[a]	390	+ 90	*12[a]*
Thorn EMI	UK	El	450[a]	1 200	– 750	*7[a]*
Ferranti–International Signal	UK	El	440[a]	580	– 140	*54[a]*
Toshiba	Japan	El Mi	410[a]	650	– 240	*1[a]*
Honeywell	USA	El Mi	400	1 800	– 1 400	*6*
NEC	Japan	El	380[a]	570	– 190	*1[a]*
Mitre	USA	El	370[a]	350	+ 20	. .[a]
Diversified						
General Motors	USA	Ac Eng El Mi	7 380[a]	7 500	– 120	*6[a]*
United Technologies	USA	Ac El Mi	4 000	4 500	– 500	*19*
Daimler Benz	FRG	Ac Eng MV El	3 920	3 420	+ 500	*7*
Mitsubishi Heavy Ind.	Japan	Ac MV Mi Sh	3 040	2 840	+ 200	*17[a]*
Litton Industries	USA	El Sh	3 000[a]	2 920	+ 80	*58[a]*
IRI	Italy	Ac Eng El Sh	2 670[a]	1 950	+ 720	*36[a]*
LTV	USA	Ac MV El	1 800	2 150	– 350	*24*
Textron	USA	Ac Eng MV	1 800	1 500	+ 300	*23*
EFIM	Italy	Ac MV El	1 710[a]	1 490	+ 220	*79[a]*
INI	Spain	Ac A MV El Sh SA/O	1 560[a]	1 290	+ 270	*9[a]*
Armscor	S. Africa	Ac A MV El SA/O	1 330[a]	1 870	– 540	*74[a]*

Production sector/company	Country	Industry	Arms sales 1991	1988	Change	Arms sales dependency
Oerlikon-Bührle	Switzer.	Ac A El SA/O	1 100	930	+ 170	*44*
VSEL Consortium	UK	MV Sh	920	830	+ 90	*100*
Gencorp	USA	Ac Eng El Mi SA/O Oth	790[a]	950	– 160	*45*[a]
Bremer Vulkan	FRG	El Sh	780	130	+ 650	*44*
Thyssen	FRG	MV Sh	770	600	+ 170	*3*
Eidgenössische Rüstungsbetriebe	Switzer.	Ac Eng A SA/O	700[a]	550	+ 150	*95*[a]
Ford Motor	USA	Ac MV El Mi	700[a]	1 200	– 500	*1*[a]
Thiokol	USA	Eng Mi SA/O Oth	690	580	+ 110	*55*
Dowty Group	UK	Ac El	450[a]	410	+ 40	*33*[a]
Ships						
DCN	France	Sh	3 710	3 030	+ 680	*100*
Tenneco	USA	Sh	2 220	1 670	+ 550	*16*
Ishikawajima-Harima	Japan	Eng Sh	860	600	+ 260	*39*
Devonport Management	UK	Sh	470[a]	450	+ 20	*94*[a]
Lürssen	FRG	Sh	400[a]	170	+ 230	*81*[a]
Other						
CEA Industrie	France	Oth	1 750	1 670	+ 80	*25*

Note: A = artillery, Ac = aircraft, El = electronics, Eng = engines, Mi = missiles, MV = military vehicles, SA/O = small arms/ordnance, Sh = ships, and Oth = other.

[a] Figures are for 1990.

Source: SIPRI data base.

The overall positive trend of reductions in arms procurement has apparently affected all sectors of industry. The exception of the shipbuilding industry is likely to be a temporary phenomenon[41] and might be a consequence of the fact that, in shipbuilding, the process of capacity reduction and concentration of companies began during the 1970s. There is no empirical evidence in the data displayed here to verify the proposition occasionally put forward that traditional arms producers, like manufacturers of platforms (tanks, ships and aircraft) and mechanical engineering skills (gun producers) are particularly affected while demand for high-technology products, electronics and software will remain stronger.[42]

[41] In shipbuilding, arms sales fluctuate greatly because of the long periods required to complete a single unit.

[42] Gummett and Walker (note 38), p. 71.

V. Conclusions

The political changes and more benign security environment have led to economic difficulties for the arms-producing industry. Companies have adopted a variety of strategies in order to stay in business in this changed arms market environment. Not all of these strategies will lead to a reduction of arms production and to the release of resources from military use. If, for instance, companies react with more aggressive marketing strategies and are successful in finding new markets both at home and abroad, and if this strategy is fostered by government assistance, as is presently the case in a number of countries, this will contribute to a slowdown of the disarmament process or could contribute to the continuation or even fostering of regional arms races.

The most favourable strategy for arms control *and* stable domestic economic conditions is the conversion of arms industries to civilian production, but even this process has experienced great difficulties and will, at best, be slow.

The reversal in the long-term trend from a dynamic arms race to gradual arms control has economic effects in five general areas. The economic effects in industry described in this chapter are but one aspect of the consequence of arms control and disarmament. Public finance, manpower (both in industry and in the armed forces), military hardware and infrastructure are important aspects of the same process. The economic consequences are both positive and negative. In the short term economic distortions are often the predominant trend. Jobs are lost, the destruction of surplus weapons or other military hardware is costly, and cleaning up the waste in military bases is not only costly and often technically complicated but also an ecological hazard. However, in the long term disarmament offers positive economic opportunities. Skills and creativity and human, financial and material resources that have been absorbed by the military can be reallocated to non-military use. Regions that have depended on military activities might suffer from the withdrawal of troops, but such regions can also make use of facilities that were previously restricted exclusively to military purposes.

The long-term positive economic effects of disarmament will, however, not emerge automatically. Resistance of industry and other economic interest groups might slow down or even halt the process that began with the end of the cold war.

Part II
The United States

2. The United States: trends in defence procurement and research and development programmes

Gordon Adams and Steven M. Kosiak

I. Introduction

The US military has traditionally tried to equip its forces with the most modern and technologically sophisticated weapon systems in the world. During the 1980s, the United States underwent a massive military buildup which emphasized new weapon acquisition. Overall, US defence budgets grew by 54 per cent between fiscal year (FY) 1980 and FY 1985. However, during this period funding for research and development (R&D), where new technologies are investigated and ultimately made ready for production, nearly doubled, while funding for weapon procurement more than doubled (for the long-term development of US national defence spending, see figure 2.1).

The boom in Department of Defense (DoD) acquisition funding (which consists of programmes in R&D and procurement) is now, however, clearly over. Driven primarily by concern over the size of the federal deficit, US defence budgets began to follow a downward course in the second half of the 1980s. The collapse of the Warsaw Pact and the dissolution of the Soviet Union during 1989–91 substantially accelerated this decline. Moreover, US defence budgets are projected to continue falling through, and perhaps well beyond, FY 1997. Funding for R&D has been cut relatively modestly to date, but is projected to continue to decline through FY 1997, even under the current Administration plan. By contrast, procurement funding has been cut dramatically over the past several years, falling by more than 50 per cent between FY 1985 and FY 1992, and is projected to drop even lower by FY 1997.

Altogether, acquisition funding fell by almost a quarter between FY 1985 and FY 1990. Under the current Administration plan, it is projected to fall by another one-third between FY 1990 and FY 1997; alternative proposals for deeper reductions in overall defence spending might well lead to even deeper cuts in acquisition funding. Adapting to these significantly lower levels of funding will prove to be a challenge for both the Defense Department and the US defence industry. The Administration's current Future Year Defense Plan (FYDP) for FY 1992–97 appears to contain a potentially significant mismatch between acquisition plans and projected funding levels. Unless that plan is substantially restructured, the US military could find its procurement and R&D

programmes in considerable disarray by the mid-1990s or even sooner. Likewise, the USA will have to substantially change the way it develops and produces weapons if it is to maintain an adequate defence scientific and industrial base in the context of declining budgets.

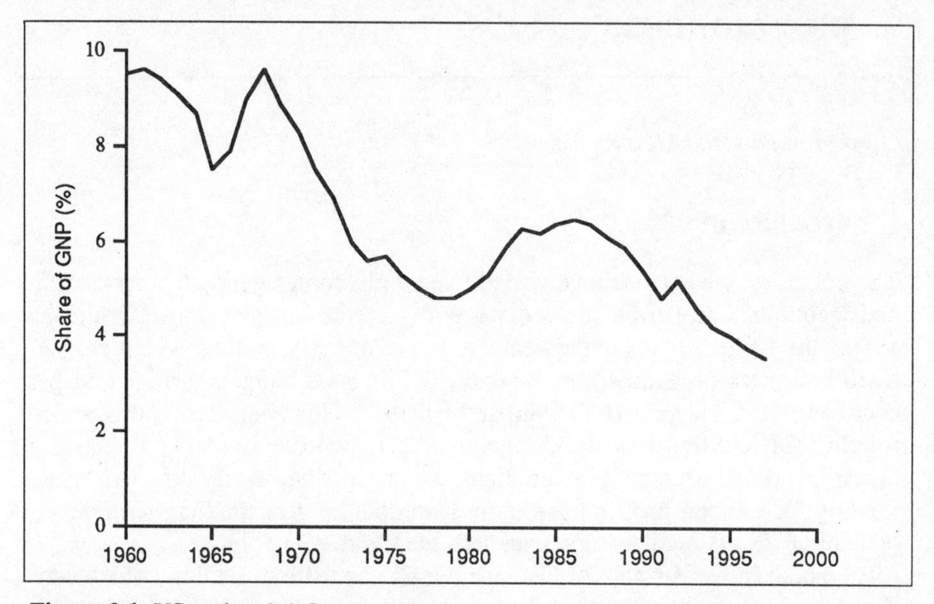

Figure 2.1. US national defence expenditure as a share of the gross national product, FY 1960–97[a]

[a] National defence estimates include outlays for the 1991 Desert Shield/Storm operations and the allied Persian Gulf War contributions.

Source: The Defense Budget Project data base.

The next two sections of this chapter briefly review the history of US military procurement and research and development funding from World War II through FY 1991, with an emphasis on modernization trends in the 1980s. Section IV describes the procurement and R&D aspects of the Administration's proposed FY 1992–97 FYDP. The last section discusses issues and problems likely to affect the Administration's procurement and R&D plans and the prospects for these two areas of weapon acquisition in the longer term.[1]

[1] Unless otherwise noted, all funding figures are 'budget authority', expressed in FY 1993 US dollars, and all growth or decline is in 'real' (inflation-adjusted) terms. *Budget Authority* (BA) is the authority granted by Congress to enter into obligations for payment of government funds. *Outlays* are the actual payments of these funds. *Total Obligational Authority* (TOA) is a DoD financial term which expresses the total 'value of the direct defense program' and is equivalent to new budget authority plus any additions or subtractions due to, for example, transfers of funding from previous years and rescissions. TOA generally differs only marginally from BA and was used in this chapter only for certain breakdowns of the research and development budget which were unavailable in BA.

II. The history of acquisition funding

Historically, US Department of Defense funding for procurement has followed a path roughly parallel to that of the DoD budget as a whole. Until the end of the 1970s, overall US defence spending followed a fairly predictable and unsurprising course; spending tended to be relatively low during times of peace and relatively high during times of war. In the first three decades after World War II, the US defence budget peaked twice, once in FY 1952, during the Korean War, and again in FY 1968 during the height of the Viet Nam War. Similarly, DoD funding for procurement peaked in FY 1952 and FY 1967. Funding for DoD R&D has followed a more stable course, less closely tied to swings in overall defence spending and less dependent on whether the country is at war or peace. In 1950–80 funding for R&D grew relatively slowly, but steadily, increasing from $3.9 billion to $22.6 billion.

With the coming of the Reagan presidency in FY 1981, the relatively predictable pattern of overall US defence spending changed and with it so did the pattern of spending for procurement and R&D. During the first half of the 1980s, the USA embarked on an unprecedented peacetime military buildup: the defence budget increased by 54 per cent between FY 1980 and FY 1985. Funding for procurement grew even more dramatically. Between FY 1980 and FY 1985, US procurement funding grew by 116 per cent. Like procurement, R&D was one of the major beneficiaries of the Reagan buildup of the early 1980s, increasing by 82 per cent between FY 1980 and FY 1985.

After peaking in FY 1985, funding for national defence entered a period of relatively modest, but steady, decline. The half decade of extraordinary growth in funding enjoyed by the Defense Department was financed in large part by reductions in domestic programmes and a growing federal deficit. By 1985, concerns about the Soviet threat began to be overtaken by concerns about the federal deficit, and between FY 1985 and FY 1990 US funding for national defence fell by 13 per cent. During this same period, DoD procurement funding fell by 30 per cent. Despite this decline, by historical standards, in FY 1990 both the overall defence budget and the DoD procurement budget remained relatively high. At $336 billion, the overall defence budget was still 34 per cent higher than it had been in FY 1980, before the Reagan buildup had begun. Likewise, at $89.7 billion, the FY 1990 DoD procurement budget was still 52 per cent higher than it had been in FY 1980. R&D funding fared better than either the procurement budget or DoD funding generally in the later half of the 1980s. Between FY 1987, when R&D funding peaked, and FY 1990, funding for R&D fell by only 9 per cent. Moreover, at $40.3 billion, in FY 1990 it was still 78 per cent above the FY 1980 level.

In FY 1991, funding for national defence fell a further 8.7 per cent, marking the largest annual reduction since the closing years of the Viet Nam War. The FY 1991 defence budget included a 21 per cent reduction in procurement funding and a 5 per cent cut in R&D. As had been the case with the five preceding defence budgets, the desire to trim the federal deficit was probably the prime driver of this reduction. However, in 1989–90 the Soviet Union's hold on Eastern Europe collapsed and, much more clearly than in earlier years, the perception of a declining Soviet threat also played a significant part in driving the US reductions. The Administration's proposed FY 1991 defence budget also marked a change from previous years in that it projected the termination of a large number of current-generation procurement programmes, including the F-15E fighter, the Apache attack helicopter, the F-14D fighter and the M-1 tank. Even more unusual was the proposal to cancel several next-generation weapon systems, including the Sea Lance anti-submarine weapon and the Advanced Short-Range Air-to-Air Missile, in addition to the Marine Corps' tilt-rotor V-22 Osprey transport aircraft which the Bush Administration had first sought to cancel in its revision of the Reagan Administration FY 1990 budget. Although not all of these proposed terminations were accepted by Congress, to a greater degree than in previous years savings in the FY 1991 defence budget were achieved by terminating rather than stretching out arms procurement.

Defense industry-related employment rose and fell during the 1980s in a pattern roughly paralleling the rise and fall of US defence spending. According to DoD estimates, in FY 1980 US defence-related industries employed about 1.99 million men and women; that figure jumped to 3.37 million in FY 1987 and fell back slightly to 3.1 million in FY 1991.

Defence industry employment is generated not only by spending on procurement and R&D, but by spending in the operations and maintenance, military construction and family housing accounts as well. DoD estimates do not specify industry employment levels related to procurement and R&D alone. However, a rough estimate derived from DoD data is that DoD procurement and R&D spending together generated about 1.4 million of the 1.99 million defence industry jobs in FY 1980, 2.3 million in FY 1987 and 2.0 million in FY 1991.[2]

III. Modernization in the 1980s

Between FY 1981 and FY 1991, the Department of Defense budgeted a total of $1.1 trillion for procurement and $400 billion for R&D. These two investment accounts absorbed, respectively, 29 per cent and 11 per cent of total national defence funding during this period. Reflecting the traditional high-technology emphasis of the US military, aircraft, missile and electronics, and communications programmes accounted for by far the largest shares of procurement and

[2] These estimates may substantially understate or overstate defence industry employment related to DoD procurement and R&D spending because they are based on the simplifying assumption that every dollar spent by the Department of Defense on non-pay purchases—whether it is spent in the procurement, R&D, operations and maintenance, military construction, or family housing accounts—generates the same number of jobs. Estimates were derived from data on DoD non-pay purchases. See *National Defense Budget Estimates for FY 1993*, Office of the Comptroller of the Department of Defense, Mar. 1992, pp. 89–90.

R&D funding in the 1980s. During the decade, aircraft programmes accounted for 37 per cent of procurement outlays; missile programmes, 16 per cent; 'other procurement' (including communications and electronics), 25 per cent; ship construction and conversion, 13 per cent; weapons, 6 per cent; and ammunition, 3 per cent. In 1981–91 the Navy (including the Marine Corps) accounted for approximately 40 per cent of total DoD procurement funding; the Air Force, 38 per cent; and the Army, only 19 per cent of that total.[3]

As concentrated as US procurement budgets have been in aerospace, electronics and communications programmes over the past decade, R&D funding has been even more heavily concentrated in these areas. Between FY 1981 and FY 1990, aircraft R&D accounted for 16 per cent of major DoD R&D prime contract awards; missiles and space systems, 36 per cent; electronics and communications, 22 per cent; ship programmes, 3 per cent; weapons, 2 per cent; ammunition, 2 per cent; and tanks and other vehicles, 1 per cent.[4] Between FY 1981 and FY 1991, the Air Force accounted for 41 per cent of total R&D funding; the Navy (including the Marine Corps), 27 per cent; the Army, 15 per cent; and DoD defence agencies, 17 per cent.

Overall, R&D spending in the 1980s strongly favoured development efforts tied to the near-term procurement of specific weapon systems. This focus came at the expense of basic R&D, which is intended to improve scientific understanding of technologies which could form the foundation for future military systems.[5] Between FY 1980 and FY 1991, when the total R&D budget grew by 65 per cent, funding for technology base programmes increased by less than 10 per cent. Throughout the period, funding for technology-base programmes

[3] Defence agencies and other programmes accounted for the remaining 2% of FYs 1981–91 procurement spending.

[4] The remaining 17% of major R&D prime contract awards (those over $25 000) were allocated to other minor programmes and services.

[5] The DoD R&D budget breaks down into five different budget 'activities.' These activities include the technology base, advanced technology development, strategic programmes, tactical programmes, intelligence and communications, and defence-wide mission support. The technology base activity consists of programmes which strengthen the scientific base for developing weapons in the future. It includes programmes in the research and exploratory development research 'categories'. Technology base research programmes—which include efforts carried out at both DoD research facilities and universities—are not directly tied to specific military applications but rather are intended to expand scientific knowledge of fields related to national defence. Exploratory development funds support R&D efforts aimed at solving specific military problems, using rudimentary hardware to attempt to establish the feasibility of the concept being explored. Such efforts include, for example, the Air Force's Aerospace Propulsion and Aerospace Flight Dynamics programmes and the Navy's Submarine Technology programme. The advanced technology development activity consists of programmes which have begun development of hardware, including prototypes, for testing and 'proof of design', but have not yet been selected to move from advanced development to engineering development—where hardware is developed according to precise DoD specifications—and eventually to production. The strategic, tactical, and intelligence and communications activities include all R&D on programmes which have passed through the earlier stages of R&D and have been selected to move through engineering development to eventual procurement in the strategic, tactical, and intelligence and communications mission areas. In addition to dividing the R&D budget into 6 different budget activities, the DoD divides it into 6 research categories: research, exploratory development, advanced development, engineering development, management and support, and operational systems development. With the exception of the 'management and support' category, these categories correspond to the historical stages in the development of a weapon programme.

hovered between $3.7 billion and $4.6 billion.[6] By contrast, R&D funding for systems already approved for full-scale development and eventual procurement (those in the tactical, strategic, and command and communications activities) more than doubled between FY 1980 and FY 1987 and was still about 60 per cent higher in FY 1991 than at the start of the decade. On average, during the period FY 1981–91, the Defense Department provided $25.1 billion annually for programmes in these latter stages of the R&D process.

Funding for programmes in advanced technology development also grew substantially during the 1980s. Between FY 1980 and FY 1991, funding for programmes in this activity grew nearly sixfold. Despite this growth, however, funding for advanced technology development, like funding for technology base programmes, fell far short of the levels provided for the tactical, strategic, and intelligence and communications activities: advanced development funding increased from $1 billion in FY 1980 to $5.6 billion in FY 1991, and averaged $4.1 billion annually between FY 1981 and FY 1991. Although generally less closely tied to specific weapon systems than programmes included in these latter activities, programmes in advanced technology development came to focus largely on one mission area in the 1980s—ballistic missile defence. Indeed, the growth in funding for this activity essentially reflects the growth in funding for the Strategic Defense Initiative (SDI) that occurred in the 1980s. From its beginning in 1983, all but a small fraction of SDI funding has been for programmes in this stage of R&D. By FY 1991, advanced technology development accounted for over 99 per cent of SDI funding and, at $3.1 billion, SDI funding accounted for approximately half of that year's advanced technology development budget. Defence-wide mission support, the last R&D budget activity, grew by about 35 per cent in FY 1980–91, averaging about $4.6 billion during the period.

IV. Acquisition in the 1990s

In the early 1990s the US defence budget, which had been declining modestly since FY 1985, began to take a much sharper downward turn, which is projected to continue through FY 1997. Of all the areas of the defence budget, procurement funding has been cut the most deeply over the past several years and there is little prospect of relief in sight. Although it has been cut much less deeply, R&D funding has also been cut consistently and significantly over the past several years and is projected also to continue declining through FY 1997.

The dramatic changes that occurred in Eastern Europe and the Soviet Union in 1989–90 were first reflected in long-range Pentagon force and budget planning in February 1991, with the presentation of the proposed new Base Force concept and the FY 1992–97 Future Years Defense Plan. Responding to the reduced Soviet threat, the Administration proposed reducing the size of the

[6] These figures and those provided in the breakdown of R&D funding by 'activity' which follows are in Total Obligational Authority (TOA).

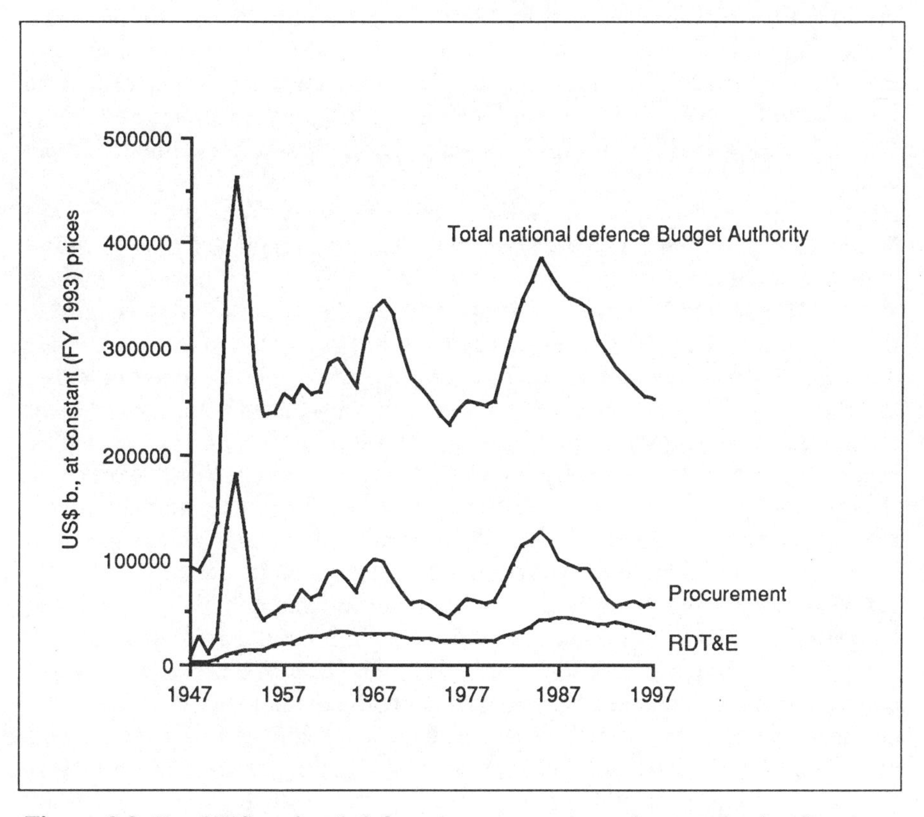

Figure 2.2. Total US national defence, procurement, and research, development, testing and evaluation budgets, FY 1947–97

Source: The Defense Budget Project, based on US Department of Defense data.

US military by 20 per cent, measured in terms of active military personnel, between FY 1990 and FY 1995.[7] The number of Army divisions would be reduced from 28 to 18 (plus two cadre divisions), the number of tactical fighter wings from 36 to 26, and the number of carrier battle groups from 13 to 12.

The Administration also expanded the list of production terminations it had begun with its FY 1991 budget submission. Among other programmes, the Administration proposed terminating procurement of the F-16 fighter, the Bradley Fighting Vehicle, the Trident submarine, the MX missile and the re-manufactured F-14D fighter, leaving only a relatively small number of current-generation systems in production. Similarly, the Administration expanded its list of proposed next-generation weapon cancellations; these additions included the Navy's A-12 attack aircraft and P-7 anti-submarine patrol plane, the naval

[7] Compared to FY 1987, when US military personnel strength peaked, the projected FY 1995 level will represent a 25% decline.

version of the Advance Tactical Fighter and the Air Force version of the A-12. Reflecting these proposed cuts in force structure and acquisition terminations, the new FYDP projected reducing the defence budget by 14 per cent from FY 1992 to FY 1997. Procurement funding was projected to remain essentially flat through FY 1997, while R&D funding was projected to decline by 18 per cent.

Less than a year after this Base Force and FYDP were presented, the Soviet Union itself dissolved, to be replaced by the loosely organized Commonwealth of Independent States (CIS). Some observers argued that the new Base Force and FYDP had rapidly become out of date. The Administration's response to the dissolution of the Soviet Union came with its presentation of the FY 1993 defence budget in January 1992. The Administration remained committed to the Base Force goals announced in 1991, arguing that the Base Force was already predicated upon a greatly reduced Soviet threat and represented an irreducible minimum to meet US security needs in the post-cold war world. The Administration did, however, revise the FY 1992–97 budget plan, including a modest additional reduction in overall defence spending and some significant cuts in a number of procurement programmes in particular.

The revised FYDP would reduce overall defence funding by 18 per cent over the period FY 1992–97, cutting a total of $50.4 billion (in current-year dollars) from the FYDP presented in February 1991.[8] The bulk of these savings would come out of procurement funding. Under the revised plan, over these years procurement funding would fall 22 per cent, to $55.6 billion in FY 1997. This would be 56 per cent below its level at the height of the Reagan buildup in FY 1985 and 6 per cent below its level at the start of the buildup in FY 1980. Similarly, under the revised plan R&D funding would decline by 18 per cent in the period FY 1992–97, reaching $31.6 billion in FY 1997. This would be 29 per cent below the R&D funding level at its peak in FY 1987 but still 40 per cent above its level in FY 1980.

In the FY 1991 and FY 1992 defence budgets, the Pentagon primarily sought savings through the termination of current-generation production and reductions in force structure—attempting to protect hardware modernization plans. By comparison, the plan presented in January 1992 achieved additional savings primarily by terminating or deferring the acquisition of next-generations systems. That plan significantly reduced acquisition funding for 10 major next-generation weapon systems. By far the most significant changes were the decisions to end production of the B-2 bomber after only 20 aircraft, rather than 75 as originally planned, and to terminate production of the Seawolf (SSN-21) attack submarine programme after only one boat. The Defense Department projected that cancelling these two programmes would result in $32 billion (current-year dollars) in savings from the FY 1992–97 budget plan presented in

[8] In current dollars, the revised FY 1992–97 plan constitutes a $63.8 billion reduction from the Feb. 1991 plan for the same period. However, $13.4 billion of these reductions are due to adjustments to last year's plan required under the 1990 Budget Enforcement Act, based on the lower than projected inflation rate.

1991.[9] In addition to these changes, the Defense Department decided to terminate R&D on the Small ICBM (SICBM) and the Navy's Advanced Air-to-Air Missile; cap production of the Advanced Cruise Missile (ACM) at 640 missiles (the number authorized through FY 1992); cancel production of the Army's new Air Defense Antitank System (ADATS); stretch out R&D on the Army's new Comanche (RAH-66) scout/attack helicopter, the Line-of-Sight Antitank (LOSAT) system, and the Block III tank; and forgo procurement of the Navy's proposed Fixed Distribution System (FDS) underwater surveillance network, while continuing R&D on the programme. Altogether the Administration estimates that these programme changes would result in $42.1 billion (current-year dollars) in savings from the previous year's FY 1992–97 plan.

The combination of next-generation weapon cancellations included in the FY 1991, FY 1992 and, especially, FY 1993 budgets has resulted in a marked decrease in the scope of the Administration's modernization plans. However, the Pentagon still plans to procure a significant number of next-generation systems. Major next-generation Air Force acquisition programmes in the revised FYDP include: the F-22, the planned replacement for the F-15 air superiority fighter; the C-17, the Air Force's new strategic airlifter; the Joint Surveillance Target and Attack Radar System (JSTARS) aircraft, which is designed to locate enemy ground forces operating behind the forward edge of the battle area; and the Advanced Medium-Range Air-to-Air Missile (AMRAAM). Major Navy programmes include: the advanced E/F version of the Navy's F/A-18 fighter/attack aircraft; and CVN-76, a new Nimitz Class aircraft-carrier. Major Army acquisition programmes in the revised FYDP include: the Army Tactical Missile System (ATACMS), designed to strike targets deep behind enemy lines; the Advanced Anti-tank Weapon System-Medium (AAWS-M); and the Sense and Destroy Armor (SADARM) system.

Potentially the most significant change in the Department of Defense's new plan is the announcement of a drastically revised acquisition strategy. Traditionally, the DoD has moved new weapon systems into production well before completing the R&D phase. Moreover, with very few exceptions, weapon systems at the advanced stages of R&D have almost invariably been put into production. In presenting the FY 1993 budget, the Administration announced that it was adopting a new acquisition strategy under which more reliance would be placed on the R&D process. Under this new strategy, the Defense Department would move new weapon systems into 'full-scale production only after verifying the need for producing the system, and after minimizing technical, manufacturing, and operational risks. The strategy will also emphasize the upgrading of existing weapons systems using proven technologies, when operational needs can be met, instead of producing new weapons systems.'[10] According to Secretary of Defense Dick Cheney, 'we can now afford to take

[9] This savings estimate assumes that the Administration is successful in its effort to rescind $3.4 billion in FY 1991 and FY 1992 funding approved by Congress for the second and third Seawolf submarines.

[10] 'DoD to slow pace of modernization, cut strategic nuclear arsenal, while maintaining essential forces', News Release, Office of the Assistant Secretary of Defense (Public Affairs), 29 Jan. 1992, p. 4.

more time to refine the capabilities of new systems, moving into production only after we are fully satisfied that we'll get what we need at a price we can afford.'[11]

V. Future defence funding prospects and potential problems

According to the Administration, the revised FYDP and the proposed new acquisition strategy in particular will accomplish three important objectives: (*a*) make equipping the Base Force affordable within projected budgets; (*b*) preserve the technological and production capabilities of the United States to introduce next-generation weapon systems in relatively short order if and when they are needed; and (*c*) ensure that the US military retains its technological superiority over the long term. However, it is far from clear that the revised plan and the new acquisition strategy, at least as applied so far, will enable the Administration to accomplish these goals.

Plans/budget mismatch: procurement

The FYDP presented in February 1991 appeared to include a potentially significant mismatch between acquisition plans and projected funding levels, and the 1992 revised FYDP may well suffer from a similar under-funding problem. The Defense Department's decisions in both the FY 1991 and FY 1992 budgets to terminate the production of a wide variety of current-generation weapon systems were partially explained as consistent with plans for a smaller force structure. However, the suddenness with which so many systems were terminated and the fact that at least some of the reductions (in helicopters, for example)[12] appeared likely to leave the smaller force somewhat short of requirements suggest that the Defense Department was struggling to fit desired systems into declining resources. The Army's endless search for a clear 'long term' aviation modernization plan and the continual downward revisions of naval aviation plans similarly seemed to indicate an uneasiness in the Pentagon over the affordability of its modernization plans. The most pointed evidence of underfunding came in a June 1991 statement by DoD Under Secretary for Acquisition Donald Yockey, contending that the services were not providing realistic projections of the funding levels that would actually be required in the FYDP for 11 specific weapon systems and 3 general areas.[13] The weapon systems cited included the Navy's T-45 trainer aircraft, the Army's Light Heli-

[11] See note 10.

[12] Some planned cuts in procurement objectives appear to go beyond the reductions implied by the decline in the force structure projected in DoD plans. Two such examples are the Army's UH-60 utility helicopter, for which the procurement objective was reduced by 36%, from 2267 to 1447, and the Army Helicopter Improvement Program (AHIP), for which the procurement objective was reduced by 58%, from 583 to 243. By comparison, Army plans call for reducing the number of divisions by 29%, from 28 in FY 1990 to 20 (including two cadre divisions) in FY 1995.

[13] 'Yockey gets tough on underfunded Pentagon programs', *Defense Week*, 10 June 1991.

copter (RAH-66), the Armored Systems Modernization plan and the Line-of-Sight Anti-Tank System, and the Air Force's Advanced Tactical Fighter (F-22).

It is far from clear that the revised FYDP presented in January 1992 has resolved the plans/funding mismatch. As part of the new acquisition strategy, the Pentagon announced that it would place increased emphasis on upgrading existing weapon systems. If equipment is upgraded and modified it need not be replaced as quickly, slowing funding requirements for next-generation weapon systems. However, the proposed FY 1993 budget (the last budget in the revised FYDP for which information on modification funding levels is available) does not seem to support this new emphasis. The FY 1993 request for aircraft modifications, for example, represents a 19 per cent decline compared to FY 1992 and a 21 per cent decline from the FY 1993 funding level projected in the February 1991 DoD plan.

In addition, concerns persist that many proposed next-generation weapon systems remain under-funded in the revised FYDP, including a number of those singled out by Yockey, such as the F-22.[14] There is also growing concern over the cost of developing the new E/F version of the F/A-18 fighter/attack aircraft. In 1991 the Navy estimated that developing the F/A-18EF would cost about $3.3 billion (current-year dollars). The official Navy estimate rose to $4.9 billion in early 1992, and some observers suggest that development costs may eventually reach $7 billion.[15] The new plan envisions building a smaller number of new weapon systems and has reduced projected funding levels to reflect those cuts, but there is little evidence that the underlying problem has been solved. In short, if the FY 1992–97 FYDP presented in 1991 was tantamount to trying to fit 100 pounds of potatoes into a 90-lb sack, the revised FYDP may change the situation only to the extent that DoD is now trying to fit 90 lbs of potatoes into an 80-lb sack.

The Administration plan may also fall short in terms of the goals of preserving the technological capability of the United States to put next-generation systems into production on short notice and maintaining the US military's technological superiority over the long term. The speed with which hardware can be moved out of R&D and into production depends in large part on the degree to which a trained work-force and appropriate industrial facilities are available. Overall, as funding for acquisition drops, the size of the work-force involved in hardware production and R&D will decline. As a very rough estimate, under the Administration's current plan, between FY 1991 and FY 1997 this work-force might be expected to decline from about 2.0 million to 1.5 million.[16]

[14] The Congressional Budget Office (CBO), for example, has estimated that F-22 average unit procurement costs could range from $80 million (consistent with Air Force estimates) to $115 million. See *Balance and Affordability of the Fighter and Attack Aircraft Fleets of the Department of Defense*, CBO Papers, Apr. 1992, p. 25.

[15] See Rosenberg, E., 'Pentagon review of billion-dollar Hornet variant is delayed', *Defense Week*, 30 Mar. 1992, p. 16.

[16] Under the Administration plan, acquisition spending (outlays) is projected to fall by 25% between FY 1991 and FY 1997. This estimate thus assumes that FY 1997 acquisition-related defence industry

Given the US military's reduced force requirements in the post-cold war world, it is appropriate that the US defence industry contract: the question is whether the contraction will be so deep or uneven that some areas of the defence scientific and industrial base will be left too weak to support DoD technological and production requirements over the long run.

Funding is projected to remain relatively high for some types of system, such as airlift, throughout the FYDP, ensuring the relative health of some sectors of the defence industrial base. In the case of other programmes, where military and civilian technologies substantially overlap—such as some types of avionics, for example—the required production skills and technologies may be adequately preserved in the civilian commercial market. As it stands now, however, the revised FYDP may not permit the retention of all important sectors of the defence industrial base. The plan includes relatively long production gaps for some types of system that depend on certain technologies with relatively little applicability to products on the civilian commercial market.

These gaps in production threaten to squeeze major platform integrators—some of the biggest names in defence industry like McDonnell Douglas, General Dynamics' Electric Boat, Lockheed and FMC—the hardest. In the aircraft manufacturing sector, where these is an acknowledged over-capacity, this squeeze is causing individual companies to 'down-size' (contract) and portends increases in 'teaming' arrangements, spreading the production of major subsections of new systems to the participating manufacturers. With procurement of the next-generation Block III tank now projected to begin in the middle to latter part of the next decade, the Defense Department may be forced to choose between low-rate production, armour plate stockpiling, modifications and new foreign sales strategies to maintain an armour production capability. Submarine construction faces many of the same options: continuing production of current-generation attack submarines—even though there is no pressing military requirement—in order to bridge the gap until an afforable next-generation submarine is developed and funded, or preserving certain unique manufacturing techniques, like high-strength welding and silent valves and pumps, by, for example, applying these processes or equipment to commercial use, albeit at uneeded performance over-specification.

Plans/budget mismatch: R&D

An emphasis on R&D programmes and funding are a key aspect of the proposed new acquisition strategy. Such funding is required both at the basic technology level—to ensure that the US military retains its technological edge over the long term—and at the advanced stages of R&D—to ensure that prototypes of next-generation systems can be fully tested and produced. However, the Administration's projected R&D budgets may not support this approach.

employment would be about 25% smaller than the FY 1991 work-force. (For an explanation of how the FY 1991 acquisition-related defence industry work-force estimate was derived, see note 2.)

Under the FYDP, overall funding for R&D would be lower than it was in the 1980s. Between FY 1981 and FY 1991, annual DoD funding for R&D averaged about $37.9 billion.[17] During the six years covered in the revised FYDP, funding for R&D would average $36.2 billion, a level 4 per cent lower. Moreover, by FY 1997, the last year covered by the FYDP, R&D funding is projected to fall to $31.6 billion, a 17 per cent reduction from the FY 1981–91 average.

In addition, R&D funding in the revised FYDP will apparently be substantially less balanced that it was in the 1980s. The FYDP projects an increasing priority on R&D for ballistic missile defences. Between FY 1981 and FY 1991 DoD spending on ballistic missile defence R&D averaged about $2.7 billion. Between FY 1992 and FY 1997 it may average as much as $5.7 billion, reaching as high as $6.1 billion by FY 1997.[18] During the 1980s, SDI absorbed an average of about 7 per cent of the total DoD R&D budget, and by 1997 it may account for as much as 19 per cent of that budget. There may well be merit in developing and eventually even deploying some type of both strategic and tactical ballistic missile defences. However, in the context of a shrinking overall R&D budget, significant increases in SDI-related R&D can occur only at the expense of other elements of the R&D budget. The priority on SDI may make it difficult to accomplish the broader technology goals of the new acquisition strategy.

While funding for SDI-related R&D is projected to rise substantially, average annual funding for non-SDI-related R&D is projected to fall from about $35.2 billion in the period FY 1981–91 to an average of $30.5 billion over the next six years, representing a decline of 13 per cent.[19] Moreover, by FY 1997 non-SDI-related R&D funding would drop to $25.5 billion, a decline of some 28 per cent. This does not necessarily mean that R&D funding for *'tactical'* (conventional) weapon systems will fall below its 1980s average. R&D funding for non-SDI related *'strategic'* programmes has declined significantly since FY 1985 and is likely to remain substantially below its 1980s average throughout the period FY 1992–97. If that is indeed the case, it might be possible to keep R&D funding for tactical programmes relatively near its 1980s level, even in the face of a significant reduction in overall non-SDI-related R&D funding.

[17] These figures and those provided in the discussion of SDI and non-SDI-related R&D which follow are in Total Obligational Authority.

[18] These projections of SDI R&D funding were derived from SDIO (SDI Organization) estimates of funding requirements presented in the spring of 1991. Congress has since directed the Administration to focus efforts on a smaller, more limited system. It is not clear that this congressional action will lead to any substantial reduction in R&D funding requirements. Indeed, if the SDIO were to attempt to meet the 1996 initial deployment date included in the Missile Defense Act of 1991, R&D funding might even have to be increased. In any case, the FY 1993 request for SDI does not suggest a dramatic reorientation of the programme. For information on the spring 1991 SDIO plan, see US General Accounting Office (GAO), *Strategic Defense Initiative: 15-Year Funding Requirements*, Feb. 1992.

[19] These estimates of future non-SDI-related R&D funding levels were derived by subtracting the projected FY 1992–97 SDI funding levels included in SDIO's spring 1991 plan from the overall R&D funding levels projected in the revised FYDP presented in Jan. 1992.

If the next generation of military hardware were to cost no more to develop than their predecessors, this level of R&D funding would presumably be consistent with the proposed new acquisition strategy—especially given that, among other things, the new strategy posits a somewhat slower approach to modernization. Historically, however, R&D costs have grown substantially from one generation to the next. Although R&D funding has not enjoyed continuous growth over the past 30 years, the overall trend has been unmistakably and consistently upward; on average, between FY 1951 and FY 1991, R&D funding grew by over 6 per cent a year. Given this trend in R&D funding requirements, it may well be that the revised FYDP includes a substantial mismatch between projected new weapon development and projected R&D funding levels. If projected budgets are too low to allow full prototyping, testing and evaluation of a broad range of systems, then clearly DoD will be limited in its ability to quickly move new weapon systems into production should the need arise.

Finally, the revised FYDP may not be entirely consistent with the goal of retaining the long-term technological superiority of the US military. The most innovative technological developments generally emerge from DoD technology base programmes. This level of R&D, not tied to the procurement of any particular system, currently receives only about 10 per cent of total R&D funding, down significantly from 1980 when it accounted for 17 per cent of the R&D budget. Given the potential technological gains from such spending, a substantial increase in technology base funding would be consistent with the new acquisition strategy, but funding for technology base programmes is projected to increase by only 1 per cent between FY 1992 and FY 1993 (the last year of the FYDP for which the DoD has provided a detailed breakdown of R&D funding).

Little room for manœuvre in other DoD accounts

If procurement and R&D accounts are indeed under-funded in the FYDP, carrying out the Administration's modernization plans on schedule and within the overall defence spending levels included in the FYDP would require that funds be shifted out of other DoD accounts—namely, military personnel and operations and maintenance.[20] However, the plans/funding fit may be no less tight in these accounts than it is for procurement or R&D. Evidence suggests that the military personnel and operations and maintenance accounts are being reduced about as deeply and rapidly as they can be without requiring deeper force structure cuts and lower readiness goals than currently projected. Even a modest shift of funding out of these accounts into procurement or R&D could necessitate reductions in the US force structure and readiness.

[20] The three remaining titles in the DoD budget—military construction, family housing and other—typically account for under 5% of the total DoD budget.

If the Administration's proposed Base Force is to be adequately manned, it is doubtful that the military personnel level can be cut below the level projected for FY 1997. Nor is it likely that substantial additional savings could be achieved by accelerating the pace of the planned draw-down. The current plan is already relatively fast-paced: 318 000 active military personnel are to be cut from the force structure over the next three years.

It would also be difficult to find substantial additional savings in operations and maintenance (O&M) accounts without undercutting the readiness of US forces. Current Administration plans call for cutting O&M funding essentially by the same proportion and at the same rate as the force structure. The number of active-duty military personnel is projected to fall by about 21 per cent between FY 1990 and FY 1997.[21] Over that same period O&M funding is projected to drop by 20 per cent. However, not all areas within O&M are likely to fall at the same rate. The O&M accounts are in many ways 'black boxes': some areas of O&M, including training, exercises, depot maintenance and logistical planning, have a clear direct link to the readiness of armed forces for combat, while other areas, such as administration, base operations support and supply operations, have a more indirect impact. Unfortunately, experience suggests that it is easier to make reductions in direct readiness funding than indirect and overhead O&M; it took some seven years for indirect and overhead support costs to adjust to the force structure cuts made by the US military after the Viet Nam War.[22]

Notwithstanding this experience, the Administration is projecting that its proportional and relatively face-paced O&M reductions will not significantly reduce the readiness of US combat units over the next five years. It is counting on management initiatives, specifically detailed in the Secretary of Defense's 1989 Defense Management Report (DMR), to substantially reduce indirect and overhead O&M costs. The goals of the DMR effort—for example, to consolidate supply depots in the United States, reduce the size of the depot maintenance infrastructure and streamline the acquisition work-force and process—are laudatory and could make an important contribution to reducing the DoD overhead, particularly by leading to deeper reductions than those currently projected in civilian DoD personnel, who account for a major portion of indirect O&M costs. Unfortunately, actual DMR savings may fall well short of DoD estimates. The US General Accounting Office (GAO) has pointed out that the DoD estimates were based primarily on management judgements rather than empirical cost data and thus may be overstated.[23] Currently projected O&M funding levels may also prove insufficient if next-generation weapon systems

[21] The FY 1997 active military personnel level will be 25% below the force level in FY 1987, when the size of the US military was at its 1980s peak.

[22] See Statement of Robert F. Hale, Assistant Director, National Security Division, Congressional Budget Office, before the House Armed Services Committee, 19 Mar. 1991, p. 17.

[23] General Accounting Office (GAO), *Observations on the Future Years Defense Program*, Apr. 1991, pp. 3–4.

turn out to cost substantially more to operate and support than their predecessors—as they often have in the past.[24]

It may be that the current Defense Department FYDP contains sufficient O&M funding to maintain today's levels of readiness over the next five years; the proposed DMR savings may materialize as projected; and the next generation of military hardware may cost no more to operate and support than current-generation systems. However, the margin for error appears to be very small. For this reason, it is unclear whether even modest additional reductions in O&M could be made over the next five years without consequences for levels of readiness. Recent DoD testimony seems to support the notion that O&M funding is being cut at as deep and rapid a pace as it can be without affecting readiness and specifically emphasizes the problems associated with eliminating indirect and overhead areas of O&M. Major General Robert F. Swarts, Deputy Assistant Secretary of the Air Force (Budget), testified in March 1992 that 'reductions to direct mission O&M since FY 1990 have kept pace with the force structure drawdown, but cuts to essential system support infrastructure have run ahead of actual savings. . . . For example, flying hours have been reduced 26.5 per cent, while numbers of major bases and square footage have been reduced by only 12.0 and 5.5 per cent, respectively.'[25] Another indication that further O&M cuts would be risky is suggested by the fact that even under the current plan the Air Force's depot maintenance backlog is projected to grow through FY 1993.[26] According to Swarts, the currently planned glide path of O&M reductions will allow the Air Force to reach its 'target without bending the gear,' but 'if the rate of decline is increased, we'll break something'.[27]

Possible effects of deeper cuts

What makes the potential under-funding problem particularly troublesome is the fact that actual FY 1992–97 funding levels are almost certain to fall short of the levels projected in the FYDP. Many inside and outside Congress are calling for substantially deeper budget reductions. Senator Sam Nunn, chairman of the Senate Armed Services Committee, has proposed cutting the FYDP by an additional $30–$35 billion (current-year dollars). Representative Les Aspin, chairman of the House Armed Services Committee, has proposed a plan which would increase this reduction to $48 billion. Others have pushed for cuts ranging as high as an additional $350 billion below the current FYDP. How much below the Administration's projections the FYDP will in fact be cut is impossible to predict, but additional reductions are likely, given the strong

[24] See Congressional Budget Office (CBO), *Operations and Support Costs for the Department of Defense*, July 1988.

[25] See Statement of Major General Robert F. Swarts, Deputy Assistant Secretary (Budget), on Operations and Maintenance, Air Force, Mar. 1992, p. 6.

[26] Swarts (note 25), p. 14.

[27] Swarts (note 25), p. 10.

pressures in the United States to increase spending for some domestic programmes, reduce the federal deficit and cut taxes.

It is possible that the revised FYDP may adequately fund the procurement and R&D accounts and that the planned pace of modernization could be carried out without making additional offsetting reductions in force structure or readiness in order to pay for it, but the revised FYDP represents, at best, a very tight fit between projected plans and projected funding. If substantial additional reductions in funding for defence spending were made over the next six years, the Defense Department would clearly be forced to alter the FYDP in a way that would require at least some further cuts in force structure, reductions in readiness or a further slow-down of modernization plans.

If current Pentagon estimates of weapon acquisition costs turn out to be roughly on the mark and the additional cuts in overall defence funding levels are modest, the changes in DoD plans might not have to be dramatic. A relatively small additional reduction in the force structure, for example, might go far towards solving under-funding problems, since reductions in the force structure can lead to savings not only in military personnel costs, but in both O&M and procurement accounts as well. When a unit is removed from the force structure it not only eliminates manning requirements but also, over the long run at least, saves O&M and equipment costs.

A decision to make major reductions to the proposed Base Force might make it possible not only to solve the plans/funding mismatch in procurement and R&D but also to make substantial additional cuts in procurement funding. Over the long run, a cut in the force structure should lead to a roughly proportional cut in procurement requirements; cutting the number of Air Force tactical fighter wings by 20 per cent, for example, would reduce long-run average annual Air Force tactical fighter procurement requirements by roughly 20 per cent.[28] By comparison, a reduction in the force structure can lead to far deeper cuts in procurement requirements in the short to medium term.

If the planning goal is to maintain a force structure of a constant size and age, the number of weapon systems of a particular type which are procured each year must be kept essentially equal to the number of such weapons retired. However, as one reduces the size of the force structure, the immediate driver for new procurement—the need to replace systems reaching the end of their service lifetimes—can disappear entirely in the short to medium term. It may be some time before the more modern weapon systems in the remaining forces reach the end of their service lifetimes and need to be replaced.

The difference between the long-run and short- or medium-term effects of a decision to reduce the force structure can be illustrated with a simple (hypothetical) example. Assuming that a destroyer has an average service lifetime of 40 years, maintaining and regularly modernizing a fleet of 120 destroy-

[28] This is not necessarily to say that a 20% reduction in force structure requirements would result in a 20% reduction in equipment *costs*. Since substantial economies of scale could be lost, at least in the short term, higher unit procurement costs could result from such a decision.

ers would, over the long run, require that an average of 3 destroyers be procured each year. By comparison, a fleet one-third smaller consisting of 80 destroyers could be maintained over the long run by procuring an average of only two (one-third fewer) destroyers each year. However, a decision to move from a 120-ship fleet to an 80-ship fleet would allow a much deeper reduction in procurement over the short to medium term. Indeed, assuming that the 40 destroyers selected for elimination from the force structure were the oldest 40 destroyers in the fleet, it might be possible to meet force structure requirements over the following 13 years without constructing a single new destroyer.[29]

In reality, the level of procurement savings resulting from a decision to further reduce the size of the Base Force would almost certainly be less significant than might be inferred from the above illustration. For many programmes, current procurement plans and projected funding levels already reflect the disproportional effect that a reduction in the force structure can have on short-term procurement requirements. This is clearly true in the case of Air Force tactical fighter aircraft and Army tanks, for example, where the FYDP calls for gaps in procurement ranging from 2 years to perhaps as many as 15 years.[30] For programmes such as these, where little or no production is now taking place, it is difficult to see how a decision to make still further reductions in the force structure could translate into any additional procurement savings in the immediate future.

Another factor that might limit the potential for additional procurement savings over the short to medium term is related to concerns about the defence industrial base. Even if no new procurement is necessary to meet force structure requirements in the short or medium term, for certain types of weapon it may be desirable to fund a minimal level of procurement in order to preserve the industrial base for such production. In some instances, funding of modifications and upgrades of existing systems may help maintain a sufficient industrial base; in others it may be necessary to continue low-rate production. There is no consensus on how much and what type of procurement is required to ensure that the overall defence industrial base remains adequate. However, some preservation is likely to be required, and, due to overhead costs, low-rate production of the kind that might be justified to preserve the industrial base tends to result in very high unit costs, further reducing potential savings.

Since many procurement programmes have already been cut to minimum levels and some areas of the defence industrial base need to be preserved, the

[29] If an average of 3 ships a year were procured in the past, then it would take 13 years from the time the decision was made to reduce the fleet from 120 to 80 ships before the oldest destroyer in the new, smaller force structure would reach retirement age.

[30] The last F-16s (the only Air Force fighter currently in production) are to be authorized in FY 1993, and the first F-22s (the replacement for the F-15) are to be authorized in FY 1996. A gap of 8 years or more may separate procurement of the last F-16 and procurement of the first Multirole Fighter, its planned replacement. The Army plans to buy the last M-1 tanks in FY 1992. In late 1991 Army plans called for procuring the first Block III tank, beginning in FY 2001. However, under the revised FYDP development of the Block III tank has be deferred until after FY 1997, suggesting that procurement may not begin until late in the decade.

extent to which further procurement funding cuts could follow from a further reduction in the Base Force is limited. Clearly, however, a decision to further cut the force structure could result in *some* additional savings; certainly, in some areas of procurement, industrial base concerns could be adequately addressed with lower levels of funding than are currently projected.

Acquisition funding after FY 1997

If the Administration's current long-range force structure and hardware plans are carried out, there will need to be some growth—possibly substantial—in procurement funding after FY 1997. A December 1991 report by the US Congressional Budget Office (CBO) estimated that carrying out the planned modernization of the Administration's Base Force would require that the total annual DoD budget be increased by $21–67 billion above its projected FY 1997 level by the middle of the next decade.[31] According to the CBO, by the year 2010 the Army could require a procurement budget of $11–15 billion; the Navy (including the Marine Corps), $34–54 billion, and the Air Force, $27–42 billion. Assuming that DoD procurement requirements outside the services (primarily defence agencies' procurement) remained at recent levels of about $3 billion a year, the total DoD procurement budget in 2010 would be expected to be $75–114 billion.[32] That would be $8–47 billion more than the DoD—under the then current plan—was projecting to budget for procurement in FY 1997. These projected budget requirements are also far in excess of the $56 billion procurement budget included in the current Pentagon plan for FY 1997.

Despite the January 1992 announcement terminating or deferring 10 major next-generation weapon systems, there is little reason to believe that this requirement for increased procurement funding after FY 1997 will disappear. Although the decision to cancel development of the SICBM could lead to significant savings after FY 1997, when it was projected to enter production,[33] for the most part the savings due to the announced changes accrue through FY 1997 and not in the years beyond. For example, the Defense Department estimates that the decisions to terminate production of the B-2 bomber after 20 planes and to terminate production of the new Seawolf attack submarine after

[31] For consistency with the rest of this chapter, CBO estimates were converted from FY 1992 dollars to FY 1993 dollars. See CBO Staff Memorandum, *Fiscal Implications of the Administration's Base Force,* Dec. 1991, p. 2. For discussions of each service's modernization plans, see CBO Staff Memorandum, *The Costs of the Administration's Plan for the Army Through the Year 2000,* Dec. 1991; CBO Staff Memorandum, *The Costs of the Administration's Plan for the Navy Through the Year 2000,* Dec. 1991; and CBO Staff Memorandum, *The Costs of the Administration's Plan for the Air Force Through the Year 2000,* Dec. 1991.

[32] The CBO's high and low estimates were based on different assumptions about cost growth. Essentially, the low-range estimates assume that unit costs of planned new weapon systems do not grow above their estimated costs and in some cases actually cost less to procure than the systems they are replacing. By contrast, the high estimates assume that, consistent with past experience, the actual procurement costs of new weapon systems substantially exceed both their estimated costs and the costs of the systems they are replacing.

[33] The Administration had not fully committed itself to moving beyond the R&D phase with the SICBM. CBO's analysis, however, assumed that procurement of the SICBM would begin in FY 1998.

only one boat will result in net acquisition cost savings of $32 billion (current-year dollars) through FY 1997—the bulk of these being procurement savings. Procurement savings from the B-2 termination will be minimal after FY 1997, since only the last 9 of the planned 75 aircraft were scheduled for authorization after FY 1997. Savings from the Seawolf decision would be minimal after FY 1997 if, as currently planned, the Navy were to begin production of a new Centurion Class boat by the turn of the century. The decisions to defer R&D and procurement of the Army's new RAH-66 Comanche scout/attack helicopter, Block III tank and LOSAT system will only increase procurement funding requirements for these programmes after FY 1997.

The announced new acquisition strategy will need to be applied methodically and vigorously in order to substantially reduce the procurement funding requirements for the late 1990s and beyond. Even this action would not, by itself, eliminate the need for substantial increases eventually. At best, it might hold procurement funding requirements to the low end of the CBO estimates, which are based on optimistic assumptions about holding down R&D and procurement cost growth.[34]

Over the long term, however, it is unrealistic to assume that overall defence budgets and R&D and procurement budgets in particular can be kept on a downward course indefinitely. New weapon systems tend to cost more to develop and produce than the systems they are designed to replace. A new acquisition strategy which allows systems to be more fully developed and tested before they are put into production and relies more on upgrades and modification of existing weapon systems will lead to substantial savings, but it is unlikely to reverse the strong historical trend towards higher weapon costs. Ultimately, if the United States is going to maintain a relatively large force structure, high levels of readiness, and a reasonable pace of modernization, the defence budget will have to stabilize and may have to include at least a minimal annual rate of increase in funding for procurement and R&D.

Restructuring modernization plans

This is a time of considerable uncertainty for US national security and defence planning. There is disagreement and uncertainty over how large the post-cold war US military should be and what missions it should emphasize. There is also disagreement and uncertainty over how much the United States should spend and is likely to spend on defence through the mid-1990s and beyond. These uncertainties make it difficult to precisely predict what will happen to Defense Department plans and budgets for procurement and R&D through the current FYDP and beyond.

[34] In an earlier report the CBO estimated that, over the long run, even keeping the proposed Base Force equipped with *current-generation* systems would require an average of $71 billion (FY 1993 dollars) in procurement funding annually. See Statement of Robert F. Hale, Assistant Director, National Security Division, Congressional Budget Office, 19 Mar. 1991, p. 24.

The Administration's current FYDP calls for procurement funding to be cut substantially in FY 1992 and FY 1993, and to hover near its FY 1993 level through FY 1997. R&D funding is projected to remain essentially flat through FY 1994 and to decline modestly over the last three years of the FYDP. Under the Administration's plan, by FY 1997 procurement and R&D funding would be, respectively, 22 and 18 per cent below their FY 1991 levels. There is reason to believe that there is a mismatch between the Administration's modernization plans and the projected funding levels included in the plan. This under-funding problem will be made doubly difficult to manage if, as seems likely, actual funding levels for defence through FY 1997 fall short of the Administration's projections. Moreover, the Administration's long-term modernization plan would seem to require that there be at least some growth in defence spending in the late 1990s and past the turn of the century. While such growth is possible, it may be dangerous to plan today based on that prospect.

The Pentagon needs to restructure its modernization plan to ensure that it is flexible enough to adjust appropriately to a range of possible threats, levels of defence spending and next-generation weapon costs. A flexible post-cold war acquisition strategy must involve three basic elements: (*a*) slowing the rate at which next-generation weapons are put into production; (*b*) maintaining high levels of R&D; and (*c*) continuing to modify and procure a significant number of current-generation weapons. The new acquisition strategy announced in January 1992 puts the Administration rhetorically on record in favour of the first two of these elements. Moreover, during the past several years the Administration has both cancelled and deferred the production of a relatively large number of next-generation systems and kept R&D funding levels relatively high. However, effectively carrying out this new approach may require that the Pentagon delay the acquisition of at least a few additional next-generation systems, such as the F-22, and reorient R&D funding away from SDI. The most critical weaknesses of the announced new acquisition strategy are its failure to live up to its rhetoric—at least to date—with regard to funding modifications of current-generation systems, its failure to include funding for the continued production of more than a handful of major current-generation systems and its potential failure to provide for an adequate level of R&D funding in the future.

3. The United States: unmanaged change in the defence industry

Judith Reppy

I. Introduction

The end of the cold war and the collapse of the Soviet Union have left the US military establishment without a major security threat to justify its weapon programmes and military budgets. This simple fact, coupled with the continuing pressure from the federal budget deficit, has set the stage for sizeable cuts in defence spending from the peak years of the Reagan Administration. In a major policy shift, the US Department of Defense (DoD) has proposed to change its arms procurement practices to fund most weapon programmes only through the prototype and testing stages, sharply reducing the number of new weapon systems carried forward into production. Thus, the private defence contractors that make up the bulk of the US defence industry face a shrinking market at home and a rupture in the historically close links between weapon development activities and arms production. In addition, they face greater competition abroad, as producers in other countries also seek to offset cuts in national defence budgets by expanding into foreign markets.

The varied responses of defence contractors to the changes in the international environment are transforming the US defence industry. Unlike earlier draw-downs in real defence spending—for example, in 1972–76 following the US withdrawal from Viet Nam—the cutbacks this time are expected to be both deep and long lasting, and defence contractors must plan their strategies accordingly. Larger contractors are teaming together to share the remaining contracts, exerting pressure on smaller sub-contractors, who are exiting the market in droves. Not only sub-contractors and vendors[1] but also some major contractors have left the market by selling off their defence units. Other contractors are taking advantage of the depressed market to buy up competitors, thus further reducing competition in their market niches. In this kaleidoscope of different responses, perhaps only one generalization is valid: there is very little interest among US defence contractors in converting their defence operations to civilian production. Instead, companies have chosen to become smaller, laying off thousands of workers and, in some cases, closing plants.

These changes in the defence industry are largely a reaction to changes in government policy, including defence spending policy. In turn, they pose fresh

[1] Vendors are companies that sell parts and services to prime contractors and sub-contractors. For a description of these terms, see section IV below.

problems to the US Government, which must worry about ways to maintain a viable defence industrial base and to encourage new technology for both military and civil purposes. The federal government has not concerned itself very much with another problem caused by the cutback in defence spending, namely, the hardships faced by the unemployed defence workers and their communities. This chapter describes the changes taking place in the US defence industry and points to some of the policy issues arising from these changes.

II. The historical context

More than at any other time since the 1950s, when new firms entered the defence market to develop and produce the new weapons of the missile age, the US defence industry today faces a market upheaval. In the 1950s the turmoil in the industry was attributable to the opportunities provided by rapid rates of technological change and the military's adoption of new kinds of weapon.[2] The idea that the US Government would be spending large amounts of money on the military when the country was not involved in an active war was still novel and took some time to settle in, but as weapon programmes multiplied, more companies were interested in entering the defence market. In 1962 Peck and Scherer listed 18 companies that entered as prime contractors for guided missile programmes between 1940 and 1960, most of them coming in after after 1945.[3] Other firms, already engaged in the manufacture of military aircraft, added missile divisions to their lines of business. Even higher rates of entry were noted for sub-contractors and vendors during this period.

Ironically, even as Peck and Scherer were completing their classic work, the tumultuous growth in the industry was coming to an end. In subsequent years, beginning in the 1960s, the US defence industry came to be characterized by considerable stability.[4] The rate of technological change did not slacken, although the areas showing the greatest changes shifted from aircraft to electronics and computer-related technologies, but the ability of prime contractors to internalize and routinize technological change grew, so that new technology did not mean new companies. As established companies forged ever-closer links to the military and built up research and development (R&D) complexes funded by the DoD and dedicated to weapon development, the barriers to entry for outside firms increased. Increasingly complex arms procurement regulations and specialized accounting requirements posed further barriers, while exit from

[2] Peck, M. J. and Scherer, F. M., *The Weapons Acquisition Process: An Economic Analysis* (Graduate School of Business Administration, Harvard University: Boston, 1962), chapter 7; Simenson, G. R., 'Missiles and creative destruction in the American aircraft industry, 1956–1961', ed. G. R. Simenson, *The History of the American Aircraft Industry: An Anthology* (MIT Press: Cambridge, Mass., 1968), pp. 233–41.

[3] See Peck and Scherer (note 2), pp. 192–93.

[4] Baldwin, W., *The Structure of the Defense Market, 1957–64* (Duke University Press: Durham, 1967); Reppy, J. V., 'The United States', eds N. Ball and M. Leitenberg, *The Structure of the Defense Industry: An International Survey* (Croom Helm: London, 1983), p. 27.

the industry was discouraged by the same factors, since firms specialized in the defence market culture have found it difficult to succeed in other markets. Thus, it is no surprise that the list of major defence contractors has varied little over the past 30 years, with most of the turnover coming from mergers of established contractors and joint ventures rather than new company entries or exits.

There has been a more cyclical pattern of company entry and exit at the sub-contractor and vendor level, particularly during turn-downs in defence spending such as occurred in the early 1970s after the USA withdrew from Viet Nam. In general, the sub-contractors and vendors neither have the close ties to the Pentagon that the major prime contractors enjoy nor do they suffer the disabilities of the prime contractors in diversifying into civilian markets. In some cases, however, these companies produce items that are essential to important weapon systems and are heavily dependent on defence contracts. The challenge of maintaining a defence industrial base at the sub-contractor level despite lower levels of defence spending is one of the difficult policy problems facing the US Government today.

Whereas the industry turnover of the early missile age was in response to increased opportunities in the defence market, the turnover of the 1990s is clearly a result of the decline in defence spending. A smaller industry, whether measured by number of contractors or the volume of resources devoted to arms production is evolving. However, it is still dominated by the institutions and practices that are the legacy of 40 years of cold war spending; whether these institutions and practices will persist despite the decline in the size of the industry or a new culture of defence procurement will emerge is an open question.

III. Defence industry employment

Throughout the cold war the primary justification for the high levels of US military spending was always the Soviet threat, but there was also a secondary argument that pointed to the jobs created by military spending and the benefits of military-funded technology for the civilian sector. The jobs argument has had political potency because of the interest of individual congressmen in maintaining employment in their district; now that the military threat has disappeared, the fear of job losses has become the major political barrier to deep cuts in defence spending.

Figure 3.1 displays defense industry employment over a 42-year period. The peaks associated with the Korean and Viet Nam Wars are evident, as well as the buildup during the Reagan Administration. Total defence industry employment has declined since its recent high point of 3.36 million in fiscal year (FY) 1987 to 2.75 million in 1992. Since defence-related employment follows from defence spending, future industry employment can be expected to be lower still. The effects will not fall equally on all states and localities, however. Military-related employment has been concentrated in a relatively few states, creating a natural lobby for continued spending among those most

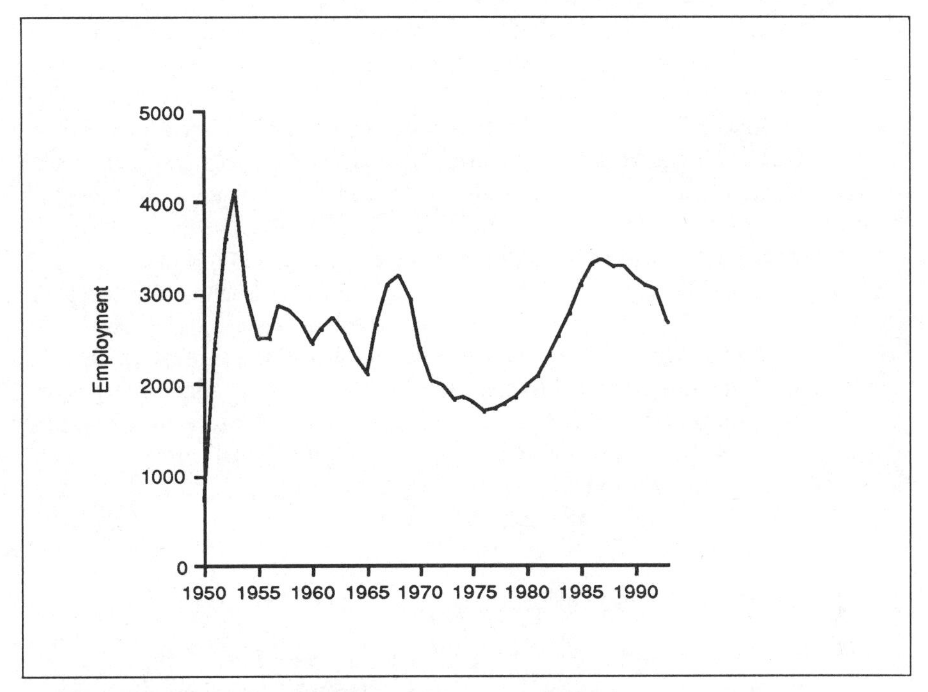

Figure 3.1. US military-related employment, FY 1950–93

Sources: *National Defense Budget Estimates for FY 1993* (US Department of Defense, Office of the Comptroller: Washington, DC, 1992), pp. 136–37; US Congress, OTA, *After the Cold War: Living with Lower Defense Spending*, OTA-ITE-524 (US Government Printing Office: Washington, DC, 1992), p. 137.

affected. In 1991 California led the country, with over 800 000 defence jobs—active military, civil service and defence-related industry—with Virginia (328 000 jobs) a distant second. Other states with relatively high levels of defence employment include Maryland, Massachusetts, Washington, Missouri and Connecticut.[5] These states will be seriously affected by the cuts in military spending; although their degree of dependence on defence spending varies, they all face considerable dislocations in their labour markets. Other states that will be affected because of their dependence on defence spending are Alaska, Hawaii and Maine.[6]

Job losses from cutbacks in military spending are more than a matter of lay-offs by defence contractors; active-duty military forces are to be reduced by at least 20 per cent between 1990 and 1995, and DoD civilian employees are being cut by about 10 per cent in the same period. Much of the draw-down will

[5] US Congress, OTA, *After the Cold War: Living with Lower Defense Spending*, OTA-ITE-524 (US Government Printing Office: Washington, DC, 1992), p. 158.

[6] Schmidt, C. S. and Kosiak, S., *Potential Impact of Defense Spending Reductions on the Defense Industrial Labor Force by State* (Defense Budget Project: Washington, DC, 1992), p. 9.

be accomplished by normal attrition, encouraged by special incentive programmes in the military, but there will inevitably be some involuntary layoffs. Whereas there is a range of services available for military and government employees who are laid off, those who lose their jobs in private industry are less well served.[7] Their situation is made more difficult by the continued high levels of unemployment in the general economy and, in particular, by the fact that manufacturing employment has declined in the USA so that alternative jobs equivalent to those disappearing in the defence economy are hard to come by.

US Government policy has left most of the adjustment in the private sector to market forces, arguing that the cuts in defence-related jobs are small relative to total US employment and well within the capacity of the economy to absorb. Thus, the actions of the private defence contractors will determine in large part how many employees are laid off and what alternatives they face. Job losses in the industry could be minimized if contractors were to expand into other markets and reallocate their existing work-force to new product lines. The companies, however, are following a different strategy.

IV. The current arms market structure

The defence industry in the United States is comprised of a heterogeneous set of private companies, large and small, more or less dependent on Department of Defense business, and manufacturing many different kinds of weapon systems and components.[8] They can be usefully divided into *prime contractors*, the companies that negotiate directly with the DoD and are responsible for the development and production of major weapon systems; *sub-contractors*, the companies that enter into contractual relationships with the prime contractors to provide sub-assemblies and components; and a third tier of *vendors*, the companies that sell parts and services to both prime contractors and sub-contractors. The industry is commonly identified with a small set of the largest prime contractors—the 25 leading companies, for example, regularly account for 50 per cent of the total value of DoD prime contracts—but the picture is really more complicated. Prime contractors act as sub-contractors on some projects, and some, highly specialized, sub-contractors and vendors may be more dependent on defence business than the prime contractors, even though they never make a direct sale to the DoD.

Given this range of circumstances, it is not surprising that individual companies have adopted widely varying strategies to cope with the changes in the defence market. These strategies can be classified under the headings of *stabilization strategies*, in which companies act to maintain or increase their market share, and *withdrawal strategies*, in which they exit from the defence market,

[7] US Congress, OTA (note 5), chapter 3.

[8] Standard descriptions of the defence industry can be found in Gansler, J. S., *The Defense Industry* (MIT Press: Cambridge, 1980); Reppy (note 4); McNaugher, T., *New Weapons Old Politics* (Brookings Institution: Washington DC, 1989); and Weida, W. J. and Gertcher, F. L., *The Political Economy of National Defense* (Westview Press: Boulder, Colo., 1987).

either through selling off assets or conversion of facilities.[9] Because of the variety of company responses, generalizations about changes in the industry—beyond the obvious one that the industry will shrink with the shrinking budget—are not possible. Instead it is necessary to look in more detail at the various segments of the defence market to try to discern the new structure that is emerging.

Prime contractors

The major US prime contractors are clustered in the aerospace and electronics industries, with the top companies concentrated in aerospace: all of the 10 leading prime contractors, accounting for 32 per cent of total defence procurement in FY 1991, are principally manufacturers of aircraft, aircraft engines and/or missiles. These companies, listed in table 3.1, have long-established ties to the Department of Defence and have adapted themselves to the specialized requirements for doing business in the defence market. Most of them are heavily dependent on DoD sales, as the figures in table 3.1 indicate, although for some that dependence is beginning to decrease with the fall in defence spending. Those companies for which DoD business is less important, such as General Electric and United Technologies, have divisions that specialize in defence production and that are typically kept separate from the rest of the company to make it easier to comply with DoD accounting and classification requirements. General Motors is on the list because of its purchase in 1985 of Hughes Aircraft, now GM Hughes Electronics, which produces missiles, electronics equipment and satellites for the Department of Defense.

Even before the current down-turn in DoD procurement there was considerable excess capacity in the aerospace sector of the defence industry, but the extent of the problem was masked by the high levels of defence spending in the 1980s.[10] Now, however, there is clearly not enough business to occupy all of the defence aircraft production lines and it is inevitable that the major producers of military aircraft will shrink in size, at least on the defence side of their business. It is even likely that one or more of the producers may exit the business entirely; a Rand Corporation official is quoted as saying that the number of aerospace divisions acting as prime contractors could drop from 10 to 5 or fewer in the next three to five years.[11]

[9] Anthony, I. and Wulf, H., 'The economics of the West European arms industry', in eds M. Brzoska and P. Lock, SIPRI, *Restructuring of Arms Production in Western Europe* (Oxford University Press: Oxford, 1992), pp. 24–29.

[10] See Gansler (note 8), p. 170; and US Congress, Office of Technology Assessment (OTA), *Redesigning Defense: Planning the Transition to the Future US Defense Industrial Base*, OTA-ISC-500 (US Government Printing Office: Washington, DC, 1991), p. 65.

[11] Smith, B. D., 'Airframe building capability loss looms for full-service defense contractors', *Aviation Week & Space Technology*, vol. 136, no. 11 (16 Mar. 1992), pp. 41–42.

Table 3.1 Defence dependence of the 10 leading US defence prime contractors, FY 1991

Figures are percentages.

Rank	Company	Arms sales as share of total sales 1988	1989	1990
1	McDonnell Douglas	60	62	55
2	General Dynamics	84	84	82
3	General Electric	12	11	11
4	General Motors	6	6	6
	(GM Hughes Electronics	61	56	57)
5	Raytheon	67	61	57
6	Northrop	90	90	86
7	United Technologies	25	21	19
8	Martin Marietta	75	75	75
9	Lockheed	79	75	75
10	Grumman	82	80	72

Source: '100 companies receiving the largest dollar volume of prime contract awards, fiscal year 1991' (US Department of Defense: Washington, DC, 1992); SIPRI data base.

The major contractors have responded to the drop in procurement orders by strenuously lobbying for the defence programmes in which they are involved and by insisting on relaxation of some of the more stringent procurement regulations introduced in the 1980s: for example, the use of fixed-price contracts[12] for development work. They are also reducing costs by laying off workers and cutting back investment and other discretionary expenses wherever possible, with the ironic result that a number of companies have improved their financial positions even though their defence sales have decreased.[13] Despite a rhetorical commitment to maintaining R&D activities at a high level, there is evidence that individual companies are cutting their non-contract R&D programmes.[14] The trend towards teaming together for large programmes continues, as, for example, in the competition to develop the next tactical fighter aircraft for the Air Force, in which General Dynamics, Lockheed and Boeing joined together in one team and McDonnell Douglas and Northrop formed another.

Virtually all the major contractors are paying more attention to the possibility of increased foreign sales, particularly in the lucrative category of commercial sales.[15] US contractors are joining European-based teams in order to compete for NATO programmes, although they have not shown the same interest in

[12] In fixed-price contracts the price is set during contract negotiations, so that the contractor bears the risk of cost overruns, whereas in cost-plus contracts the government pays the contractor's costs plus a fixed fee.

[13] Velocci, A. L., Jr., 'Greater defense firm discipline seen as key to weathering budget storm', *Aviation Week & Space Technology*, vol. 136, no. 11 (16 Mar. 1992), pp. 43–44.

[14] Finnegan, P., 'Companies sacrifice r&d to survive cuts', *Defence News*, 2 July 1990, pp. 1, 20.

[15] For a detailed analysis, see Anthony, I., 'The United States: arms exports and implications for arms production', chapter 4 in this volume.

acquiring European companies as Europeans have in buying up US companies—a difference that can be explained by the greater attraction of the large US market.[16] It should be noted, however, that increased foreign sales are not possible for every company, since the number of potential suppliers has grown, military spending is falling in most countries of the world, and even an increased market share may not translate into higher sales in absolute terms.

One strategy that could sustain company sales and employment—namely, conversion of defence production to civilian goods and services—has not seemed attractive to defence contractors. Many of these same contractors tried unsuccessfully to enter civil markets in the aftermath of the Korean or Viet Nam Wars: Grumman's unhappy experience with manufacturing buses and Boeing-Vertol's failure with light railway cars are just two examples of the pitfalls that lie waiting in commercial markets for companies whose corporate culture has been shaped by the defence market, with its emphasis on high performance over cost and reliability and its absence of marketing and service networks.[17] However, a number of prime contractors have pursued a kind of conversion by increasing their non-defence business with the federal government, especially the National Aeronautical and Space Administration (NASA).[18] This strategy allows the companies to capitalize on their technological strengths without changing the way in which they do business; that is, they are still selling high-technology products and services, customized to a particular buyer's requirements in a market governed by extensive procurement regulations.

The dominant strategy for the defence contractors has been to reduce costs by laying off workers. Over 100 000 jobs were lost in the aerospace industry in 1991, and the defence industry as a whole is expected to lose perhaps 1.4 million jobs by 1995.[19] In part this is a reaction to the large expansion of the industry in the 1980s, but the current draw-down is not expected to be followed by a cyclical up-turn. By reducing employment through permanent layoffs, companies are able to economize on the costs of health insurance and pension plans, which represent an increasing share of total labour costs. In the absence of successful strategies for conversion to civilian production, defence companies will be permanently smaller and their former employees will be forced to find jobs in other industries.

Because individual companies have selected different combinations from the set of strategies, the outcomes differ considerably. The defence-dependent contractors are more vulnerable than the more diversified companies and less able to contemplate replacing lost defence contracts with civilian sales. Instead, they

[16] Reppy, J. V. and Gummett, P., 'Changing military industrial structures: implication for national technology capabilities', paper presented at the International Political Science Association, Buenos Aires, 21–25 July 1991; MacPhail, M. 'Europe role awaits US defense firms', *Defense News*, 24 Sep. 1990, p. 33.

[17] US Congress, OTA (note 5), pp. 206–14.

[18] US Congress, OTA (note 5), p. 203.

[19] Velocci, A. L., Jr., 'Aerospace jobs decline as firms seek new efficiencies during down-turn', *Aviation Week & Space Technology*, vol. 136, no. 3 (20 Jan. 1992), pp. 57–58; US Congress, OTA (note 5), p. 59.

are trying to position themselves to compete effectively in the smaller defence market. General Dynamics, which is in the unique position of being a prime contractor across the board for aircraft, missiles, tanks and ships, has adopted a strategy of concentrating on its defence business. This involves becoming a smaller company: employment peaked in 1987 at over 105 000, but the company expects to employ only 63 000 by the end of 1994; and some of its operations—for example, the Electric Boat Yard at Groton, Connecticut, and the tank factory at Detroit—are expected to shut down.[20] Spending on investment and R&D has been cut to less than half the levels of the late 1980s.[21] As part of its strategy of concentrating on its core defence businesses, General Dynamics has sold its missile division to Hughes rather than continue to compete in a segment of the market where there are numerous competitors. It has also sold Cessna Aircraft, which it had purchased in 1986 in an attempt to diversify into the civilian market for small planes, and it has put its other civilian divisions up for sale.[22]

Other US firms following a strategy of concentrating on the defence market include Northrop, Grumman, Martin Marietta and Loral. Whereas Northrop and Grumman are shrinking as their major programmes are completed or cancelled, Loral has pursued an aggressive campaign of defence acquisitions while divesting itself of non-electronics businesses, thereby increasing both its defence dependence and its share of the defence electronics market. In 1989 it acquired Honeywell's Electro-Optics Division and in 1990 it purchased Ford Aerospace; subsequently it sold the BDM division of Ford Aerospace and spun off the satellite business to a joint venture with a European consortium.[23] In these acquisitions it was taking advantage of the weak market and the desire of the sellers to get out of the defence industry: Honeywell and Ford are no longer defence contractors.

Martin Marietta has also been busy acquiring former competitors as it seeks to maintain a balanced portfolio of defence contracts. Its CEO, Norman Augustine, is quoted as saying, 'There may be some opportunities to essentially buy up the defense industry . . . There probably will be a few companies that do that and they will be the survivors.'[24] In a move outside its main areas of specialization in defence electronics and space launch vehicles, Martin Marietta teamed up with Lockheed to bid for the aerospace and missile divisions of the LTV Corporation. They were, however, out-bid by a combined offer from the French electronics company, Thomson-CSF, and the Carlyle Group of Washington, and congressional opposition to a sale to foreign interests forced Thomson to drop out of the competition. The final sale was to a combined bid from Loral Corporation, the Carlyle Group and Northrop Corporation, with

[20] US Congress, OTA (note 5), p. 199.

[21] Stevenson, R. W., 'Dynamics set to trim 27,000 jobs', *New York Times*, 2 May 1991, pp. D1–D2.

[22] Dickson, M., 'Slimming to fit a shrinking business', *Financial Times*, 20 May 1992, p. 20.

[23] Tracey, P., 'Loral's CEO hitches his fortune to firm', *Defense News*, 16 July 1990, p. 38–39; Betts, P., 'Loral goes on the attack in defence', *Financial Times*, 12 Mar. 1992, p. 20.

[24] Finnegan, P., 'Martin Marietta considers acquisition of defense firms', *Defense News*, 2 Dec. 1991, p. 32.

Loral taking the missile business and Carlyle and Northrop sharing
business.[25]

Lockheed is an example of a major defence contractor that is f
strategy of increasing the non-defence side of its business, largely through join
ventures overseas and contracts with non-defence government agencies. It has expanded its maintenance business that supplies services to commercial jet airliner fleets.[26] As one of the successful team to develop the Air Force's advanced tactical fighter, the F-22, it is assured of defence contracts for the next several years. However, production levels for the F-22 are in doubt, and whatever happens with that project, Lockheed will be a much smaller company: in 1989-90 it laid off over 8000 workers at its Marietta, Georgia, plant, and in 1990 it closed its Burbank, California, manufacturing facility, laying off an additional 9500 employees.[27]

Within the more diversified companies, mixed strategies are possible. General Electric is an example. GE's aircraft engine division already serves both civilian and military customers, so a shift to emphasizing civilian business requires little adjustment in company organization and practices. Lower military sales with commercial demand holding steady necessarily implies a shift in the customer mix. GE Aerospace, by contrast, is pursuing a strategy similar to that of General Dynamics: stick to defence and shrink. GE Aerospace reports that it has halved the number of R&D projects that it supports with company funds and hopes to stay in business by pursuing a large number of smaller projects.[28]

Raytheon, another diversified company, with civilian interests ranging from home appliances to Lexington Books, is planning to reduce its reliance on the military market still further.[29] Raytheon will not exit the market; it has a strong position in defence electronics and missiles, and it expects to increase its sales overseas in the next five years (it is the prime contractor for the Patriot missile, which has a bright future after the 1991 Persian Gulf War). However, its total defence business is shrinking, and like other contractors it has been laying off employees and closing defence plants.

McDonnell Douglas has, like Boeing Corporation, been able to compensate in part for declining defence sales with sales of commercial jet airliners manufactured by its Douglas division, although it has been forced to postpone development of a new jumbo jet because of lack of orders.[30] McDonnell has been the top defence contractor for a number of years on the strength of its contracts for a range of aircraft and missile systems, but it has lost out in three recent Pen-

[25] Hayes, T. C., 'Two offers for LTV unit submitted', *New York Times*, 4 Aug. 1992; Velocci, A. L., 'Market focus', *Aviation Week & Space Technology*, vol. 137, no. 10 (7 Sep. 1992), p. 27.

[26] US Congress, OTA (note 5), pp. 201–202; Polsky, D., 'With growth outside defense, Lockheed will hold its own in '91', *Defense News*, 17 Dec. 1990, p. 30.

[27] US Congress, OTA (note 5), pp. 76, 104.

[28] See the interview with John Rittenhouse, Senior Vice President, General Electric Aerospace, 'One on one', *Defence News*, 3 Feb. 1992, p. 38.

[29] Stevenson, R. W., 'Raytheon is looking beyond the Pentagon', *New York Times*, 20 Feb. 1992, p. D1.

[30] 'McDonnell jettisons the jumbo', *New York Times*, 21 June 1992, Section 3, p. 2.

tagon decisions—namely, the cancellation of the Navy A-12 attack plane, with which it was teamed with General Dynamics; the Army's light helicopter programme; and the competition for the Air Force advanced tactical fighter programme, in which it and Northrop made up the losing team. In addition, the company has faced serious cash flow problems that led it to seek relief from the Pentagon.[31] It has responded with large-scale layoffs on both the military and civil side of its business; it leads the aerospace industry in the number of jobs lost.[32]

This review of the highest-ranked prime contractors confirms the diversity of strategies adopted in response to the declining defence budget. For some companies, exit from the market has seemed the best way to protect their shareholders' interests, and, as noted above, defence contractors seeking to consolidate their position in particular defence markets—for example, Martin Marietta and Hughes—have been interested buyers. However, the sell-off of defence divisions to other companies has been impeded in some cases by the poor prospects for the defence industry, so that a number of firms have found no buyers for their defence units. Instead, they have spun them off into free-standing companies, with the shares of the newly created company distributed to shareholders of the parent company on a pro rata basis. Honeywell sold part of its defence business to Loral Corporation, but it was driven to spin off other divisions to form Alliant Techsystems, after it failed to receive acceptable bids for them. Similarly, Emerson Electric also sold off its defence units in 1990.[33] The result is an increase in the set of defence contractors heavily dependent on DoD contracts—Alliant ranked number 25 in DoD prime contracts in FY 1991—and a reduction in the group of diversified companies among the prime contractors.

The net effect of these changes on the structure of the defence industry is a smaller number of contractors, many of which have themselves shrunk in size since the late 1980s. Some of these companies are more dependent on the defence market than they were in the past because of a conscious decision to concentrate on their defence business, while others have shifted their mix of business towards commercial markets, if only because the defence side of their business has declined. The various buy-outs and spin-offs have created further incentives for shrinkage, as the new owners seek to eliminate duplication and consolidate production lines.

[31] Morrocco, J. D., 'McDonnell Douglas sought $1 billion advance from US to ease cash crunch', *Aviation Week & Space Technology*, vol. 135, no. 5 (29 July 1991), p. 23.

[32] 'Production cutbacks, program losses bring McDonnell Douglas to forefront in job cuts', *Aviation Week & Space Technology*, vol. 136, no. 3 (20 Jan. 1992), p. 58.

[33] Tracey, P. 'Many defense firms shun buyer's market, opt to spin off units', *Defense News*, 27 Aug. 1990, p. 25.

Sub-contractors and third-tier vendors

Sub-contractors and vendors constitute an important part of the defence industry, both in terms of numbers of companies involved and in terms of the importance of their contribution to the final product. Major prime contractors frequently serve as sub-contractors to other prime contractors—a phenomenon that is probably increasing with the increased prevalence of prime contractor teaming together on large programmes in order to share the available business—and companies that are usually thought of as sub-contractors, such as Loral Corporation and Alliant Technosystems, also appear on the list of the 100 leading prime contractors. In addition to these major players, there are literally thousands of other companies that are part of the defence industrial network. Surveys of the sub-contracting networks of major defence programmes regularly report hundreds of companies involved at the sub-contracting level and hundreds more at the third tier.[34]

These companies, especially the suppliers that are specialized in defence business, face even harder times than the prime contractors when defence spending falls. As the number of new defence programmes shrinks, major contractors tend to bring in-house business that they would previously have contracted out; their 'make-or-buy' decisions shift to 'make' in order to soak up some of their excess capacity. In addition, the prime contractors are now competing in the markets for defence maintenance and service contracts that in the past were left to smaller companies. Moreover, sub-contractors and vendors do not have the bargaining power *vis-à-vis* the prime contractors that the prime contractors have enjoyed with the Pentagon; as a consequence, they are unlikely to gain relief from onerous contract provisions even when the prime contractors succeed in wresting concessions from the Defense Department.

As in the case of the prime contractors, however, there is a range of company situations and strategies. Many of the specialized firms may be forced out of business, but other companies with a diversified customer base will be able to survive. They may lack the financial resources to diversify through acquisition, but they do not face the same cultural barriers to conversion that the large prime contractors do. Unlike the prime contractors they are actively pursuing new civilian business. Their difficulty lies in the short term, in expanding the commercial side of their business in the face of a serious recession.[35] The long-term fate of this sector of the defence market will depend heavily on whether the government succeeds in shifting defence procurement towards dual-use technology and the use of commercial components, in place of the current requirement that parts and components meet specialized military specifications. To the extent that military demand becomes less specialized, the DoD will be like any other customer and there will be less need for an extensive defence sub-contractor base.

[34] For example, US Congress, OTA (note 10), p. 42.
[35] US Congress, OTA (note 5), pp. 217–23.

V. Government policy responses

The ongoing changes in the structure of the defence industry and in the strategies of the individual defence contractors pose serious problems for the US Government. There is an inherent conflict between the government's desire to spend less on arms procurement and its desire to maintain a healthy defence industrial base as a source for current requirements and a hedge against possible future needs. Moreover, the shrinking defence budget threatens to erode government support for the civil technology base, since much of the funding for such technology, for example, advanced manufacturing processes for semiconductors, has been funneled through the Department of Defense. Recommendations to rely less on dedicated domestic defence companies leave open the risk of increased reliance on foreign suppliers.

In principle, the US Government could have a policy of encouraging conversion of defence production and maintaining employment, at least for the technically skilled work-force, through support of civilian technology projects accompanied by protection of a core defence industrial base. In practice, the government's response to its dilemma has been predictably schizoid, with moves to relax some procurement regulations and to apply *ad hoc* measures to help certain defence companies, even while proclaiming a faith in the ability of free market forces to guide the process of shrinking the defence base.[36] Rhetorical calls for increased reliance on civilian technology to supplement the defence technology base have been accompanied by cuts in DoD funding for dual-use technologies.[37] Federal support for industry conversion to civilian production has been limited to modest programmes for counselling and retraining displaced workers, programmes that are generally viewed as inadequate for the size of the problem.[38] These contradictory policies are a reflection of conflicting policy goals as well as different beliefs about the proper role for government in the economy.

From an analytical point of view, defence procurement can be viewed as a principal/agent problem. In the US system, which relies on privately owned companies to develop and produce weapons, the government must motivate its agents, the defence contractors, to develop and produce—in a timely manner and at a reasonable cost—the weapons that the government wants and to remain available for future needs. Given the non-competitive structure of the market, the government has to rely on procurement regulations and the incentives incorporated into defence contracts rather than on market competition to achieve its ends. In general, fixed-price contracts provide an incentive for contractors to minimize costs, but not to pursue high-risk technological developments, whereas cost-plus contracts have the opposite effect.[39]

[36] Baker, C., 'Officials label industrial base study short-sighted', *Defense News*, 2 Dec. 1991, p. 1.
[37] Leopold, G., 'Congress may restore some proposed cuts', *Defense News*, 3 Feb. 1992, p. 28.
[38] See US Congress, OTA (note 5), chapter 3.
[39] See note 12.

The relative bargaining power of the government and its defence contractors shifts over time, both over the lifetime of a programme and with broader political cycles. Thus, during the design competition phase of a programme the government can require that companies invest their own money in R&D and production facilities; the companies will, in general, be willing to do so, knowing that in the production phase of the programme, when other contractors have been dropped from the competition, the government will be dependent on them to produce a needed weapon, and they will be in a position to recover their costs. During repeated waves of acquisition reform, different policies with respect to contract type and procurement regulation have risen and fallen in popularity, but the basic pattern of defence contractors recovering their investments in the production phase of the cycle has been a constant factor.[40]

This cycle of behaviour was particularly clear during the second half of the 1980s, when, in response to congressional concerns about the scandals that marked the early years of the Reagan defence buildup, the Department of Defense instituted tougher rules governing types of contract, contractor investment in plant and equipment, and progress payments. The government insisted on fixed-price contracts, even for development work; on wider use of competitive procurement practices, including second sources[41] for production contracts; on lower rates for progress payments; and on contractor investment in new production facilities.[42] In the atmosphere engendered by record high defence budgets and the expectation that the new weapons under development would be produced in quantity, contractors were willing to agree to these more onerous contract provisions. When defence spending turned downwards, however, many of them were forced to take losses on programmes that were cut back or cancelled, and as the cuts grew deeper the contractors successfully lobbied to have the procurement regulations relaxed.[43] Since 1989 the government has quietly dropped requirements for second sources for production contracts, raised the rate on progress payments, and removed restrictions on independent R&D programmes. Under pressure from contractors and the Council on Competitiveness, the DoD has cancelled its policy of extending recoupment of R&D costs to cover commercial spin-offs as well as foreign military sales.[44]

In addition to generalized relief through relaxation of some procurement practices, the Pentagon has the power to rescue firms that are in particular trouble. There is some evidence of the follow-on imperative at work in the award of the advanced tactical fighter to the Lockheed team; without this contract Lockheed would have been without a major military aircraft programme.[45]

[40] McNaugher (note 8).

[41] Second sourcing refers to the practice of qualifying a second contractor, in addition to the company that won the original contract, to compete for follow-on production contracts.

[42] Stevenson, R. W., 'New risks in military deals', *New York Times*, 24 Feb. 1987, pp. D1, D7.

[43] Scott, W. B., 'Stunned defense industry scrambles to recoup in era of shrinking budgets', *Aviation Week & Space Technology*, vol. 134, no. 11 (18 Mar. 1991), pp. 58–60.

[44] Baker, C., 'Pentagon halts r&d recoupment policy', *Defense News*, 20 Jan. 1992, pp. 4, 28.

[45] Kurth, J., 'Aerospace production lines and American defense spending', ed. S. Rosen, *Testing the Theory of the Military–Industrial Complex* (Lexington Books: Lexington, Mass., 1973), pp. 135–56; Shoop, T., 'The top 20 government contractors', *Government Executive*, vol. 23, no. 8 (Aug. 1991), p. 63.

McDonnell Douglas apparently benefited from unusual payments that amounted to a cash advance from the Pentagon when it faced a cash flow crisis in late 1990, and both McDonnell Douglas and General Dynamics were allowed to defer payments that they owed in connection with the cancellation of the A-12 attack aircraft in December 1990.[46] These actions can be justified in terms of maintaining the contractor base; they also respond to pressures from members of Congress to avoid plant closings in their districts. However, they are evidence of the ambivalence surrounding defence acquisition policy, since they run directly counter to the goal of reducing excess capacity in the industry.

Still more conflicting signals are embedded in the policy changes announced by Secretary of Defense Dick Cheney in January 1992.[47] The DoD intends to shift to a strategy of developing new military technology to the prototype stage without quantity production. Devised as a way of maintaining the technology base in an era of shrinking budgets, the new policy, if implemented, will have the effect of altering the basic structure of incentives that has driven weapon development and acquisition since the early 1950s. Not surprisingly, the contractors have attacked the plan, maintaining that without production contracts the industry, including the sub-contractors, will lose the skilled labour and capacity necessary to maintain leadership in world aerospace markets and respond to future defence needs.[48] From the government's perspective the new policy is a way of shaping the inevitable decline in the size of the industry by favouring R&D activities, while eliminating expensive excess capacity. Deprived of the prospect of profits in the production phase, however, it is not clear that defence contractors will choose to remain in the industry, so the expected benefits for the technology base could prove elusive.

Even while it has announced its plans to maintain the defence technology base and to increase its reliance on civilian technology wherever possible, the DoD has cut its spending for dual-use technology.[49] During the 1980s a number of technology programmes were initiated by the government and placed in the DoD in order to avoid conflict with the prevailing free-market ideology, which ruled out government assistance to specific sectors or industries except under the rubric of national security. Such dual-use technologies as the design and manufacture of semiconductors, high-definition television, and computer software development were all funded through the DoD budget, primarily in the Defense Advanced Projects Research Agency (DARPA).[50] With the cuts in

[46] Morrocco, J. D., 'Pentagon officials linked with plan to bolster McDonnell Douglas cash flow', *Aviation Week & Space Technology*, vol. 13, no. 13 (30 Mar. 1992), p. 26; Finnegan, P., 'Pentagon allows A-12 firms to delay repayment', *Defense News*, 11 Feb. 1991, p. 3.

[47] Schmitt, E., 'Military proposes to end production of most new arms', *New York Times*, 24 Jan. 1992, pp. A1, A15.

[48] Brown, D. A., 'Defense base erosion could hurt military aircraft exports', *Aviation Week & Space Technology*, vol. 136, no. 9 (2 Mar. 1992), p. 66; Silverberg, D., 'Acquisition rule irks industry', *Defense News*, 10 Feb. 1992, p. 10. See US Congress, OTA (note 10) for a full discussion of the proposed acquisition strategy.

[49] Holzer, R. and Leopold, G., 'Pentagon to curtail dual-use research', *Defense News*, 10 Feb. 1992, p. 4.

[50] Broad, W. J., 'Pentagon wizards of technology eye wider civilian role', *New York Times*, 22 Oct. 1991, pp. C1, C11.

defence spending, however, these programmes have become vulnerable because they are not viewed as essential to the military mission. Congressional efforts to push the development of civilian technology have repeatedly foundered on the shoals of executive branch resistance, and the funding available for the small programmes that have been put in place is offset by the cuts in DoD spending for dual-use technology. Lists of 'critical' technologies have been prepared, but there is no hint of an effective programme to co-ordinate government support for those technologies.[51]

VI. Conclusions

The reduction in US defence spending is leading inevitably to a reduction in the large US defence industry, which was sustained for many years by cold war budgets.

Individual defence contractors are reacting to the changes in their business environment with a variety of strategies, but all of them point towards a smaller, more concentrated industry. Despite its strong interest in the future of the defence industrial base, the US Government's ability to 'manage' the transformation of the defence industry is limited by its reluctance to intervene directly in the market. Instead, it has adopted a mixed set of policies that sends conflicting signals to the industry and is unlikely to achieve the goals of eliminating excess capacity while maintaining a balanced defence industrial base.

The future defence industry may, then, look much like a scaled-down version of the past, with a few large defence prime contractors dependent on the DoD for most of their sales and locked into a tight network of regulation and special arrangements. Alternatively, the market may evolve towards a more open structure, as called for by those who believe that the best procurement policy for the DoD is greater reliance on the civilian sector.

In a time of rapid change and uncertainty in the international situation, it is not surprising that the future direction of US Government policy for the defence industry is also uncertain. The conflicting voices reflect different judgements about what the future holds, as well as the different interests at stake. Similarly, the varied reactions of the private defence contractors can be read as a set of bets on the course of future events. In the absence of a new major military threat, it is likely that the current confused situation will continue for a long time to come.

[51] See, for example, US Office of Science and Technology Policy (OSTP), *Report of the National Critical Technologies Panel* (OSTP: Washington, DC, 1991); US General Accounting Office (GAO), *Defense Industrial Base: Industry's Investment in the Critical Technologies*, GAO/NSIAD-92-4 (GAO: Washington, DC, Jan. 1992).

4. The United States: arms exports and implications for arms production

Ian Anthony

I. Introduction

US arms exports took on a new importance after 1989 for three reasons. The United States emerged as the world's predominant arms exporter as well as the source of a high percentage of new military technology development. In both of these areas the degree of US predominance is expected to grow in the near term. While the exact scale of near-term reductions in US military expenditure is not known, it is clear that procurement for the US armed forces will decline over the next few years and that this prospect has heightened the relative importance of the economic and industrial aspects of US arms transfers, at least in public statements by producers.[1] As Judith Reppy notes, '[v]irtually all the major contractors are paying more attention to the possibility of increased foreign sales'.[2] The third reason for heightened attention to US arms exports is the dramatic changes in the international environment after 1989, which have left the United States in a unique position to influence the prospects for arms transfer control.

Although the USA will be the single most important actor in the global arms market, there is no consensus in the United States about how to approach the issues raised by arms transfer policy. Early statements by advisers close to President-elect Bill Clinton suggest that removing barriers which restrict export opportunities to US companies will be high among the list of priorities for the new Administration.[3]

This chapter addresses questions related to the changing patterns of US arms exports. How important are US arms sales politically and economically? To what extent is the US arms industry export-dependent? Is the level of export dependence likely to rise or fall? How are levels of dependence distributed across different industrial sectors? While tentative answers to these questions are offered, this is by no means a systematic or exhaustive treatment of US arms exports in the 1990s, a subject which deserves book-length treatment.[4]

[1] For example, the article by Joel L. Johnson (Vice President of the Aerospace Industries Association), 'In search of a sensible US arms transfer policy', *Military Technology*, Oct. 1991.

[2] Reppy, J., 'The United States: unmanaged change in the defence industry', chapter 3 in this volume.

[3] 'Clinton policy would loosen export controls on high-tech items', *Defense Marketing International*, 30 Oct. 1992, pp. 1–2.

[4] At least one book on this subject is in preparation, by William Hartung of the World Policy Institute in New York. Hartung has recently published two articles discussing US arms transfers: 'Breaking the arms-

Although statements about the future technological direction of US arms production and the political orientation of export policy are currrently still speculative, some conclusions are suggested by the discussion contained in this chapter.

1. Limits on the degree of US dominance of much of the international arms market will be self-imposed. In other words, US producers will take as much of the international market as their government permits them to do.

2. The economic importance of international sales to US companies, while limited, is likely to grow.

3. Within these international sales, exports are likely to become less important than sales by US-owned companies located overseas.

II. Measuring the US share of the global arms trade

There is no comprehensive quantitative measure of the value of the global arms trade. All of the data which are available can be challenged from one or another perspective, and their utility lies primarily in the fact that they fill what would otherwise be an information vacuum.

While the available sources are prepared by different people in different places using different methodologies, they offer independent confirmation for some trends in the international arms trade. Therefore, while the caveats about the data are important, there is reason to believe that these general trends are accurate.

Essentially, the data suggest that the market share accounted for by US arms exporters is growing while the overall market is shrinking.

Measuring the US share of aggregate value

According to SIPRI estimates, in 1991 the United States accounted for 51 per cent of total deliveries of major conventional weapons—compared with 30 per cent in 1987. The USA has inherited this dominant position as a result of the dramatic decline in the volume of arms exports from the former Soviet Union rather than by increasing its own sales dramatically. In 1991 the Soviet Union accounted for 18 per cent of total deliveries—compared with 39 per cent in 1987.[5]

Estimates of the aggregate size of the international arms market from other sources suggest a similar trend.

sales addiction', *World Policy Journal*, winter 1991; and 'Curbing the arms trade: from rhetoric to restraint', *World Policy Journal*, spring 1992.

[5] For a discussion of exports from the former Soviet Union, see Wulf, H., 'The Soviet Union and the successor republics: arms exports and the struggle with the heritage of the military–industrial complex', chapter 7 in this volume.

The Arms Control and Disarmament Agency (ACDA) recorded a 27 per cent reduction in the value of US arms delivered—from $15.4 billion in 1988 to $11.2 billion in 1989.[6]

More recent official data also indicate a significant decline in the value of US commercial arms sales. The value of equipment deliveries fell from $8.4 billion in 1989 to $5.6 billion in 1990 and to $3.8 billion in 1991.[7] While there is a massive increase in the value reported for commercial sales in 1992 (to over $28.7 billion), this figure represents the value of licences approved and not contracts signed or items delivered. In other words, these reflect the estimated value of contracts which US companies might or might not be awarded. Normally, only a small percentage of licences lead to firm orders after discussions with foreign governments.

These figures for commercial sales exclude the value of US security assistance programmes, chiefly the Foreign Military Sales (FMS) programme. Again, the figures are difficult to interpret. As indicated in table 4.1, the value of FMS agreements has grown considerably in the period after 1986, while the value of FMS deliveries has not.

Table 4.1. Value of US Foreign Military Sales agreements and deliveries, 1982–91

Figures are in US$ million, at current prices.

	1982	1983	1984	1985	1986	1987	1988	1989	1990	1991
Agreements	16 422	14 339	12 784	10 482	6 540	6 449	11 739	10 747	13 948	22 982
Deliveries	8 765	10 790	8 198	7 520	7 268	11 103	8 844	6 991	7 389	8 845

Source: Defense Security Assistance Agency (DSAA), *Foreign Military Sales, Foreign Military Construction Sales and Military Assistance Facts*, 30 Sep. 1991 (DSAA: Washington, DC, 1991).

Part of the explanation for this is that equipment agreed in the aftermath of the war against Iraq in 1991 has not yet been delivered, and therefore the value of FMS deliveries could be expected to begin to grow significantly over the next three or four years. However, Foreign Military Sales is not a synonym for arms sales since it includes the value of all support services provided along with weapon systems. A substantial (but unspecified) percentage of the value of FMS agreements is for items such as teaching maintenance routines for equipment or English-language training for foreign military personnel. Moreover, the value of the agreements is not fixed but represents an estimate. Items transferred under FMS are managed by the Department of Defense (DoD), and contracts are between private sector arms producers and the DoD. The DoD negotiates FMS agreements with foreign governments on the basis of its own estimate of what equipment will cost and not on the basis of the real price,

[6] Arms Control and Disarmament Agency, *World Military Expenditures and Arms Transfers 1990* (US Government Printing Office: Washington, DC, Nov. 1991), p. 127.

[7] Department of State and Department of Defense data, reproduced in 'US commercial arms sales', *Defense & Economy World Report*, 24 June 1991.

which may not be known for several years. When it comes to negotiating a price with a manufacturer, the DoD will combine foreign orders with orders for the US armed forces in order to achieve economies of scale.

In order to avoid having to re-open negotiations with foreign governments because it has been unable to secure equipment within the agreed ceiling, the DoD makes its cost estimates deliberately high and then refunds money to the purchasing government at the end of the transaction. Therefore, the value of FMS agreements declared by the Defense Security Assistance Agency (DSAA) in its annual data is usually higher than the actual monies received.

Looking at the provisional data for 1991, the Bush Administration agreed $23 billion worth of FMS sales in 1991, of which $16.6 billion were for Middle Eastern countries (over $13 billion for Saudi Arabia alone).[8] These sales agreements are conditional on congressional approval, on whether the recipient countries take all of the equipment agreed. Some of the FMS sales agreed in recent years face significant obstacles before being translated into deliveries. Agreements with Pakistan worth $2.8 billion (6 per cent of the total value of FMS agreements between 1986 and 1990) are currently embargoed under the 1985 Pressler Amendment to the 1961 Foreign Assistance Act. The amendment mandates that 'no military equipment or technology shall be sold or transferred to Pakistan' unless the President is able to certify that Pakistan 'does not possess a nuclear explosive device'.[9] This is something the Bush Administration has felt unable to do since September 1990, and at the time of writing (October 1992) it is an open question whether the equipment agreed with Pakistan (including 71 F-16 fighters and 10 attack helicopters) will be delivered.

Data produced by the US Library of Congress Congressional Research Service (CRS) records a decrease in the value of US arms exports to the Third World. The value of new US arms agreements with the Third World fell from $19.1 billion in 1990 to $14.2 billion in 1991. However, the US share in the value of all new agreements with Third World countries increased from 44.3 per cent in 1990 to 57.4 per cent in 1991. This largely reflected the dramatic fall in Soviet export agreements. In tems of deliveries, the value of US arms deliveries to Third World countries remained constant—at just over $5 billion. However, this represented a significant increase in the US share of total deliveries to Third World countries from 17.8 per cent in 1990 to 29.2 per cent in 1991.[10]

While the scale of the increase in the US market share can be disputed, the trend is fairly clear and confirmed by several sources. The emergence of one predominant supplier in the arms market reverses the trend of the previous 15

[8] Statement of Representative Lee Hamilton in *Congressional Record*, 24 Jan. 1992. A brief description of FMS and Direct Commercial Sales is contained in US Congress, Office of Technology Assessment, *Global Arms Trade*, OTA-ISC-460 (US Government Printing Office: Washington, DC, June 1991), p. 11.

[9] It is currently a source of dispute between the Department of State and the Senate whether the embargo applies only to FMS deliveries or to all US arms sales.

[10] Grimmett, R. F., *Conventional Arms Transfers to the Third World, 1984–91*, CRS Report for Congress (92-577F), 20 July 1992.

years during which a growing number of countries became active arms suppliers. In 1990 ACDA observed that 'since the late 1970s, almost all major arms have been offered by several exporters, and over two dozen countries offer munitions and simple support equipment'.[11] In 1990 a SIPRI study noted the extent to which West European suppliers had made inroads into markets such as Saudi Arabia and Turkey, which had been almost exclusively US clients for much of the 1970s.[12] Developments in 1991 tended to underline the predominant European view that the net effect of the 1991 Gulf War would not be to increase the size of the global market. Rather it would be to restore the strong dominance over the arms market that the US enjoyed prior to the late 1970s.

Whether or not this happens partly depends on whether European countries are prepared to combine their financial, scientific and production resources. Were they to do this, there is no technical reason why Western Europe could not compete with the United States—although development of a competitive industrial base would take many years.[13] Many independent observers have supported a restructuring of European political institutions in order to bring about a common approach to arms industrial questions. However, efforts at rationalization have stumbled over the hurdle of government unwillingness to relinquish national sovereignty.[14] The gridlock in European decision making has meant that restructuring is taking place according to commercial and industrial logic and not as part of a political or strategic agenda.[15]

The emergence of a common European approach to defence industrial and arms export questions is therefore something which can only be expected to occur over the long term if at all.

The other major current centre of production and export for advanced weaponry is the countries of the Commonwealth of Independent States (CIS), in particular the Russian Federation.[16] However, exports from the former Soviet Union declined dramatically in 1991 and the volume of sales recorded before 1989 is unlikely to be recovered.

Therefore, it is probable that the United States will not simply retain its considerable edge as the predominant exporter of advanced weapons but will continue to increase its lead over the other industrialized countries.

[11] Arms Control and Disarmament Agency, 'Diversification of arms sources by Third World nations', *World Military Expenditures and Arms Transfers* (ACDA: Washington, DC, 1990), p. 25.

[12] Anthony, I. and Wulf, H., 'The trade in major conventional weapons', *SIPRI Yearbook 1990: World Armaments and Disarmament* (Oxford University Press: Oxford, 1990), p. 233.

[13] Kolodziej, E. A., 'Europe as a global power: implications of making and marketing arms in France', *Journal of International Affairs*, summer 1988.

[14] Brzoska, M. and Lock, P. (eds), SIPRI, *Restructuring of Arms Production in Western Europe* (Oxford University Press: Oxford, 1992).

[15] The changing nature of West European arms industry is discussed in Wulf, H., 'Western Europe: facing over-capacities' and Sköns, E., 'Western Europe: internationalization of the arms industry', chapters 8 and 9, respectively, in this volume.

[16] The changing nature of the arms industry in the CIS is discussed in Cooper, J., 'The Soviet Union and the successor republics: defence industries coming to terms with disunion', chapter 5 in this volume.

Regional distribution of US arms exports

The regional distribution of the recipients of US arms exports is highly concentrated in three areas: Western Europe, the Near East and East Asia. These regions account for more than 95 per cent of US arms exports.[17]

Within the regions indicated in table 4.2, sales to formal allies of the United States—Australia, Japan, the NATO countries and South Korea—make up one-third of the total. Sales to what might be called 'quasi-allies' account for another 45 per cent of the total. These countries (to which the USA has security commitments that fall short of formal alliances) include Egypt, Israel, Kuwait, Pakistan and Saudi Arabia.

Table 4.2. Regional distribution of US Foreign Military Sales agreements, 1982–91
Figures are percentages.[a]

	1982	1983	1984	1985	1986	1987	1988	1989	1990	1991	Average 1982–91
East Asia	29	14	12	23	29	29	21	15	23	11	21
Near East and South Asia	54	34	31	35	30	30	61	63	56	72	47
Europe and Canada	13	50	53	37	34	33	14	18	19	12	28
Africa	1	0	1	1	2	1	1	0	0	0	1
American republics	3	1	2	3	4	5	3	3	2	1	3
International organizations	1	0	1	1	1	2	1	1	1	1	1

[a] 0 = less than 1 per cent; columns may not add up to 100 due to rounding.

Source: Defense Security Assistance Agency (DSAA), *Foreign Military Sales, Foreign Military Construction Sales and Military Assistance Facts*, 30 Sep. 1991 (DSAA: Washington, DC, 1991).

This distribution is important because it underlines that few US exports are discretionary. Most transfers are made to countries that the United States is very unlikely to turn down. As Under Secretary of State Reginald Bartholemew noted in testimony to Congress, the Bush Administration could not support any policy that prohibited sales necessary for the security of friends and allies.[18]

The distribution of FMS agreements is not a perfect proxy for the regional distribution of US arms exports. Apart from the fact that not all of these sales will materialize, as noted above, the exclusion of commercial sales probably exaggerates the importance of the Middle East and underestimates the importance of European countries in particular. Under commercial sales agreements, US companies deliver equipment directly to a foreign government or corporation without the Department of Defense playing a management role in the

[17] The regional aggregates used by the US Departments of State and Defense in compiling arms export data do not include the 'Middle East' but refer instead to a belt stretching from the Maghreb to Pakistan under the heading the Near East and South Asia. SIPRI data for the same countries are aggregated into three regions: North Africa, the Middle East and South Asia.

[18] Prepared statement to the Subcommittee on Arms Control, International Organizations and Science of the Foreign Affairs Committee of the House of Representatives on 24 Mar. 1992, reprinted in *Department of State Dispatch*, 30 Mar. 1992.

transaction. These kinds of transaction are more likely within NATO, where the recipient has no need for logistical assistance and there are fewer foreign policy issues at stake. Nevertheless, this measure is sufficient to suggest that US policy towards the Middle East and whether and how intra-alliance defence markets will be regulated will both be important determinants of the scale of US arms exports. These questions are discussed more fully in section IV below.

III. The economic dimension of US arms sales

Since commodities move from country to country against payment, there is inevitably an economic dimension to any discussion of arms transfers. However, the weight assigned to economic factors varies greatly.

At the broadest macro-economic level, ACDA commented in 1990, 'the role the arms trade plays in world trade continues to diminish'.[19] While the international arms trade in 1990 represented around 1.3 per cent of total trade (at most), it may be that it is of greater importance to the United States than to most other advanced industrial countries. Joel Johnson, Vice President of the Aerospace Industries Association, has argued that the arms industry provides the largest positive trade balance of any US industrial sector. Moreover, he expects that the proportion of US arms production which is exported may grow from 15 per cent to over 20 per cent by the end of the 1990s.[20]

The economic costs and benefits of US arms sales can be evaluated both directly and indirectly. Most directly, the economic benefits of overseas sales for national and company revenue and employment can be quantified and analysed. Indirect economic benefits could be claimed if US sales contribute to enhanced political stability and security in regions where the United States has important economic interests. In recent times the need to secure access to oil supplies from the Middle East is the most common variant of this argument. However, the argument that US economic development rested on the security of Europe formed one rationale for the massive gifts of military assistance to Western Europe in the 1950s and 1960s.[21]

Discussion of the economic costs and benefits of arms transfers has been a common undercurrent in the debate but has waxed and waned along with US procurement expenditure. In the mid-1970s—following the end of the US involvement in Viet Nam—a rise in the value of US exports to Middle Eastern countries coincided with downward pressure on the defence budget. Sales to Iran and Saudi Arabia were seen in part as 'a critical safety valve for ensuring the viability of US military industries'.[22]

[19] ACDA (note 5), p. 22.

[20] Johnson, J. L., 'In search of a sensible US arms transfer policy', *Military Technology*, Oct. 1991. Johnson assumes that US exports will rise to between $12–13 billion while the procurement budget shrinks to $60–70 billion.

[21] Pach, C. J., *Arming the Free World: The origins of the United States Military Assistance Program 1945–50* (University of North Carolina Press: Chapel Hill, 1991).

[22] Klare, M., *American Arms Supermarket* (University of Texas Press: Austin, 1984), p. 33.

During the 1980s—when the level of US arms procurement spending increased significantly—the economic dimension of US arms sales was rarely highlighted. In 1988 William Bajusz and David Louscher wrote: 'economic considerations traditionally have not figured prominently (if at all) in deliberations over specific arms transfers, which instead have been shaped principally by foreign policy considerations'.[23]

In 1978 the value of US arms exports was equivalent to 12 per cent of total US military expenditure for that year. In 1989, by contrast, the value of US arms exports was equivalent to less than 4 per cent of US military expenditure.[24] Moreover, during the 1980s a growing number of customers asked for direct or indirect offsets as a condition for awarding a contract, further complicating the calculation of economic benefit. In exceptional cases the value of the offset can be higher than the value of the arms sale. The size of US arms exports must rise dramatically or the value of US military expenditure must fall precipitously before exports can be argued to be of the same economic importance as they obtained in the 1970s.

Nevertheless, by 1991 economic and industrial base arguments were advanced in support of specific arms transfers—notably sales to Saudi Arabia. In April 1991 Bajusz and Louscher argued that refusing to sell arms to Saudi Arabia would cost more than 1.5 million jobs and $19.8 billion in lost revenue.[25] In the 1970s the true level of commercial and employment benefits derived from arms exports was challenged by economists.[26] This area remains a methodological minefield. As a simple example of the potential for confusion, while Bajusz and Louscher suggest that the cancellation of one arms transfer agreement alone (albeit a major one) may cost more than 1.5 million jobs, Johnson estimates total employment in the US arms industry as 1.3 million. In 1992 economists use similar arguments to question new assessments of the value of arms exports.[27]

In 1992 a study by the Congressional Budget Office (CBO) based on a survey of the situation in 1990 concluded that limits on the arms trade 'would not significantly affect the vast majority of exporting industries in the country'. The CBO found that even a reduction of $20 billion per year in annual defence sales

[23] Bajusz, W. D. and Louscher, D. J., *Arms Sales and the US Economy* (Westview Press: Boulder, Colo., 1988), pp. 2–3.

[24] These percentages are derived from ACDA estimates of US arms exports and military expenditure in *World Military Expenditures and Arms Transfers 1990* (US Government Printing Office: Washington, DC, 1991).

[25] Bajusz, W. D. and Louscher, D. J., *The Contribution of Arms Exports to the US Economy* (Foresight International for the American League for Exports and Security Assistance: Washington, DC, 17 Apr. 1991).

[26] For example, Fried, E., 'An economic assessment of the arms transfer problem', in ed. A. Pierre, *Arms Transfers and American Foreign Policy*, Council on Foreign Relations (New York University Press: New York, 1979).

[27] Greg Bischak and James Raffel argue that the figure of $19.8 billion assumes that all possible sales are actually realized and that the methodology used to calculate employment consequences artificially inflates the numbers of jobs created. Bischak, G. and Raffel, J., *Economic Conversion and International Inspection* (National Commission for Economic Conversion and Disarmament: Washington, DC, Oct. 1991).

would have significant effects on only a small number of companies in 6 out of 420 major industrial sectors.[28] Since orders from the Department of Defense to US arms-producing companies are falling, the impact of losing export sales will become proportionately greater. This is likely to be especially true for sectors where US procurement is expected to drop sharply (such as tanks and other armoured vehicles). Is it possible to identify either the types of company or the specific companies whose future is likely to be most dependent on arms exports?

IV. Export dependence in the US arms industry

Table 4.3 contains a list of US companies derived from the SIPRI arms production data base. The list indicates the extent to which these arms-producing companies are dependent on (*a*) arms sales and (*b*) foreign sales.

This listing of companies is by no means comprehensive. It reflects the data available to the author at the time of writing. Moreover, the figure for foreign sales refers to all income derived overseas, including direct exports and sales by overseas subsidiaries in both civilian and military product areas. In cases where a company has both a significant reliance on foreign sales and a large defence division, it is possible that most foreign sales are in civilian product areas. This is certainly the case for Boeing, for example. Conversely, a diversified corporation—such as Textron—has a limited arms sales dependence. However, these arms sales are concentrated in an offshore subsidiary (Bell Helicopter in Canada) with a history of international sales. It is probable that a high proportion of Textron's foreign sales are military items. Nevertheless, even if the table has only a suggestive value it is interesting in the absence of any better measure.[29]

Of the 31 companies listed, 13 depend on foreign sales for 30 per cent or more of their turnover. Of these 13 companies, 8 have a relatively low level of dependence on arms sales (20 per cent or less) and 4 depend on arms sales for less than 10 per cent of total sales. These four—ITT, Honeywell, Motorola and IBM—all make and sell computers or other electronic goods. Of the 10 companies in the list that depend on arms sales for 50 per cent or more of total sales, 6 generate 20 per cent or less of their sales overseas.

These two sets of companies conform to the result one would expect. Most arms-producing companies are heavily dependent on sales to the Department of Defense, while the difficult regulatory environment for arms exports deters many companies—especially those engaged in 'dual-use' product sectors—from seeking military contracts. As the Office of Technology Assessment (OTA) concluded in 1991: 'US Government policy may be the single most

[28] Congressional Budget Office (CBO) of the Congress of the United States, *Limiting Conventional Arms Exports to the Middle East*, Sep. 1992, p. 57.

[29] A detailed company-by-company approach to US arms exports is taken in Ferrari, P. L., Knopf, J. W. and Madrid, R. L., *US Arms Exports: Policies and Contractors*, (Ballinger: Cambridge, Mass., 1988).

Table 4.3. Dependence of selected US arms-producing companies on foreign sales and arms sales, 1990

Figures are percentages.

Company	Foreign sales as share of total turnover	Company	Arms sales as share of total turnover
Motorola	54	Esco Electronics	96
IBM	52	Alliant Tech Systems	92
Boeing	45	General Dynamics	82
FMC	40	E-Systems	75
General Dynamics	40	Lockheed	75
Honeywell	33	Grumman	72
Control Data	32	Oshkosh Truck	64
Tenneco	32	Litton Industries	58
United Technologies	32	Raytheon	57
ITT	32	McDonnell Douglas	55
TRW	32	TRW	37
Harris	30	Sequa	36
Alliant Tech Systems	30	Rockwell International	33
Texas Instruments	28	Texas Instruments	32
Hercules	28	FMC	28
Textron	28	Harris	26
Litton Industries	27	Hercules	25
Ford Motor	26	Textron	24
Allied Signal	24	Control Data	20
McDonnell Douglas	21	United Technologies	19
Sequa	20	Boeing	18
Esco Electronics	18	Westinghouse Electric	18
General Electric	16	Tenneco	15
Rockwell International	16	Olin	13
Westinghouse Electric	12	General Electric	11
Olin	8	Allied Signal	11
E-Systems	7	ITT	8
Lockheed	6	Honeywell	6
Raytheon	6	Motorola	6
Grumman	5	IBM	2
Oshkosh Truck	4	Ford Motor	1

Source: SIPRI arms production data base.

important factor influencing the international prospects of US defense companies'.[30]

Thirteen companies derive more than 20 per cent of their sales overseas and also depend on arms sales for more than 20 per cent of sales. These companies are Alliant Tech Systems, Control Data, FMC, General Dynamics, Harris,

[30] US Congress, Office of Technology Assessment (OTA), *Global Arms Trade*, OTA-ISC-460 (US Government Printing Office: Washington, DC, June 1991), p. 48.

Hercules, Litton Industries, McDonnell Douglas, Rockwell International, Sequa, Texas Instruments, Textron and TRW.

V. The impact of domestic arms procurement on arms exports

The future prospects for US arms producers in the international market depends to some extent on choices made in the United States about future military technology development.

The domestic debate in the United States about the pattern of procurement for the armed forces is considered at greater length in chapter 2 and chapter 3 in this volume. The discussion here is confined to considering how the outcome of the domestic debate will impact on the US share of international sales.

Some politicians (notably in Congress) have suggested continuing production of current-generation equipment improved through incremental changes to electronic, propulsion and weapon systems. If this was to be the preferred policy, then it is logical to expect increased US dominance of certain important sectors of international production. This is particularly true for aircraft and aerospace products, where the USA has many advanced systems in a mature stage of production. Conversely, if the United States moves into a new generation of even more sophisticated equipment, then over the long term companies may have increasing difficulty operating in the international market.

The United States has always exported many of its front-line weapon systems—although strategic delivery systems and dedicated long-range attack aircraft (such as the F-111) have been transferred only to close allies such as the United Kingdom and Australia. Systems such as the A-6 and A-10 attack aircraft have never been exported, and no exports are made that might undermine the capacity for US military intervention in regions where vital interests are seen to exist. However, as indicated in table 4.4, if an item is cleared for export, then sales have often been made only two or three years after the systems was first produced for the US armed forces. In other words, US attitudes to arms export control are not related to 'newness' or levels of technology but are derived principally from the desire to prevent certain military capabilities from spreading beyond close allies.

In compiling table 4.4 it proved impossible in some cases to distinguish between exports of new systems and exports of equipment from US inventories. The numbers given for M-48, M-60 and M-113 armoured vehicles probably include significant numbers of vehicles from the US Army which were refurbished before being transferred under a security assistance programme. Therefore, the figures for armoured vehicles in the column 'total exports' should be treated as estimates close to the maximum number of new platforms that could have been exported. Until 1985, when the drawdown in US defence spending began, systems transferred in this way would have been replaced in US service with new equipment. However, the practice of transferring equipment out of inventory without replacement has now started and is likely to

Table 4.4. US export sales and total production for selected systems

System name	Year first produced	Total produced	First year of export	Total exports	Exports as share of total production	Year last produced
Aircraft						
C-130	1954	1 904	1957	780	*41*	–
A-4	1956	2 960	1966	212	*7*	1979
F-4	1960	5 195	1964	2 196	*42*	1981
F-14	1972	497	1974	80	*16*	–
F-15	1973	1 569	1976	319	*20*	1993
F-16	1975	3 699	1975	1 464	*40*	–
F/A-18	1980	1 256	1980	445	*35*	–
AV-8B	1982	422	1982	141	*33*	1993
Helicopters						
Model-204/5	1958	10 986	1964	3 450	*31*	1984
CH-47	1961	1 498	1964	404	*27*	–
UH-60	1978	2 139	1984	485	*23*	–
SH-60	1983	579	1984	135	*23*	–
AH-64	1983	807	1990	86	*11*	–
Armoured vehicles						
M-47	1950	8 676	1950	8 000	*92*	1953
M-48	1952	11 703	1952	8 000	*68*	1959
M-113	1959	93 844	1960	56 000	*60*	–
M-60	1960	15 914	1961	3 500	22	1987
M-2	1980	6 132	1988	600	*10*	–
M-1	1981	8 558	1987	920	*11*	–

Note: Export figures exclude overseas licensed production. A dash in the final column indicates that the system is still in production.

Sources: *Jane's All the World's Aircraft*, editions *1966–67, 1967–68, 1971–72, 1978–79, 1981–82, 1990–91, 1991–92* (Jane's Information Group: Coulsdon); *World Military and Civil Aircraft Briefing* (Teal Group: Washington DC); *World Helicopter Systems*, editions *1983, 1985* (Interavia: Geneva); *General Dynamics Annual Report 1992*; *Jane's Armour and Artillery*, editions *1981–82, 1991–92*; *World Military Vehicle Forecast* (DMS).

become more common as the size of the armed forces shrinks through the remainder of the 1990s.

In these cases the US Government is to some extent setting itself up in competition with its own manufacturers. In one case—the 'cascading' of recently modernized M-60-A3 tanks from US Army stocks in Europe to Greece, Portugal, Spain and Turkey—the equipment transferred was not considered surplus by the Department of Defense but was transferred in line with arms control commitments.[31]

[31] The 'cascade' is the NATO Equipment Transfer and Equipment Rationalization Programme, which is a programme of arms transfers between member countries undertaken in the context of the 1990 CFE Treaty. Recipient countries import TLE more modern than that they currently deploy and thereby take on the reduction commitments of donor countries. For a discussion, see *SIPRI Yearbook 1992: World Armaments and Disarmament* (Oxford University Press: Oxford, 1992), pp. 288–91.

Table 4.5. New agreements for sales of fighter aircraft, 1988–92

	1988	1989	1990	1991	1992 (Jan.–June)	**Total 1988–92**
USA	116	89	66	204	138	**475**
UK	68	0	30	0	0	**98**
France	0	0	0	0	0	**0**

Source: SIPRI arms trade data base.

From the point of view of military utility, US systems remain extremely competitive even though of the systems listed in table 4.4 only a few are currently considered to represent the 'state of the art' in the USA itself. Probably only the AH-64 attack helicopter and the M-1 tank—both of which are criticized for being too sophisticated to be operated efficiently—would fit this description. Nevertheless, with the exception of the M-47 tank and the Model-204 utility helicopter, all of the systems listed are still in demand internationally. Aircraft and armoured vehicles no longer in production (such as the A-4, F-4, M-48 and M-60) are being rebuilt or repaired for export.

Given that the producing companies are so far along the production curve for the F-14, F-15, F-16 and F/A-18 fighter aircraft, these aircraft could be offered at a comparatively low cost to new customers and would be more than sufficient for most air forces in the world to manage their allocated tasks. Moreover, there would be a very strong international interest in an upgraded 'F-15 plus', F-16 Agile Falcon or F/A-18 Super Hornet if these offered new capabilities without requiring a completely new logistical and maintenance system to support them.

The availability of these upgraded aircraft would strengthen the US dominance in international fighter aircraft sales.

In August 1992 the President of the US Aerospace Industries Association complained that 'the government puts impediments on the US aerospace industry that our competitors do not have. . . . We have a lot of very competent competitors out there that are also going after the export market.' At the same Executive Roundtable meeting the Chairman of McDonnell Douglas noted that 'the government really doesn't have a strategy on what it should do to support foreign sales'.[32] However, as suggested by table 4.5, the most severe competition faced by US companies in the recent past has usually come from other US companies. This has created something of a dilemma for the US Government in that active export promotion would have to avoid giving competitive advantages to one US company over another.

In recent years McDonnell Douglas and General Dynamics have contested several fighter aircraft contracts overseas. In Japan, South Korea and Kuwait

[32] Speaking at the Senior Aerospace Executive Roundtable, reprinted in *Aviation Week & Space Technology*, 24 Aug. 1992, pp. 35–39.

the General Dynamics F-16 and the McDonnell Douglas F/A-18 were the final two aircraft in competition. The F-16 was successful in Japan and Korea and the F/A-18 in Kuwait. In 1977 the F-15 (also made by McDonnell Douglas) was selected over the F-16 as the basic air superiority fighter for the Japanese Air Self Defense Force. The same two aircraft will make up the bulk of the Israeli Air Force once aircraft currently on order are all delivered.

Western European aircraft have not fared well in competition with US fighters. The French Mirage-2000 and the Swedish JAS-39 lost competitions in Switzerland and Finland (in both cases to the F/A-18) while the Anglo-German-Italian Tornado fighter-bomber was rejected by South Korea and Japan.

While US fighter aircraft represent an extremely cost effective procurement choice from the perspective of most foreign armed forces, there is much less interest in these upgraded versions of existing equipment within the US Air Force and Navy. While upgrading aircraft through their service lifetime is already standard practice, this is not regarded as a substitute for the development of an entirely new generation of technology—represented by the F-22, Multi-Role Fighter and Advanced Strike Aircraft programmes. It is certain that new generations of equipment will incorporate technologies which are subject to strict export control or, in some cases, exports of which are prohibited. In addition the unit procurement costs, maintenance costs, manpower and training requirements, and changes in force structure needed to operate such aircraft effectively would be a major obstacle to foreign sales.

As table 4.4 makes clear, many of the systems which have formed the backbone of the US armed forces have historically been exported from very early in their production runs. However, it will be many years before an F-22 fighter becomes available for export assuming a buyer could be found.[33]

The United States does not have the same kind of dominance in other production sectors, notably naval vessels. Looking at foreign sales of large surface warships, the pattern in table 4.5 is largely reversed, with West European countries (especially Germany) being the dominant suppliers. Table 4.6 lists agreements for transfers of surface warships from the United States and European Community (EC) countries between 1988 and 1992.

Moreover, this table underestimates the dominance of European suppliers in this product category since virtually all of the vessels supplied by Germany and France were new construction. The United Kingdom, the Netherlands and the United States, by contrast, have largely transferred major surface combatants being retired from their navies. Twenty of the 29 US ships listed in the table are ex-US Navy vessels, many of which are at least technically still US property. These have been leased to foreign navies by the United States rather than sold—although in reality it is difficult to imagine the US Navy accepting any of the ships back.

[33] According to Air Force General Phil Nuber, it would be the year 2002 at the earliest before any foreign request for the aircraft would even be considered. *Defense News*, 22–28 June 1992, p. 22.

Table 4.6. New agreements for transfers of corvettes, frigates and destroyers, 1988–92

	1988	1989	1990	1991	1992 (Jan.–June)	**Total 1988–92**
USA	11	12	6	0	0	**29**
FR Germany	4	16	2	1	0	**23**
France	1	3	2	6	0	**12**
UK	2	0	3	2	2	**9**
Netherlands	0	2	0	0	0	**2**
EC	7	21	7	9	2	**46**

Source: SIPRI arms trade data base.

The US Navy has already followed the path of demanding continuous increases in capability from its equipment suppliers. This has translated into increases in the unit size, capability and cost of the vessels bought. As a result, almost all of the ships operated by the US Navy are 'unexportable' while successful export designs (notably the FFG-7 Class) are accepted only reluctantly by the US Navy.

In spite of its current market dominance, designs such as the F-117 short-range bomber and the B-2 long-range bomber suggest that the USA may go down the same path in terms of aircraft design and development. Alternatively, however, the US companies may seek to re-enter production sectors which they had previously left. During the period 1988–92 the Federal Republic of Germany agreed to transfer 10 diesel submarines—systems no longer in production in the United States. In 1992 members of both the US Senate and the House of Representatives advocated the production of diesel submarines for export after cutbacks in procurement of nuclear-powered submarines for the US Navy led to the closure of the Electric Boat shipyard in Groton Connecticut. The plan is opposed by the Navy.[34]

VI. Political aspects of US arms transfer policy

The future volume and geographical distribution of US arms exports will also depend to some extent on the nature of the restructuring which takes place in the international arms industry. The decisive political dimension in this will be the nature of intra-alliance political relations fostered under the new administration. Perhaps more important, however, will be the nature of US arms transfer policy.

[34] *Arms Sales Monitor*, no. 16 (June–July 1992), p. 1; *Defense News*, 10–16 Aug. 1992, p. 9.

Intra-alliance relations

A second factor which will determine the future pattern of US arms exports is the extent to which US companies participate in the restructuring of the international arms industry which seems to be under way in West Europe. After the disintegration first of the Warsaw Treaty Organization and then the Soviet Union Western Europe is the second largest market in the world for advanced military systems. While Western Europe has a major defence industrial capacity which it intends to retain (albeit on a smaller scale) the US has historically supplied a significant percentage of European military equipment. Under the new political conditions, however, it is not clear that European governments will in future grant the same level of access to their markets for US companies as they did during the cold war.

While the multinational military operations in the Persian Gulf illustrated the advantages of having common equipment—a factor which might favour standardization on US equipment—European governments are also interested in issues of economics and technology development.

In the post-cold war period greater attention will inevitably be paid to issues of industrial competitiveness and technology development in the United States. As one group of authors put it, 'the passing of the postwar military and economic order requires a fundamental reassessment of the wellsprings of American military and economic power'.[35] This is equally true for other countries, and the outcome of these assessments will have consequences for the overseas sales possibilities of US companies.

European industries are asking governments not to remove existing barriers faced by US companies doing business in Europe while the disparity in size between the arms industries on either side of the Atlantic is so great. As one industrialist expresses the problem, 'The US market is 2.2 times as large as that in Europe, only 0.5 per cent of its equipment is imported. In addition, the European military uses a far greater percentage of American equipment. . . . The facts clearly show that a complete liberalisation of procurement would only benefit the stronger supplier'.[36]

As underlined by the procurement choices for fighter aircraft listed above, however, an increasing number of European governments are not responsive to the arguments of industry, preferring to seek better value for procurement expenditure to the benefit of the taxpayer. While they have no interest in competing with US companies, European arms manufacturers are interested in importing specific US technical assistance which will raise their competitiveness with one another. Regardless of the political debate about burden-sharing, advantages in price and technology seem certain to allow US companies increased access to European markets unless European governments take a con-

[35] Alic, J., Branscomb, L., Brooks, H., Carter, A. B. and Epstein, G., *Beyond Spinoff: Military and Commercial Technologies in a Changing World* (Harvard Business School Press: Boston, Mass., 1992), p. 3.

[36] Frédéric Puaux, Thomson Group, speaking on Defence Procurement in the Nineties, Centre for European Policy Studies Business Policy Seminar, Brussels, 5 June 1992.

scious decision to obstruct their activities. Under present circumstances this seems very unlikely; on the contrary, efforts have been made within NATO to reduce barriers to intra-alliance arms transfers.[37]

At the same time as assessments in possible recipient countries may reduce overseas sales prospects, US decisions may also place some limit on the ability of companies to supply advanced technologies. As the perceived need for co-operation against a Soviet military threat fades, decision makers in the US Congress are questioning the wisdom of selling advanced production process technologies to allies in Western Europe and the industrialized countries of Asia. Whereas arguments of military necessity would have made such sales routine during the cold war, the question is now being raised whether selling Japan and South Korea licences to manufacture advanced aerospace products will undermine the future position of US firms in civilian markets. This issue of technology transfer is likely to become an increasingly contentious element in the management of US alliance relationships.[38]

US arms transfer policy

Three successive Republican Administrations have consistently argued that arms transfers are an effective foreign policy instrument. Bush Administration officials continued to make this argument in testimony before various congressional committees in hearings held after the coalition war fought against Iraq in 1990–91.

While there are certainly many congressmen who share this view, there are also powerful individuals and committees which reject it. In 1992 correspondence between Congress and the White House outlined the basic dispute. In May 1992 a letter cosigned by 237 congressmen put forward the view that the overriding US interest in the Middle East was the success of the arms control element of the Arab–Israeli peace process. Responding for the Administration, National Security Adviser Brent Scowcroft replied that Saudi Arabia was a vital factor for the success of the peace process but that the support of the Saudis was more likely if its ability to defend itself in the face of significant potential threats was secure.[39]

Under Secretary of State Reginald Bartholemew has stated that the belief that US arms transfers enhance US influence among key regional decision-makers

[37] The USA has bi-lateral Memoranda of Understanding (MoU) with almost all of its allies regarding the trade in military equipment. However, the effectiveness of these MoUs in creating open market conditions has been widely questioned. See, for example, Government Accounting Office (GAO), *International Procurement: NATO Allies Implementation of Reciprocal Defense Agreements*, GAO/T-NSIAD-92-29, 29 Apr. 1992 (GAO: Washington, DC, 1992).

[38] Aspects of this issue are discussed in Sköns (note 14).

[39] Letter of 9 Apr. from Representative Mel Levine and 236 co-signatories to President Bush and reply of 15 May by General Brent Scowcroft, National Security Adviser. On 21 May the same Congressmen wrote to British Prime Minister John Major, expressing similar sentiments and received a similar response.

in the Middle East as one 'basic assumption' underlying the policy.[40] If it could be demonstrated that market dominance gave the United States the possibility to use arms transfer policy in this way it would increase incentives to continue or even expand arms transfers. This temptation would be especially strong in the Middle East, where regional governments have attached particular importance to the procurement of major weapon systems and where the United States has enduring policy interests.

On this basis it seems unlikely that the United States will voluntarily reduce the level of its arms exports but more probable that it will seek the maximum economic and political benefits from this aspect of foreign policy. The factors which may restrict US export performance are more likely to stem from choices made in the USA about domestic arms procurement and technology development.

[40] Prepared statement to the Subcommittee on Arms Control, International Organizations and Science of the Foreign Affairs Committee of the House of Representatives on 24 Mar. 1992, reprinted in *Department of State Dispatch*, 30 Mar. 1992.

Part III

The former Soviet Union and the successor republics

5. The Soviet Union and the successor republics: defence industries coming to terms with disunion

Julian Cooper

I. The historical background

In the autumn of 1991, in the aftermath of the failed August coup, the Soviet Union ceased to exist. The Baltic republics became fully independent, and most of the remaining republics agreed to form the fragile Commonwealth of Independent States (CIS). Communist rule came to an end and, with Russia to the fore, with varying degrees of commitment, the newly independent countries embarked on radical economic transformation. The institutions and practices of central planning were rapidly dismantled, opening the way for market forces and private ownership.

The former Soviet Union (hereafter, FSU) possessed a defence industry of formidable proportions (see table 5.1).[1] From the beginning of the 1930s the Soviet Communist Party and state strove to create conditions for its priority development. The system of central planning and non-market allocation of material resources was managed in such a manner as to ensure that the enterprises and research organizations of the defence industry were adequately supplied with materials and equipment of high quality. Privileged conditions of employment, not only in terms of pay and bonuses but also in terms of above average provision of housing and social amenities, ensured that the defence sector was able to attract and maintain a labour force of relatively high quality. Both at a national and local level the Communist Party apparatus played an active role in securing these privileged conditions of work. With its 'first departments' in all defence sector enterprises and institutes, the KGB maintained a regime of extraordinarily strict secrecy, which not only protected militarily sensitive information but also served to conceal the sector's privileges from the population at large.

The defence industry of the FSU had a highly centralized system of management. The state-owned enterprises and research and development (R&D) organizations were subordinate to a set of powerful industrial ministries of Union scope, in recent years eight or nine in number,[2] the activities of which were co-

[1] The principal features of the former Soviet defence industry are described in the author's Chatham House Paper: Cooper, J., *The Soviet Defence Industry: Conversion and Reform* (Pinter/Royal Institute of International Affairs [RIIA]: London, 1991).

[2] During most of the 1970s under Brezhnev, there were 9 ministries; under Gorbachev for a while there were 8.

Table 5.1. Economic indicators of the arms industry in the former Soviet Union, 1991

Indicator	Figure	Comment/source
Total military expenditure	96.6 b. roubles	Official figures
Procurement expenditure	39.7 b. roubles	Official figures
R&D expenditure	10.4 b. roubles	Official figures
Industrial employment	7.5 m. (1988), of which 4.7 m. in military production	20% of industrial labour force; author's estimate
R&D employment	1.7 m., of which approx. 1.3 m. engaged in military R&D and 0.1 m. in military R&D outside the defence complex	Author's estimate
Gross output	140 b. roubles (1988), of which 88 b. roubles was the military component	16% of gross industrial output

Source: Based on Julian Cooper, *The Soviet Defence Industry: Conversion and Reform* (Pinter/ Royal Institute of International Affairs: London, 1991).

ordinated by a high-level Military–Industrial Commission and a special defence industry department of the State Planning Committee, Gosplan. In the development and supply of weapons, the agencies of the defence sector worked closely with the arms procurement directorates of the armed forces, but the customer–supplier relationship was not typical of most Western countries. The development and supply of weapons took place according to party–government decrees, payment being made not directly by the customer from the Ministry of Defence budget but by financial transfers to the relevant industrial organizations from the state budget. This system limited the customer power and choice of the armed forces which, during the last two to three decades of the Soviet state, felt themselves increasingly subject to the dictates of industry. Another feature of this centralized regime was the extraordinarily distorted price system. The high-quality inputs (raw materials and production equipment) used by the defence sector were systematically under-priced, leading to artificially low prices for end-product weapons and a substantially understated defence budget.

Even under President Mikhail Gorbachev, the scale of the former Soviet defence industry was shrouded in secrecy, and the situation has not improved much in the post-Soviet period. In the absence of proper definition, the limited data available are extremely difficult to interpret. According to information presented by the Central Research Institute for the Economics and Conversion of Military Production, the Soviet defence complex (the group of eight specialized defence sector industrial ministries) had about 1300 associations and enterprises and almost 1000 organizations engaged in R&D. However, each association probably included on average three or four enterprises, giving a total of approximately 5000 enterprises. The total industrial labour force was over 8 million, plus 1.7 million R&D personnel.[3] The breakdown of industrial em-

[3] *Moscow News*, no. 7 (1992).

ployment in the defence complex in 1985 was as follows: Russia, 71.2 per cent; Ukraine, 17.5 per cent; Belarus, 3.2 per cent; Kazakhstan, 1.7 per cent; Uzbekistan, 1.4 per cent; Armenia, 1.1 per cent; Kyrgystan, 0.6 per cent; Azerbaijan, 0.5 per cent; Georgia, 0.4 per cent; Moldova, 0.4 per cent; Tajikistan, 0.2 per cent; Turkmenistan, 0.1 per cent; Latvia, 0.8 per cent; Lithuania, 0.8 per cent; and Estonia, 0.1 per cent.

However, the total of almost 9.8 million industrial employees excludes other categories of workers, including those employed in specialist construction organizations, farms, social facilities and other services attached to the ministries. Only part of this vast labour force was directly engaged in military-related production and R&D, and by 1990 almost half the output of the enterprises of the defence complex took the form of civil goods. Some military production was also undertaken by enterprises of nominally civil industrial ministries. This suggests a total direct employment in military-related production of at least 6 million people. Taking into account all those owing their living to the work of the defence complex and those employed in activities indirectly linked to military production, claims by Russian public figures that some 40 million Soviet citizens were dependent on the defence industry are probably not far off the mark. There is no dispute that the Soviet economy was militarized to an extraordinary degree.

During the Gorbachev period (1981–85), the once highly privileged and protected defence industry began to experience reduced levels of funding, an inept conversion drive, loss of prestige and open public criticism. As reform efforts intensified, the centrally planned economy began to break down and even the military sector began to encounter supply difficulties. While state enterprises in the civil sector were able to exploit the possibilities of greater autonomy offered by partial reforms, including more liberal regulations for pay, the defence industry enterprises remained subject to strict central control. While in industry as a whole wages grew by almost 40 per cent between 1985 and 1990, in the defence sector they increased by only 10 per cent.[4] This loss of privilege, coupled with the uncertainties generated by defence cuts and conversion, led to an outflow of workers from the industry, especially of younger, educated personnel. In these increasingly difficult circumstances it is not surprising that some of the more conservative administrators and managers of the defence complex sought to reverse the reform process and played an active role in the August 1991 attempted coup.[5] Others, however, favoured marketization and democratization. This is the essential background to the fate of the defence industry in the post-Soviet era.

The highly centralized management of the Soviet defence industry was manifested in questions of location policy, and this has profoundly shaped the post-Soviet inheritance of the newly independent states. For each ministry, facilities were located and developed according to a single Union-wide division of labour. As a general rule, with the exception of the shipbuilding industry which

[4] *Voprosy Ekonomiki i Konversii*, no. 1 (1992), p. 74.

[5] The political involvement of the defence industry is discussed in Cooper (note 1), chapter 6.

was centred on St Petersburg (then Leningrad), most of the central research and design organizations of each ministry were located in or near Moscow. The mayor of Moscow, Yurii Luzhkov, has claimed that 75 per cent of Russia's military R&D is concentrated in the capital of the Russian Federation.[6] With the partial exception of the missile industry, strategically sensitive arms production was concentrated overwhelmingly in Russia, including all facilities for the development and production of nuclear weapons. The other republics had on their territories diverse and usually rather arbitrary sets of enterprises and research organizations, often of a highly specialized character, all answering to Moscow rather than the local authorities, who had very little, if any, control over them.

Since the aborted coup the centralized structures of the Soviet defence industry have disintegrated. The powerful Union ministries, the Military–Industrial Commission and Gosplan exist no longer. Each independent state has been forced to review its inheritance, to create appropriate structures of management, and to devise policies appropriate to the new circumstances. To make matters even more difficult, each of the states is attempting to formulate its own national security policy. Strategic systems remain under the control of a unified CIS command, but in each of the states new national conventional forces are emerging. Requirements for military equipment remain unclear; budgetary finance for procurement and R&D is severely constrained. In these extraordinary circumstances any review of the post-USSR situation cannot be other than highly provisional.

II. The Russian Federation

To a considerable extent the defence industry of the former USSR was Russian. Of the organizations of the former Soviet defence complex, 67 per cent of associations and enterprises, and 74 per cent of R&D establishments were located in Russia. For those facilities involved heavily in military work the proportions were even larger. Of the total labour force of the defence complex, 71 per cent of industrial employees and 85 per cent of personnel in R&D were found in Russia. However, the Russian R&D organizations performed no less than 90 per cent of all military research undertaken in the FSU. The newly independent Russian Federation has a very large defence complex, employing some 7 million people. Approximately 4.5 million industrial and R&D personnel are engaged directly in military production in and outside the defence complex.[7]

Within Russia there are certain regions with extremely high concentrations of defence industry enterprises. In terms of the absolute number employed in industrial facilities of the defence complex, the top 10 locations appear to be Ekaterinburg (Sverdlovsk), St Petersburg (city), Moscow (city), Nizhnii-Novgorod (Gor'kii), Moscow region, Perm, Samara (Kuibyshev), Novosibirsk, Tatarstan and Udmurtiya. It is striking that the economies of some of the for-

[6] *Nezavisimaya Gazeta*, 11 July 1992.

[7] *Moscow News*, no. 7 (1992); *Inzhenernaya Gazeta*, no. 51 (Apr. 1992).

mer autonomous republics of Russia—Tatarstan, Udmurtiya, Bashkortostan and Mariiel—are dominated by military-related production, to the extent that the first three named have larger defence industries than now possessed by any of the CIS member countries other than Russia, Ukraine and Belarus. Military-related R&D in Russia is heavily concentrated in and around Moscow, St Petersburg and Novosibirsk.

Whereas other of the newly independent states have had to create new administrative structures for oversight of the defence industry facilities on their territory, in the case of Russia it has been a question of transforming the pre-existing Union structures. The Union ministries have been abolished, but some of their personnel now constitute the staff of departments of the Russian Ministry of Industry. The Ministry has eight departments for branches of the defence sector (aviation industry, general machine-building [missiles and space technology], shipbuilding, conventional munitions and military chemicals industry, defence industry, radio industry, communications equipment industry and electronics). Unlike the former ministries which intervened actively in the affairs of enterprises, these departments are supposed to be concerned with policy issues only. In fact they appear to have steadily accumulated power since their formation, acting increasingly like the ministries from which most of their staff originated. Also within the Ministry of Industry is a department of the defence complex and conversion, which has effectively taken over functions previously exercized by the former Military–Industrial Commission.

Parts of the former industrial ministries of the defence complex have transformed themselves into new commercial, or quasi-commercial, concerns, joint stock companies and state corporations, frequently occupying the old ministerial buildings and staffed by former ministerial officials. In principle, enterprises and R&D organizations are members of these new structures on a voluntary basis. Examples include the state corporation Rosobshchemash, which includes most of the facilities of the missile–space industry, Aviaprom, a commercial, co-ordinating union of the Russian aviation industry, and a set of concerns and associations to which belong the organizations of the former shipbuilding ministry. These new bodies are forming commercial banks, commodity exchanges and other commercial structures appropriate to a market economy. Outside the formal structures of the defence industry, some leading figures of the old order have re-emerged as prominent figures of the burgeoning non-state sector. In particular this applies to the Military–Industrial Investment Company (VPIK), a billion-rouble commercial venture raising finance for conversion projects. As marketization and privatization proceed, the organizational arrangements of the defence sector are likely to evolve further and, as noted below, new forms may emerge, including multi-national corporations.

Before looking at some of the policy issues now facing the Russian defence sector, its main components are briefly reviewed.

Nuclear weapons

A virtual monopoly of Russia is the development and production of nuclear warheads and other devices. Under the Ministry of Atomic Energy are enterprises and R&D organizations employing 1.1 million people, of whom some 130 000 are engaged directly in research, design, manufacture and dismantlement in the nuclear weapons complex. The facilities of the nuclear weapon industry are located in 10 'closed towns' in relatively remote locations, a system originally developed under the overall leadership of Lavrenty Beria, chief of the security system under Stalin. These towns, belonging wholly to the Ministry and subject to extraordinarily rigorous security regimes, have a total population of almost 800 000 people. While some of the severe restrictions on visitors and other contacts with the outside world were relaxed somewhat at the end of 1991, the towns remain to a considerable extent isolated from the rest of the country, and while not as privileged as in the past in terms of supplies of food, consumer goods, housing and social provision, they still have a degree of protection from the economic and social problems currently endemic in Russia.

Two towns, Arzamas-16 in the Nizhnii-Novgorod region and Chelyabinsk-70 in the Urals, are devoted to research on nuclear weapons and the design, manufacture and dismantlement of nuclear charges. Arzamas-16, located on the site of the ancient Sarov monastery, was the first centre for nuclear weapon development. Since its foundation it has been under the scientific leadership of Academician Yulii Khariton. It was here that the first Soviet atomic and hydrogen bombs were made, the latter with Academician Andrei Sakharov's prominent involvement. Three of the towns have been engaged in the production of plutonium and other nuclear materials for weapons: Chelyabinsk-40, Tomsk-7 and Krasnoyarsk-26. Uranium enrichment is undertaken at Krasnoyarsk-45 and Sverdlovsk-44 (Verkh-Neivinsk). The remaining three towns (Sverdlovsk-45 (Nizhnyaya-Tura), Penza-19 and Zlatoust-36) have facilities for incorporating the nuclear charges into warheads, bombs and other devices.

The Russian nuclear industry has many other research organizations and enterprises concerned with the extraction and processing of nuclear materials, the design and building of reactors, and instrumentation for the military and civil nuclear industries. It also manages the country's civil nuclear power stations. Uranium mining in Russia is now concentrated in the Chita region of southern Siberia. Some nuclear materials are imported from Kazakhstan and the Central Asian republics.

In the FSU the Ministry for the Nuclear Industry, for many years the Ministry of Medium Machine Building, was without doubt the most powerful of all the industrial ministries in terms of priority and political influence. It is no surprise that now, in new circumstances, it is only the nuclear industry that has managed to retained almost intact its specialized ministerial structure. For Russia it is likely to long remain a legacy and reminder of the Soviet past.

Missiles

In the FSU, Russia possessed the majority of facilities for the development and production of strategic ballistic missiles, but Ukraine also played an important role. In or around Moscow are found most of the missile–space industry's principal research and design centres, including the country's main organization for space research technology, the Kalingrad Energiya association (the former Korolev design bureau), the Reutov Machine Building association for missiles and launchers (the former Chelomei design bureau), the Khimki Energomash association, the leading organization for liquid propulsion rocket engines, and the former Nadiradze organization, the Institute of Thermal Processes, for the development of solid propulsion missiles (including the SS-20 and SS-25 missiles). An important exception to this Moscow-centred location pattern is the design organization for submarine-launched missiles formerly headed by Academician Viktor Makeev, the Miass machine-building design bureau in the Chelyabinsk region of the Urals.

The building of ballistic missiles is undertaken at several, widely dispersed, locations. The SS-19 was built at the Moscow Khrunichev works (better known for its production of the Proton space launcher), solid-fuel ballistic missiles (including the SS-25 and the SS-20 missiles, abolished under the terms of the 1987 INF Treaty) are built at the Votkinsk works in Udmurtiya, and submarine-launched ballistic missiles at machine-building plants in Krasnoyarsk in Siberia and Zlatoust in the Urals.

Other locations for missile building are Samara (Kuibyshev), Orenburg, Perm, and, in the Moscow region, Kolomna and Serpukhov. Centres for the building of rocket engines include the vast Polet complex in Omsk, Voronezh, and St Petersburg. Facilities for the development and manufacture of missile control and navigation systems include, in Moscow, the association of automatics and instrument building (the former Pilyugin organization), Rotor and Geofizika, and in Ekaterinburg, the Avtomatika association.

The principal organization of the FSU for the development of missile launch silos, the design bureau of special machine building, is located in St Petersburg.

Transporter–erector–launchers for mobile missiles are built at the Volgograd Barrikady works and rail-based launchers at the Yurga machine-building plant in the Kemerovo region of Siberia.

Anti-ballistic missile systems have been developed by the Moscow-based Almaz association of the radio industry, at first under the leadership of the late Academician Aleksandr Raspletin, currently of Academician Boris Bunkin, who led the development of the Russian equivalent of the Patriot missile system the S-300 (SA-10). A major developer of surface-to-air missiles is the radio industry's Antei science-production association based in Moscow, with general designer Veniamin Efremov responsible for a range of SAM systems (including the SA-8, SA-10, SA-11 and SA-15). Air-to-air, anti-ship and other short-range missiles are developed by design organizations in around Moscow (including the Vympel and Zvezda bureaus), but built by enterprises dispersed more widely throughout the country.

The design and building of military satellites involve many organizations, including some of the above-mentioned facilities of the missile–space industry, the Applied Mechanics association located in the closed nuclear weapon town of Krasnoyarsk-26 and headed by Academician Mikhail Reshetnev, and the Kometa association in Moscow headed by Academician Anatolii Savin.

Aircraft

From the 1920s the development of aircraft design and production was a matter of the highest state priority of the Soviet leadership, with the result that today Russia possesses an extremely large and capable military aviation industry. Its research and design capability, including the Illyushin, Mikoyan, Sukhoi, Tupolev and Yakovlev design organizations, is heavily concentrated in and around Moscow; the production plants are widely dispersed throughout the country. Heavy bombers are built at Kazan (Tatarstan) and Samara (formerly Kuibyshev), although both plants have now ended their building of strategic bombers (the Tu-160 (Blackjack) and Tu-95, respectively). Major plants building other fixed-wing combat aircraft are Moscow (the MiG-29), Nizhnii-Novgorod (MiG-31), Novosibirsk (Su-24), Komsomolsk-na-Amure in the Far East (Su-27), and Irkutsk (MiG-29 and Su-27). Helicopters are built at Kazan, Rostov, Ulan-Ude, Arsen'ev (Far East) and Kemertau in Bashkiriya. The principal plants for building aero-engines for combat aircraft are located in Moscow, St Petersburg, Perm, Rybinsk, Omsk, Samara, Kazan and Ufa. Centres for the manufacture of aviation armament include Moscow, Tula, Sverdlovsk, Smolensk and Kirov. Organizations supplying avionic systems include not only the aviation industry's own central organizations in Moscow, but also the St Petersburg Leninets of the radio industry.

Ground forces equipment

Tank building formerly took place at four locations in Russia. The principal developer and producer is the Nizhnii Tagil Uralvagonzavod in the Urals, which has always been, simultaneously, the country's largest producer of rail freight wagons. The plant is now building the T-72 main battle tank although in greatly reduced numbers. The St Petersburg Kirov factory has ceased production of the T-80. It also built self-propelled guns and tracked chassis for military equipment. Other tank production has been located at Omsk and Chelyabinsk, but the latter appears to have ended its involvement. Prior to the programme for conversion of military facilities to civilian production, the Ekaterinburg transport machine-building works was a major producer of self-propelled howitzers. Personnel carriers and other armoured vehicles are built at several plants: locations include Kurgan in the Urals, Arzamas in the Nizhnii-

Novgorod region, Izhevsk, Kirov and Volgograd. Manufacturers of military trucks and transporters include enterprises located in Moscow, Nizhnii-Novgorod, Bryansk and Miass in the Urals. Tank transporters are built at Chelyabinsk.

Artillery systems, including multiple rocket launchers, are built by several plants, including the vast Perm Lenin works, Ekaterinburg, Kovrov and the St. Petersburg Arsenal plant, which makes naval artillery systems. The principal centres for the manufacture of infantry weapons are Tula, Kovrov and Izhevsk, the latter being the location of the famous Kalashnikov design team. Important research centres for conventional explosives and munitions are located in Moscow and St Petersburg, but production facilities are widely dispersed. Especially significant are plants at Krasnoyarsk and Biisk in Siberia. Major manufacturers of fuse mechanisms are located at Samara, Ioshkar-Ola and in the Penza region.

Optical equipment and lasers

The development and manufacture of military-related optical equipment are strongly developed in St Petersburg, Kazan in Tatarstan, Krasnogorsk in the Moscow region, Novosibirsk and Ekaterinburg. Before the collapse of the USSR, several organizations were involved in the development of laser weapon systems as part of the Soviet equivalent of the US Strategic Defense Initiative (SDI) programme. Prominent was the Moscow Astrofizika association, formerly headed by Nikolai Ustinov, son of General Secretary Leonid Brezhnev's defence minister, the projects of which included a mobile laser weapon.[8]

Naval ships

Naval shipbuilding in Russia is heavily concentrated in St Petersburg. Here are found 70 per cent of the research and design organizations of the former Soviet shipbuilding industry, including the bureaus responsible for the design of most large naval surface vessels and submarines. The city also possesses a number of major shipyards, including the Admiralty, Baltic, Severnaya Verf and Almaz associations, the latter building air-cushion military craft. In the Far East, surface ships for the navy are built at shipyards in Vladivostok and Khabarovsk. Several yards at other locations build smaller surface ships including Kaliningrad, Zelenodolsk (in Tatarstan), and, until the early 1990s, the Sretenskii works in the Chita region of Siberia. St Petersburg is also important for the development and production of naval armament: the Uran association, for example, leads all work for the development and manufacture of torpedoes.

The principal design centre for nuclear submarines is the Rubin organization of St Petersburg, currently headed by Academician Igor Spasskii. Another submarine design bureau, Lazurit, is located at Sormovo near Nizhnii-

[8] *Novoe Vremya*, no. 26 (1992), pp. 41–42.

Novgorod. Reactors for submarines and nuclear surface ships are developed by the nuclear industry's Research–Technological Institute at Sosnovyi Bor near St Petersburg, as well as by the Special Design Bureau of Machine Building at Nizhnii-Novgorod. Four yards have participated in building nuclear submarines, the most important being the vast Severodvinsk complex in the Arkhangelsk region of northern Russia. The economy of Severodvinsk, a town of 250 000 people, is completely dominated by submarine building, refitting and repair. In the Far East nuclear submarines are built by the Komsomolsk-na-Amure yard. Some have also been built by the inland Krasnoe Sormovo works near Nizhnii-Novgorod and by the St Petersburg Admiralty association. Over the period 1993–95 nuclear submarine building is to be concentrated fully at Severodvinsk, which is being given the status of the State Centre of Nuclear Submarine Building.

Command, control and intelligence

A wide range of military radar, early-warning and control systems is produced by organizations formerly under the Soviet Ministry of the Radio Industry. This ministry possessed an extremely powerful R&D capability, employing 300 000 people, many of the facilities of which are located in Moscow and St Petersburg. The vast Vympel concern embraces institutes and enterprises concerned with over-the-horizon radar and missile early-warning systems. The Moscow Skala association creates radar systems for the air defence forces, and Kibernetika, also based in Moscow, is prominent in the development of control systems of military application. Electronic components are supplied by enterprises of the former Ministry of the Electronics Industry, its principal research centre being located at Zelenograd just outside Moscow.

Even before the dissolution of the USSR, the Russian defence industry had experienced reduced levels of procurement and of R&D funding. With Russian independence and the adoption of a policy of rapid economic transformation, the defence budget was further reduced and procurement cut for 1992 by more than 60 per cent. With limited possibilities for quickly expanding civil production, by the summer most defence plants were in serious difficulty. Large-scale unemployment has only been avoided by resort to inter-enterprise and bank credit, and also by putting employees on short-time work or obliging them to take long, often unpaid, leave. Some enterprises have attempted to maintain their military output, producing for stock in the hope that policy will be reversed or that the authorities will permit them to export. The collapse of military production means that the total output of the defence complex has fallen sharply and that the share of civil work has grown. The Minister of the Economy, Andrei Nechayev, has forecast that civil production will represent 80 per cent of total defence complex output in 1992.[9]

[9] BBC, *Summary of World Broadcasts*, SU/1423 C3/3, 3 July 1992.

The Russian Government has made clear its desire to reduce the scale of military production and to convert many facilities wholly or in part to civil work. However, it is very difficult to formulate a coherent conversion policy in the absence of an agreed military doctrine and an associated lack of clarity on future requirements for military technology. The civilian First Deputy Minister of Defence, Andrei Kokoshin, who is now in charge of procurement and R&D policy for the Russian armed forces, believes that once requirements have been established it will be possible to determine which enterprises will be retained as state concerns of the defence complex, allowing others, possibly as many as 800, to transfer fully to civil work and be privatized. It is envisaged that a further set of enterprises will manufacture dual-use equipment and that such plants could have mixed forms of ownership with some state participation.[10]

Deputy Defence Minister Kokoshin has indicated that once policy has been clarified, the Russian armed forces will establish a procurement system of a type familiar in the West.

A major problem in achieving a rapid civilianization of the defence industry is an acute lack of finance to fund restructuring. The presidential adviser on conversion, Mikhail Malei, has been a vigorous advocate of an active arms export policy designed to earn hard currency to finance restructuring. In his view conversion will cost at least $150 billion, part of which could be raised by annual sales of arms to a value of some $5 billion.[11]

Some partial decentralization of arms exports has taken place, but the Ministry of Foreign Economic Relations has been endeavouring to establish a strict system of regulation and licensing, retaining a considerable measure of central control. Since 1990 arms exports from Russia have declined quite sharply and, notwithstanding the adoption of a more active policy, it is difficult to envisage any substantial reversal of this trend in the short to medium term.

Enterprises of the Russian defence industry are dependent on suppliers in other states of the FSU for some of their materials and components. Efforts have been made to devise new procedures for maintaining traditional links while simultaneously protecting the national interests of the countries concerned. In late July 1992, the Russian Government adopted a resolution establishing a new system of licences to regulate deliveries and purchases of inputs for military production to and from enterprises of CIS member states.[12] This desire to create a stable framework for continuing established forms of co-operation has been echoed in moves to restore, on a new basis, some of the traditional organizational links characteristic of the defence industry. Meeting early in August, representatives of aviation industry facilities located throughout the CIS decided to form an inter-state joint stock company, Aviaprom. The creation of such inter-state corporations has the active backing of Arkady Vol'skii, the powerful leader of the Russian Union of Industrialists and Entrepreneurs. It cannot be ruled out that some measure of re-integration of the

[10] *Izvestia*, 20 July 1992.
[11] *Izvestia*, 31 March 1992; BBC, *Summary of World Broadcasts*, SU/W0230 A/16, 15 May 1992.
[12] *Rossiiskaya Gazeta*, 6 Aug. 1992.

fragmented former Soviet defence industry could take place. Such a development could have great significance not only for Russia but also for the other member countries of the CIS.

Having inherited the bulk of the former Soviet defence industry it is not surprising that Russia has been experiencing acute problems in attempting to scale it down. A serious problem for President Boris Yeltsin and the Russian Government is that the enterprise directors and senior administrators of the defence sector constitute a powerful political force, able to lobby for a moderation of the policy of radical economic transformation. Some of the leading personnel of the industry have formed a League of Defence Enterprises, headed by Andrei Shulunov, director of the Pleshakov association of the radio industry. This body is committed to reform but could change its stance if difficulties become more acute. There is no doubt that it will take several years for Russia to establish a relatively stable defence industry of a scale appropriate to the country's new security requirements. Given the inheritance, it is likely that, in international terms, this industry will still be one of considerable size and capability.

III. Ukraine

The Ukrainian Republic cannot be said to possess a defence industry in any normally understood sense. On the territory of the republic, however, are located major elements of a defence industry, in the form of largely unrelated components of the formerly integrated Soviet armaments industry. From the point of view of a newly independent Ukraine, this is an ambiguous inheritance. On the one hand the facilities of the defence sector undoubtedly represent the most capable part of the country's industry; on the other hand they do not form a coherent industrial system, and substantial, costly restructuring will be required before the civil economy reaps benefit from them. There is another possibility: if the country's leadership so decides there is a sufficient inheritance to permit the creation of a relatively large and diverse defence industry, but such a course of action would also necessitate substantial restructuring and investment, difficult to realize without foreign involvement.

On the territory of Ukraine are located approximately 700 factories formerly affiliated to ministries of the Soviet defence complex. In 1991 these factories employed some 1.2 million people—17 per cent of the total industrial labour force. However, weapons and military-related equipment represent only a part of their output (40 per cent in 1991), and not all the plants are involved in military work. In March 1992 it was reported that 344 of the 700 factories produced directly for the military, and of these 54 per cent (185 factories) produced 80 per cent of the military-related output.[13] Thus, prior to Ukrainian independence, total military-related employment in defence complex enterprises must have amounted to approximately half a million, or 7 per cent of the total industrial labour force. To this figure must be added an unknown number of personnel

[13] *Golos Ukrainy*, 14 Mar. 1992; document of conversion proposals of Ukrainian Ministry of Machine Building, Defence Industry and Conversion, Kiev, 1991.

employed in military-related R&D. For Ukraine it can be estimated that as many as 800 000 industrial and R&D personnel might be directly engaged in military production inside and outside the defence complex.

All the defence industry facilities on the territory of Ukraine now belong to the independent republic, and new administrative arrangements have been devised for their management and oversight. Overall policy for the defence industry is now in the hands of a Ministry for Machine Building, the Military–Industrial Complex and Conversion. The Minister, Viktor Antonov, a Russian, was for many years director of the Mayak association, one of the largest concerns of the radio-electronics industry in Ukraine,

The enterprises and R&D facilities formerly under Union ministries are affiliated to new Ukrainian commercial corporations and concerns. Thus, more than 50 enterprises of the shipbuilding industry belong to the Nikolaev-based Ukrsudstroi corporation; 73 enterprises of the electronics industry to the Ukrelektor corporation; and 15 enterprises, headed by the Antonov organization of Kiev, form the association of aviation industry enterprises of the Ukraine. Facilities of the nuclear industry located in the Ukraine have a more centralized structure of management in the form of a State Committee, Ukratomenergoprom.

In terms of major end-product weapons the inherited production base is somewhat narrowly focused: on ballistic missiles, surface naval ships and tanks. From a military strategic point of view the most important facility is the Dnepropetrovsk Yuzhnoe design bureau and associated Yuzhmash plant for the development and production of ballistic missiles and space launchers and vehicles. Founded in the early 1950s and associated during its formative years with the missile designer Mikhail Yangel, this complex, employing some 50 000 people, was responsible for the SS-9 and SS-18 intercontinental ballistic missiles (ICBMs), the SS-4 and SS-5 intermediate-range missiles (scrapped under the terms of the INF Treaty), and also the Scud missile. It has also built a range of space launchers, including, the Kosmos, Tsiklon and Zenit systems, and also designed the heavy-lift Energiya rocket. In its missile building, Yuzhmash has been heavily dependent on Russian suppliers, not only for basic materials but also for such key items as liquid-propellant rocket motors. It has also drawn on the highly centralized research system of the former Union Ministry of General Machine Building, and has been totally dependent on Kazakh and Russian launch facilities.

The Yuzhmash plant has always had well-developed civilian production, its basic civil good being the YuMZ small tractor. During recent years the share of military output has steadily declined, from approximately 85 per cent in 1988 to barely 30 per cent in 1991.[14] In early 1992 the Ukrainian Government decided to halt the building of missiles and to reorientate Yuzhmash and its design bureau to civil work. While attempting to retain involvement in the space industry, efforts are being made to diversify. The manufacture of wind-power

[14] *Financial Times*, 24 Jan. 1992.

units and trolley buses has been organized, and it is likely that the plant will participate in the building of Antonov aircraft.

Pavlograd in the Donetsk region is the location of facilities for the building of strategic missiles with solid-fuel propellant motors, including most recently the SS-24 ICBM in both silo- and rail-launched modes. This missile was designed by the Yuzhnoe design organization of Dnepropetrovsk.[15] Since Ukrainian independence, there has been no information on the future of the Pavlograd facility. Ukraine is also important for the manufacture of electronic control systems and other instrumentation for missiles and space vehicles, with major facilities located in Kharkov and Kiev.

Ukraine possesses one of the pioneer tank-building plants of the FSU, the Kharkov Malyshev factory association. During the pre-war years this historic plant was responsible for the development of the legendary T-34 tank and also pioneered the building of diesel engines for tanks and other combat vehicles. In terms of scale of output during the post-war years, output it has been second only to the vast Nizhnii-Tagil Uralvagonzavod plant in Russia. Products have included the T-54 and T-64 tanks and tracked military transporters. Tank guns are supplied from a plant in the Urals. During recent years the volume of output has fallen sharply, creating increasingly difficult conditions for the plant which has been seeking to increase its civil work. By the summer of 1992 the future of tank building at Kharkov was in doubt.[16]

A field in which Ukraine has always been strong is shipbuilding. Shipyards and enterprises for ship machinery and instrumentation in Nikolaev, Kherson and the Crimea have played a major role in supplying the former Soviet Navy with surface ships. Facilities in Nikolaev include the Chernomorskii zavod, one of the largest shipyards of the FSU, responsible for building a series of major surface combatants including Kiev Class aircraft-carrying cruisers and the first Soviet full-scale aircraft-carriers, the *Admiral Kuznetsov* (originally the *Tbilisi*), which entered service with the Russian Navy, and the partly built *Varyag* and *Ulyanvosk*, the latter scrapped to make way for civil work. Also at Nikolaev is the 61 Kommunar yard, which has built a series of missile destroyers and cruisers. Shipyards in the Crimea, Kherson and Kiev have also contributed to the former Soviet Navy, building smaller surface vessels. Submarines have not been built in Ukraine. The republic has also been important as a supplier of ship gas-turbines (the Nikolaev Mashproekt and Zarya associations) and electronic and other systems manufactured by plants in Kiev, Nikolaev and the Crimea.

In building vessels for the Soviet Navy the Ukrainian shipbuilding industry was heavily dependent on Russia. The principal design bureau for heavy surface combatants, including aircraft-carriers, is located in St Petersburg, together with most of the industry's central research institutes. While 30 per cent of the surface ships of the Black Sea Fleet were built in Ukraine, all were designed in Russia. In addition, Russia supplied many systems for the vessels, including most of the armament.

[15] *Krasnaya Zvezda*, 7 Dec. 1991.
[16] *Golos Ukrainy*, 24 July 1992.

During the first half of 1992 the Ukrainian shipyards were striving to reduce rapidly their involvement in naval shipbuilding, seeing their future in civil work for foreign clients. Partially built ships intended for the Soviet Navy are being scrapped. It seems unlikely that Ukraine will continue to build major surface combatants, but the commander of the new Ukrainian naval forces, Rear-Admiral Boris Kozhin, has declared that the republic's new navy will obtain coastal defence craft from the local shipbuilding industry.[17]

Fixed-wing combat aircraft are not built in Ukraine, but the country does possess the FSU's principal design organization for heavy transport planes, the Antonov design and production complex in Kiev. Aircraft developed and built in Kiev included the An-124 Ruslan and An-225 Mriya heavy transports. Products of the Kharkov aircraft plant include the An-74 passenger plane. At Zaporozhe is one of the world's largest plants for the manufacture of aero-engines, the Motor Sich association employing some 37 000 people, with its associated Progress design organization. Engines developed and built at Zaporozhe include those installed on the Ruslan and other Antonov aircraft, plus engines for helicopters, including the Mi-24 and Ka-32. In the Trans-carpathian region of western Ukraine, near the Romanian border, there is a plant building helicopters. The models built have not been identified but must have been developed by either the Kamov or Mil design organizations in Russia.

On Ukrainian territory are several plants for the production of conventional munitions and explosives, military optical equipment (in particular the Kiev Arsenal and Chernovtsy Kvarts works) and laser systems of military application (including the Lvov Polyaron association). The Dneprovsk machine-building factory is one of the largest builders of radar systems of the FSU and remains linked to its Moscow research base as a member of the inter-state Vympel concern. Ukraine possesses a substantial electronics industry, the plants of which were important in meeting the needs of producers throughout the FSU, and has a well-developed industry for the manufacture of computer, radio and communications equipment. A near monopoly of the republic is the supply of silicon crystals for micro-electronic components—until the break-up of the Soviet Union, plants at Zaporozhe and Svetlovodsk met most of the requirements of the Russian electronics industry.

Near the town of Zheltye Vody in the Dnepropetrovsk region is an important centre for uranium mining and processing, supplying approximately one-fifth of the FSU's total primary uranium output. However, the republic does not possess any facilities for the production of weapon-grade nuclear materials.

The future of military production in Ukraine remains highly uncertain. There has been evidence of concern to retain existing supply and technological links between the Ukrainian and Russian defence industries. At the June 1992 meeting of President Yeltsin and President Leonid Kravchuk, there was agreement to set up a joint commission to regulate economic links, including those

[17] BBC, *Summary of World Broadcasts*, SU/1414 C4/2, 23 June 1992.

between defence sector enterprises.[18] What remains unclear is the extent to which the new Ukrainian armed forces will attempt to reduce dependence on Russia by developing domestic production. Representatives of the country's armed forces, including Defence Minister Konstyantyn Morozov, have suggested that this may indeed be the desired policy option.[19] As noted above, the navy intends to obtain coastal defence craft from local shipyards. However, the defence industry minister, Viktor Antonov, has noted that Ukraine does not produce machine-guns or other types of infantry weapons. In his view these arms should be purchased from Russia.[20]

Statements by government representatives, including Antonov, indicate that Ukraine intends to an active export policy, at least in the short term, as a means of generating export earnings to finance conversion and keep defence plants in work.[21] This export orientation forms part of the active conversion policy being pursued by the Ministry of Machine Building, the Defence Industry and Conversion. As in Russia, the development and implementation of a coherent conversion policy are being hampered by a lack of clarity on military requirements and by an acute shortage of finance for civil projects. Nevertheless, the ministry has developed over 200 national programmes with a particular focus on equipment for the food and consumer goods industries and the health service, and technologically complex consumer goods. Antonov, the Minister, has stated that in the absence of adequate international support for conversion to production of civilian goods, it will be necessary to produce weapons and military equipment for export to raise finance for conversion projects.

Ukraine has embarked on a privatization policy involving the removal from state ownership of some of the enterprises of the defence sector. Several such plants, mainly those with a strong civilian orientation, have been included in the official list of enterprises to be privatized in 1992.[22] Antonov believes that over half of the 700 enterprises of the former Soviet defence complex located in Ukraine should be privatized, including many of the radio-electronics plants, but thinks that such large-scale enterprises as Yuzhmash and Arsenal should be retained in state hands.[23] It is too early to say whether Ukraine will possess a sizeable defence industry with plants in the private sector. Much will depend on the future course of relations with Russia and progress in transforming the economy into a market system.

IV. Belarus

The Republic of Belarus has little if any end-product arms production but nevertheless played an important role in the defence industry of the FSU. The machine-building industry of the republic is relatively modern, with a distinct

[18] *Pravda Ukrainy*, 27 June 1992.
[19] BBC, *Summary of World Broadcasts*, SU/1430 C3/2, 11 July 1992.
[20] *Pravda Ukrainy*, 23 June 1992.
[21] *Pravda Ukrainy*, 9 June 1992.
[22] *Delovoi Mir*, 23 July 1992.
[23] *Pravda Ukrainy*, 23 June 1992.

bias towards electronics and precision engineering. There are over 30 enterprises of the electronics and radio industries, now grouped under the Belradprom association. They include important producers of micro-electronic components such as the Integral association in Minsk, Monolit in Vitebsk and Korall in Gomel, which formerly played an important role in meeting the requirements of the defence industry of the entire USSR. Belarus also possesses significant producers of radar, military control and communications systems. The Gomel radio plant forms part of an inter-state concern, Vympel, based in Moscow, for the development and building of radar systems. Other suppliers include the large Minsk Agat science-production association and the Minsk association formerly named after Lenin. Plants in Minsk and Brest supply computers. Military-related optical equipment is manufactured by the Belarus optico-mechanical association based in Minsk. Belarus also has one of the FSU's principal producers of military trucks and transporters, including those for mobile missile systems (including the SS-25 ICBM), the Minsk automobile works (MAZ). It can be estimated that approximately 150 000 industrial and R&D personnel might be directly engaged in military production inside and outside the defence complex.

Given the traditional close relations between Russia and Belarus, it is likely that established supply links and co-operative arrangements will be maintained and that Belarus will rely to a considerable extent on Russian sources of supply in meeting the future equipment needs of the republic's own armed forces. There appears to be no intention to develop a more broadly based domestic defence industry, and, if some military-related production is retained, the specializations traditionally present in the republic are well-suited to the development of dual-use technologies.

Belarus is not in a position to export new weapon systems, but regulated trade in surplus weapons of the former Soviet armed forces, especially tanks and combat aircraft, has been declared official policy.[24]

V. Moldova

The small republic of Moldova, in 1992 torn by inter-communal strife, played a relatively small role in the Soviet defence industry, restricted almost exclusively to the supply of electronic systems, including radio equipment and computer control systems for military use. Plants involved include the Kishinev Alfa, Mezon and Topaz works. In addition, factories at Bendery and Beltsy supplied equipment for the shipbuilding industry.

[24] BBC, *Summary of World Broadcasts*, SU/1445 C3/3, 29 July 1992.

VI. Kazakhstan

The Republic of Kazakhstan inherited approximately 50 disparate enterprises of the former Soviet defence complex.[25] Almost 40 of them have been grouped under the Kempo state concern (Kazakh state corporation of electro-technical and machine-building enterprises of the defence complex) while others now come under a state agency, Kazatomenergo, charged with the management of facilities of the nuclear industry.

The principal centres of the Kazakh defence industry are Ural'sk and Petropavlosk in the north of the country, and the capital, Alma-Ata, in the south. Products of enterprises in Ural'sk include heavy machine-guns for tanks (the Metallist works) and anti-ship missiles (the Zenit works, formerly of the Union shipbuilding ministry). In Petropavlovsk the heavy machine-building plant produces transporter–launchers for missiles, including the SS-23, and may also build armoured vehicles. Petropavlovsk also has two enterprises of the shipbuilding industry, one of which builds equipment for minesweepers. Torpedoes and other naval armament are manufactured by the Kirov works in Alma-Ata, which also has enterprises producing radio equipment. It is characteristic of former Soviet industrial specialization policy that one of the major military products of this large, effectively *land-locked*, republic is *naval* armament. It is estimated that 75 000 industrial and R&D personnel might be directly engaged in military production inside and outside the defence complex.

Kazakhstan had active involvement in the nuclear industry of the FSU. The Pricaspian mining–metallurgical combine on the Mangyshlak peninsula was one of the largest producers of uranium, and the Ust-Kamenogorsk Ulba works supplied beryllium and other materials used in reactors. The town of Kurchatov near the former nuclear test site of Semipalatinsk has a major nuclear industry research centre, the Luch association, the activities of which include work on the development of nuclear rocket engines.[26] This facility has been taken over by the Kazakh Academy of Sciences as a national research centre.

The Semipalatinsk nuclear test site was not the only military test facility on Kazakh territory. Others include the Baikonur cosmodrome, still being used as a CIS launch centre, and the Sary-Shagan centre for testing air defence and anti-missile systems, in use before the end of the USSR for testing the S-300 anti-missile system.[27] According to Kazakh claims, there are eight test sites in all, occupying 7 per cent of the country's territory, equal to 40 per cent of the total area occupied by test sites in the former USSR.[28]

Associated with these test facilities are a number of formerly closed towns: Leninsk (Baikonur), Kurchatov and Priozersk (Sary-Shagan); the latter, with a population of 35 000, now experiencing major difficulties as funding, has been

[25] *Kazakhstanskaya Pravda*, 31 July 1992.
[26] *Kazakhstanskaya Pravda*, 25 July 1992.
[27] *Nezavisimaya Gazeta*, 5 June 1992.
[28] *Poisk*, no. 50 (1990), p. 8; *Pravda*, 11 Jan. 1991.

withdrawn.[29] The republic also has a very serious legacy of environmental pollution arising from weapons testing over many years.

Notwithstanding the relatively good relations between Kazakhstan and Russia and an agreement concluded in March 1992 for the preservation of defence industry supply links, difficulties have appeared leading to measures to reduce import dependence. Explosives for industrial use, for example, are not produced in the republic, and the prices charged by Russian suppliers in the first half of 1992 rose by 50–65 times.[30]

In the future, Kazakhstan is likely to rely substantially on Russia for the supply of weapons and other military equipment for the republic's new armed forces. In co-operation with Russian enterprises some military production may well be maintained. In the short term, severe problems are being encountered and efforts are being made to convert some of the industry's plants to civil purposes. Enterprises have been lobbying President Nursultan Nazarbaev for assistance. A draft law on conversion has been drawn up providing for arms exports as a means of generating finance.[31] In early 1992 a stir was caused when the Ural'sk Metallist works was reported to have sold 500 large-calibre machine-guns to Germany to help pay for equipment for sewing-machine manufacture, the plant's main direction of conversion.[32]

Kazakhstan undoubtedly has the potential to remain a significant participant in military production.

VII. Georgia

Before the dissolution of the USSR, the Caucasian Republic of Georgia had over 40 enterprises affiliated to the ministries of the Soviet defence industry, employing some 30 000 people. The most important is the Tbilisi aviation factory building the Su-25 close support fighter. In producing combat aircraft this plant is very heavily dependent on external sources of supply for materials and major systems, including engines, avionics and armament. In the absence of traditional supply links with Russia, it is difficult to see how Georgia can remain in the aircraft industry. In early 1992 the Georgian Government declared its intention to seek export markets for the Su-25 missile, but once existing stocks have been exhausted this is unlikely to prove a viable policy option.

Georgia also possesses a number of enterprises of the shipbuilding industry, including shipyards at Batumi and Poti, but these are not significant builders of military craft. Electronic components and communications equipment are manufactured in the republic, which also has two research institutes of the nuclear industry: the Tbilisi institute of stable isotopes and the Sukhumi Vekua physical and technical institute, but neither is concerned directly with nuclear weapons. If the Georgian Government decided to create a modest domestic armaments

[29] *Nezavisimaya Gazeta*, 5 June 1992.
[30] *Kazakhstanskaya Pravda*, 15 July 1992.
[31] BBC, *Summary of World Broadcasts*, SU/1387 B/11, 22 May 1992.
[32] *Trud*, 21 Jan. 1992.

industry, the existing inheritance and the general level of development of the republic's civil machine-building and electronics industries would certainly be adequate.

VIII. Armenia

Armenia's principal contribution to the former Soviet defence industry was the production of electronic components and end-products as well as electrical equipment. The radio-electronics industry of the republic grew rapidly from the 1960s, and many plants became involved in military-related work: by 1991 nearly 20 enterprises were involved in conversion activity. Major facilities of the radio-electronics industry include the Razdan mashine-building works (Razdanmash), with approximately 10 000 employees, the Erevan works (Neiron), the Kamo works (Dipol) and the Abovyan works (Pozistor).

Systems for the aviation and shipbuilding industries are also produced in the republic.

IX. Azerbaijan

Azerbaijan was not a very important participant in the former Soviet defence industry. It has plants producing electronic systems, radio equipment and computers: the Radiostroenie and Elektronika plants in Baku and facilities of the shipbuilding industry (the Nord association in Baku). However, the republic does possess a quite strongly developed civilian machine-building industry with particular strengths in the development and manufacture of equipment for the oil and gas industries. It cannot be ruled out that the prolonged, bitter conflict with Armenia could lead to efforts to develop some domestic arms production, possibly in co-operation with foreign partners.

X. Uzbekistan

The Central Asian Republic of Uzbekistan has grouped facilities of the former Union industrial ministries into a number of state concerns, including one for enterprises of the aviation industry, a state concern of mashine building (Uzkontsernmash), and a concern of radio-electronics, electrical engineering and instrument making (Uzradioelektrontekhpribor).[33] The most important aviation industry plant is the Tashkent Chkalov association building transport aircraft, in particular the Il-76 used extensively as a military transport aircraft and as a base for airborne warning and control systems (AWACS). Now the plant is co-operating with the Moscow aviation production association, building the MiG-29, in developing the production of the Il-114 short-haul passenger aircraft. Close co-operation with the Russian aviation industry will be essential if the Tashkent plant is to continue as a major builder of modern aircraft.

[33] *Zhizn' i Ekonomika* (Tashkent), no. 6 (1992), p. 11.

The republic also has a number of relatively modern plants of the electronics industry supplying components and systems. The Tashkent Uzelektromash association produces communications equipment for the air defence forces but has been severely hit by a collapse of orders.[34] Uzbekistan is also a major producer of uranium. The industry's main centre, the Navoi mining–metallurgical combine, is now attempting to diversify, increasing its already established extraction of gold and silver.

XI. Kyrgyzstan

The largest industrial plant of this small, southern republic of Central Asia is the Bishkek (formerly Frunze) machine tool factory association, formerly under the Soviet Ministry of the Defence Industry and directly involved in military production. Established during World War II on the basis of a munitions plant evacuated from Lugansk in Ukraine, the plant employs some 15 000 people, the overwhelming majority of them Russians.[35] The fate of this plant is unclear. The republic also has a small electronics and radio industry, and two plants in Bishkek supply equipment to the shipbuilding industry. Uranium mining is another important activity of the republic: the Kara Balta mine supplied the uranium for the first Soviet atomic bombs. Like its neighbour Kazakhstan, Kyryzstan was also involved in the former Soviet defence industry as a provider of territory for weapon testing, in particular a test range at Lake Issyk-Kul'. Even before the dissolution of the USSR there was a vigorous local campaign for its closure.[36]

XII. Tajikistan

The involvement of this small republic in the former Soviet defence industry was limited. The mining and enrichment of uranium, but not to weapons grade, is undertaken by the Tchkalov mining-chemical combine. Shortly before the end of the USSR, the first steps were taken in the development of an aviation industry: in 1990 a branch of the Mikoyan design organization set up in the capital of Dushanbe.

XIII. Turkmenistan

Turkmenistan, the southern-most and poorest republic of the FSU, played virtually no role in the defence industry. The only plant which has been identified by the author is a facility of the shipbuilding industry located in the capital Ashkhabad.

[34] *Rabochaya Tribuna*, 31 Jan. 1992.
[35] *Trud*, 15 Jan. 1992.
[36] *Izvestia*, 7 Mar. 1991.

XIV. The Baltic republics

Now fully independent and outside the structures of the CIS, the three Baltic states made a modest, but not unimportant, contribution to the former Soviet defence industry. Their strength lay not in the production of end-product weapons but in the supply of high-precision systems and components. It does not seem likely that efforts will be made to develop domestic defence industries. In the summer of 1992 there were reports that Lithuania was buying infantry weapons from Russia on a normal commercial basis and this may set the pattern for the future.[37]

Estonia

Having formerly had a small uranium industry (at Sillamae on the north coast), Estonia possesses facilities which formed part of the Soviet nuclear industry. One of the largest enterprises of the republic, the Dvigatel works of Tallinn, manufactured equipment for the nulcear industry, while the Baltiets works of Narva produced radiation meters and other instrumentation. Estonia also possesses strengths in the electronics and radio industries.

Latvia

Latvia has a strong radio-electronics industry which used to supply components and systems used by defence sector producers throughout the FSU. Radio and communications equipment was supplied by the Radiotekhnika, VEF, Straume and Kommutator associations in Riga. The Alfa association is a very large-scale producer of electronic components, including some for specialized military applications. A number of these large associations are now being dissolved and the separate elements privatized.

Lithuania

Like Estonia and Latvia, Lithuania has particular strengths in electronics and in the manufacture of radio and communications equipment. The Vilma association of Vilnyus was the sole Soviet producer of 'black boxes' for civil and military aircraft. In addition it has a small shipbuilding industry: the Baltiya yard in Klaipeda had some experience of building transport ships for the Soviet Navy. Prior to independence approximately 20 enterprises of the republic had some involvement in military work, but the military share of total industrial output was less than 5 per cent.[38]

Again, like the other Baltic states, Lithuania seems unlikely to attempt to develop a domestic arms production capability.

[37] *Baltic Independent*, 7–13 Aug. 1992, p. 1; FBIS-SOV-92-150, 4 Aug. 1992, p. 84.
[38] *Ekho Litvy*, 12 Oct. 1991; BBC, *Summary of World Broadcasts*, SU/W0206 C1/3, 22 Nov. 1991.

6. The Soviet Union: arms control and conversion—plan and reality

Alexei Izyumov

I. Introduction

The defence industry is by far the most advanced and efficient remnant of the former Soviet system. Although it possesses a huge potential, putting it to civilian use is a formidable task. Since the first announcement of the conversion effort at the end of 1988, conversion in the former Soviet Union has developed from a pretentious old-style 'campaign' into a battleground for hard-line communists and liberal reformers, and finally into the crucial element of the transition to a market economy. The role of the defence complex in the socio-economic structures of the former USSR remains tremendous. This is why developments in the sphere of conversion of military industry to civilian production will continue to influence the course of events both within and outside the Commonwealth of Independent States (CIS) for many years to come.

II. The Gorbachev plan for conversion and its results

The policy of massive conversion of the Soviet defence industry to civilian production was first announced by General Secretary Mikhail Gorbachev in his famous 'new thinking' speech held at the United Nations General Assembly in New York on 7 December 1988. Previously, in 1987–88, the Soviet Union had already made limited attempts at conversion, but it was not until 1989 that the massive conversion of the Soviet military economy gained real momentum.

In January 1989 the Soviet Government announced a dramatic unilateral reduction in the Soviet armed forces, the military budget and arms production. Over a two-year period (1989–90) the armed forces were to be cut by 12 per cent (0.5 million), the military budget by 14 per cent and arms production by 19.5 per cent.

While significant for disarmament, the plan was also critical for the Soviet economy, which was facing a wide array of serious problems created in the era of General Secretary Leonid Brezhnev and was in dire need of new resources for development. Conversion of the huge Soviet military–industrial apparatus thus seemed to be a logical solution to such problems as under-investment in housing and health services, education and ecology, obsolete capital stock in industry, backwardness in agriculture, low pensions for the elderly, and so on. It was also thought that conversion to civilian production would alleviate the

Soviet budget deficit which, at 17 per cent of the gross national product (GNP) in 1989, was much higher than that of most industrialized countries.[1]

From the very outset, the major thrust of conversion in the USSR has been defined in terms of a shift of manpower and resources from the defence sector to the long-ailing agro-industrial complex and chronically under-funded consumer durables sectors.

To initiate the conversion process, quotas for military hardware were revised downward. Thus for the period 1991–95, orders for military aircraft have been reduced by 12 per cent, ammunition by 20 per cent, tanks by 50 per cent and helicopters by 60 per cent. In 1991 alone, compared to production in 1990, the production of long-range missiles dropped by 40 per cent, military aircraft by 50 per cent and tanks by 66 per cent.[2] At the same time, the volume of defence-related research also declined: according to Victor Smyslov, Deputy Head of the Ministry of Economy, for the period 1988–91 defence-related research was cut by 15 per cent.[3]

In 1989–90 conversion had begun in more than 400 enterprises and 200 research institutes and design bureaus of the defence industry. The scale or depth of conversion differs among various groups of enterprises. According to figures from the state planning authority, Gosplan, of all the military enterprises undergoing conversion in 1990, 124 were to cut their main line of production by less than 10 per cent, 118 by 10–20 per cent, 56 by 20–30 per cent and 124 by over 30 per cent.[4] Interestingly, some military enterprises have been asked not only to step up civilian production but also to take over the least efficient civilian factories in the hope of making them more efficient. Therefore, in 1988–89 all enterprises belonging to the Ministry of the Food Processing Equipment were brought under the control of the defence ministries. The share of civilian goods in the total output of the military industries rose from 42.6 per cent in 1988 to 50 per cent in 1990 and to 54 per cent at the end of 1991.[5]

Conversion to civilian production has resulted in a shift in employment patterns. In 1990 alone, according to a statement by former Prime Minister Nikolai Ryzkov, over 500 000 employees in the defence sector were re-assigned from military to civilian production.[6]

Conversion was not limited to only the military industry. Decommissioned hardware and stocks which could be used for civilian purposes were sold by military depots. In 1989 dual-use military supplies and equipment (automobiles, small ships, radio equipment, fuel, etc.) at a value of 365 million

[1] Hosyaistvo, N., 'SSSR v 1989 gody' [USSR economy in 1989: annual statistical abstract], *Finansy i Statistika* (Moscow), 1990, p. 612.

[2] *Izvestia*, 26 and 29 Sep. 1989; *The Economist*, 14 Dec. 1991

[3] National Conversion Program for the Soviet Defense Industry, Report of Mr. V. Smyslov, Deputy Chairman of Gosplan, at the United Nations Conversion Conference, Moscow, 1990, p. 9.

[4] Reduction of the Military Expenditures and Approaches to Conversion in the USSR, Report of Mr Y. Glybin, Head of Gosplan Department at the United Nations Conversion Conference, Moscow, Aug. 1990, pp. 3–4.

[5] *Izvestia*, 17 Oct. 1991.

[6] *Pravda*, 30 Dec. 1989.

roubles were sold to the civilian sector.[7] In 1989 the Soviet Air Force established a special permanent service for transporting civilian cargo on military aircraft. The Soviet Navy established a special department responsible for selling out-of-service navy ships to domestic and foreign buyers. In 1989 this department sold 17 old submarines and a cruiser to foreign buyers for scrap metal.[8] In December 1991 the Ministry of Defence created a unified commercial centre for the sale and lease of its vast properties. The Defence Ministry has also begun to turn over to local governments and collective farms some of the 42 million hectares of land (an area about the size of Sweden) that it controls for agricultural purposes.[9]

At the end of 1990 the Soviet Government adopted a national conversion programme for 1991–95. It provided for a sharp increase in civilian production by military ministries. Each ministry was assigned one of 12 priority areas of military–civilian conversion: consumer durables; farm machinery; equipment for light industry and food-processing, trade and public catering; medical technology; electronics; computers; communications; TV and radio broadcasting; civilian ships; civilian aircraft; space technology for peaceful purposes; and new materials and technology. The aim of the programme was to co-ordinate the conversion efforts of the individual enterprises.

Although the government conversion plans were extensive, they did not yield the desired results. However much the volume of civilian production in the military industries increased, the implementation of conversion initiatives was lagging far behind initial plans. By the end of 1989 the defence industry had managed to begin production of only 23 of the planned 120 civilian goods, and only 15 per cent of the new products met international quality standards. The defence industry also failed to meet the 28.3 billion roubles mark for its planned output of consumer goods for 1990, under-fulfilling it by 20 per cent. In 1991, according to the Vice-President of Russia, Alexander Rutskoi, the increase in the civilian output of the military industries of the former USSR has made up for only one-third of the decline in the volume of the military production of these enterprises.[10]

In fact, conversion has produced more problems for the defence industry than it has solved. From the very beginning of the conversion process, the converting factories were experiencing great difficulties in finding investment and supplies for new lines of production, in creating adequate technology for new lines at acceptable costs, and in maintaining salary levels for their personnel.

The decision to start the conversion process was not preceded by serious preparation. It came as a surprise to defence industry managers, many of whom learned about the reduction or cancellation of large military orders only three to six months before they were to begin to fill them. Orders for civilian goods,

[7] *Krasnaya Zvezda*, 14 Apr. 1990.
[8] *Izvestia*, 14 Aug. 1989.
[9] *Krasnaya Zvezda*, 12 Jan. 1990.
[10] *Arguments and Facts International*, Moscow, 1992, p. 11; *New Times*, no. 10 (1990), pp. 30–32; *Sozialisticheskaya Industria*, 21 Sep. 1990.

passed down from above, simply exacerbated the problem. Since no well-considered conversion plan had been drafted, orders for civilian goods were often unmatched by funds and raw materials, and little account was taken of the technical capabilities of the enterprise involved.

Some military facilities have had to simultaneously absorb drastic cuts in defence orders and major changes in their activity. For example, the famous Mikoyan and Tupolev aircraft-design bureaus lost 20 per cent and 50 per cent of their 1990 orders, respectively, and had to compensate for them with such projects as the design of spaghetti machines and tomato-canning equipment.[11] At a time when Soviet airlines were turning down up to 20 per cent of their potential customers due to a shortage of passenger planes, the Ministry of Aviation was assigned additional quotas for the production of consumer goods.

Most military enterprises have suffered financial losses through conversion, primarily because of the lower profitability of civilian products as compared to military products. According to defence industry estimates, military orders bring a volume of sales that is two to six times higher per 'normative' labour-hour than civilian sales.[12] New civilian products, even those in high demand, can rarely be produced by a converted factory at an acceptably low cost. As a rule, military factories cannot take advantage of the economies of scale since most products were produced in short series. In addition, the factories have much higher overhead costs than their civilian counterparts. The inability to convert enterprises to cover lost defence orders has led many of them to seek support through subsidies. In 1989–90 the government earmarked 570 million roubles just for maintaining the level of salaries in the defence industry.[13]

The worsening economic situation in the defence sector has forced thousands of highly trained workers, engineers and research scientists to leave their jobs. Most of them look for better paying jobs in newly opened enterprises in the non-state sector of the economy. National statistics on this phenomenon are still not available, but the scale of the exodus is substantial, although it is slowed down by the limited employment opportunities in other industries. For example, in the Sverdlovsk region—one of the major arms-producing areas in the Urals—in 1989–90 an 8 per cent fall in military orders led to the loss of 30 000 jobs.[14] In principle such a shift from employment in military-related sectors to civilian sectors of the economy is a welcome development, but under the present conditions it often proves to be counter-productive. In many instances it results in the breakup of highly efficient teams of designers, researchers and engineers who are unqualified when they turn to seek simple jobs in the service, trade and small-industry sectors of the economy.

[11] Interview with managers of Mikoyan and Tupolev aviation design bureaus in *Christian Science Monitor*, 13 Dec. 1990.

[12] The difference resulted from the peculiarities of the Soviet price system under which artificially high prices were paid by the state for military products as compared to civilian products of comparable technological level (for example, tractors and armoured personnel carriers).

[13] National Conversion Program for the Soviet Defense Industry, Report of Mr V. Smyslov, p. 13.

[14] *Moscow News*, no. 21 (1990).

Thus the rigid structure of the Soviet defence industry has proved to be incapable of efficiently digesting the economic impact of the disarmament process begun by Gorbachev's foreign policy initiatives. The required rate of change has been too rapid for them. Conversion—initially intended to help pull the Soviet economy out of a crisis—faces a crisis of its own. The solution has become a problem.

III. Political struggle over the direction of conversion

The failure of the initial plans for speedy and efficient conversion and its crippling effect on the military industries have aroused strong opposition in the Soviet military–industrial complex. Having come to the conclusion that its hopes of 'sitting out conversion', or accomplishing it at the expense of the state budget, are not going to materialize, traditional managers, especially those involved in the ministerial bureaucracy, began to resist conversion efforts—often aligning themselves with like-minded members of the military and politicians.

Opposition to conversion manifested itself most explicitly in early 1991, with demands to curb defence cuts, limit conversion programmes, increase prices for military hardware and preserve the privileged position of defence industries with regard to supplies and funding. The opponents also demanded that the government compensate them for the economic losses they incurred because of conversion and to provide them with investments for all major conversion-related projects. Some of the more radical spokesmen of that group went as far as to call the entire conversion effort a 'subversive act' or 'treason'.

Not all of the complaints and demands of conservative critics can be attributed to a defence of their privileged position in the economy and society. Many of their concerns are reasonable and legitimate. The government conversion policies were badly prepared and did not take into account the situation of military industries and their capacity to absorb the 'disarmament shock'. Efforts to impose conversion on defence industries, on short notice and with little regard for technological and economic factors, could only backfire, especially under the stressful conditions of economic crisis and reform.

Government policies in 1989–91 failed to take into account not only the economic but also the social and psychological aspects of conversion. No one has considered the impact of conversion on the occupational and professional personnel structures of the defence factories. Many workers and engineers have expressed great resentment when being reassigned to simpler, low-prestige jobs, even when their wages were not affected. This was especially true for chief designers and executives: for generations they were told that they were performing an important duty for their country, but then their work was downgraded and they were even called 'parasites' by the radical press. Such pressure sometimes became unbearable. In January 1990 the General Director of the Votkinsk enterprise, which was famous for producing the now banned SS-20 missile, committed suicide after 20 years as Director. In his last two years he had the unenviable task of overseeing the destruction of the results of a good part of his professional life.

Demands by military managers for the right to review the going prices on military products were also reasonable. Since 1990, suppliers, no longer fearful of the Gosplan and state Military–industrial Commission ministries, have been constantly increasing prices, pushing their partners—final assembly plants—into a corner. On the other hand, the government could not satisfy the military industry's unrealistic requests to keep the command priority system of supplies at fixed low prices as it was under General Secretary Brezhnev.

As conversion problems have mounted, so has the military industry's opposition to the conversion process. Initially, this took the form of verbal criticism and passive resistance. Managers of defence enterprises and research facilities tried to minimize conversion plans and impose quotas on the production of civilian goods. By the autumn of 1990, the resentment of military industry leaders had come out into the open. On 6 September 1990 *Pravda* published an open letter from the Supreme Soviet blaming the government for undermining the defence industry and calling for the restoration of the industry's 'special place' in the Soviet economy.[15] Later, Soviet television produced a highly publicized programme on conversion entitled 'The treason'.

Towards the end of 1990 the conservative offensive developed into a vigorous campaign against the radical '500-day' economic reform plan jointly sponsored by Gorbachev and then Chairman of the Russian Supreme Soviet Boris Yeltsin. The plan called for drastic cuts in military production and expenditures and was viewed as a great threat to the military–industrial complex. In their efforts to derail the 500-day plan, defence industry leaders joined forces with conservative communist *apparatchiks* and the military. Organizations such as the Russian Communist Party and the anti-reform Soyuz faction in the USSR Supreme Soviet became the major centres of resistance to Gorbachev's radical plans. Attacking the reformers through party channels, the opposition in the Supreme Soviet and the press pressured Gorbachev to replace the 500-day plan with an alternative pro-status quo programme, developed by Prime Minister Ryzkov.

The concerted efforts of the conservative allies brought the desired result: the fragile Gorbachev–Yeltsin alliance broke down and the 500-day programme was abandoned. No less important was the fact that the source of the conversion trouble—disarmament negotiations with the West—was slowed down, and the driving force behind these negotiations, Foreign Minister Eduard Shevardnadze, had to leave his office in December 1990.

Taking into account the role of the military and the military industry in Soviet society, it is no wonder that voices of opposition to conversion have been listened to. The Government of Prime Minister Valentin Pavlov, after assuming its duties in early 1991, immediately pledged its support for the military–industrial complex. The procurement prices for various types of military equipment

[15] *Pravda*, 6 Sep. 1990.

such as armoured vehicles were increased by 60–100 per cent and the size of the military budget was revised upwards.

However, these measures were not able to effectively protect the military–industrial complex against the accelerating disintegration of the centralized state and the mounting pressure of the democratically elected governments of the former republics, in particular that of the Russian Federation, to scale down military expenditures. Not surprisingly, the conservative faction of the military–industrial complex leadership whole-heartedly supported the August 1991 orthodox communist *coup d'état*. Characteristically, four of eight members of the short-lived junta were representatives of the military–industrial complex.

The failure of the coup has dealt the military–industrial complex the hardest blow. Now that its main pillar—the centralized state—has ceased to exist, its fate has been entrusted to the governments of republics that were not willing to continue Moscow's support for the swollen armed forces and massive military production. In the situation of dire economic crisis, it was clear that the military budget would collapse, making conversion a matter of life and death for hundreds of defence enterprises throughout the CIS states.

IV. Conversion dilemmas after the breakup of the Soviet Union

In January 1992 the Parliament of the Russian Federation, which has approximately three-quarters of the defence industries and research and development (R&D) establishments of the former USSR on its territory, has made public the figures for the first 'post-*perestroika*' military budget. Because of the rampant inflation, figures were released for only the first quarter of 1992. The overall size of the budget is defined at the level of 50 billion roubles, which, at 4.5 per cent of the Russian GNP, is about one-fifth the 1991 spending level. The structure of the budget has also been dramatically changed. According to the Chairman of the Russian State Committee on Conversion, Mikhail Bazhanov, allocations for procurement have been cut by 85 per cent and for military R&D by two-thirds.[16] Estimates of the effects on employment vary greatly. According to estimates of Russian defence industry officials, up to 1 million employees of the military industries might be unemployed by the end of 1992,[17] although social and economic compensatory measures have so far prevented passive lay-offs.[18]

At the same time the fate of the Soviet armed forces and defence industries remains unclear because of the conflicts between its member states. All of them, with the exception of four Central Asian states, are creating their own armies and intend to take over major components of the former Soviet military–industrial complex based on their territories. Thus Ukraine is trying to appro-

[16] ITAR–TASS, Moscow, 6 Feb. 1992.

[17] *Business in the ex-USSR*, June 1992, p. 45.

[18] See Cooper, J., 'The former Soviet Union and the successor republics: defence industries coming to terms with disunion', chapter 5 in this volume.

priate the Black Sea Fleet; Latvia, Lithuania and Estonia claim the military facilities of the North-Western Group of the Soviet Armed Forces in the Baltic states; and Kazakhstan plans to nationalize space and military installations in and around the Baikonur space-launching complex. So far only a tentative agreement to share the assets of the former Soviet conventional armed forces has been reached. According to President Yeltsin's military adviser, Konstantin Kobets, the agreement provides for 54.4 per cent of these funds to be allocated to Russia, 21.8 per cent to Ukraine and 6.6 per cent to Belarus.[19]

In early 1992 members of the CIS were able to work out a preliminary formula for sharing the military budget of the former USSR. According to this formula, Russia agreed to pick up 62.3 per cent of the bill, Ukraine 17.3 per cent, Kazakhstan 5.1 per cent, and the remaining republics smaller fractions of the total.[20]

The governments of Russia and other CIS states that have sizeable defence industries are aware of the magnitude of the problems resulting from the demilitarization of their economies. To deal with conversion, Russia and Ukraine have already created special government bodies. In Russia a special Committee on Conversion (headed by Mikhail Bazhanov) was created by a decree of President Yeltsin in October 1991. Conversion has also been assigned to one of Yeltsin's State Counsellors (Mikhail Malei). The Russian Government has also absorbed all the ministries of defence sectors that formerly were part of the All-Union Government and put them under unified control.

In order to cushion the effect of drastic budget cuts on the military industries, the Russian Government allocated over 40 billion roubles of credits and subsidies for conversion and promised, in the words of Economic Minister Yegor Gaidar, that half of the saving on procurement would go to social assistance for defence industry employees.[21]

In April 1992 the long-discussed special law on conversion was passed by the Russian Supreme Soviet. The purpose of this law, which covers only Russian defence industries, is to provide legal guidelines for the organization of conversion procedures at the national, local and enterprise levels as well as to provide guarantees of employment and retraining for personnel in the converting enterprises.

None the less, most observers believe that neither the new conversion law nor government measures can prevent the shut-down of many defence establishments and resulting massive unemployment. Taking into account the fact that the governments of Russia and other members of the Commonwealth lack resources and are overburdened with the more pressing agenda of social and economic reforms, they are hardly in a position to be of much help in conversion. Under such circumstances the only real hope for conversion lies in the military enterprises themselves.

[19] Radio Freedom/Radio Free Europe *Daily Report*, 14 Feb. 1992.
[20] *Moscow News*, no. 52 (1991).
[21] *Moscow News*, no. 7 (1992); *Business in the ex-USSR* (note 17), p. 46.

From the very outset of the disarmament era the more liberal, better educated and more insightful members of military industry management perceived conversion as their inevitable future. Instead of opposing the process, they began to work on ways and means of adjusting to conversion in the complicated environment of the overall transition from a planned to a market economy. This market-oriented approach to conversion does not rely on government assistance, which proved to be inefficient, but rather on individual initiatives.

With the relaxation of centralized controls, defence enterprises led by liberal management have been looking for potential civilian partners and customers in the former USSR and abroad. Their goal is to meet the conversion challenge by establishing all sorts of new 'horizontal' ties instead of using the old 'vertical' ties that have proven to be inefficient for conversion purposes.

This market-oriented approach is taking several directions. One direction involves the creation of associations and amalgamations of defence enterprises. In 1990–92 several such 'self-help' associations were founded in regions with a high concentration of military production and research facilities. At present, such associations exist in Moscow, St Petersburg, Ekaterinburg (formerly Sverdlovsk) in the Urals region, the Kiev region in Ukraine, Novosibirsk in Siberia and the Udmurtia republic of the Russian Federation. The associations are forming their own data banks, seeking customers and partners, and working out ways to cut costs and raise the quality of their civilian output. They are also actively co-operating with regional governments, organizing trade fairs and exhibitions of converting industries, such as the 'Ural-Konversia' fair held in Ekaterinburg in October 1991. This type of activity is very important in regions where one-fifth to three-fourths of industrial production is military-related.

Another direction of the market-oriented approach has been the creation of mixed military–civilian amalgamations in the form of joint-stock companies. Several such companies were created in 1990–92. The largest of them—the Rhythm shareholders society—includes 300 full-member enterprises and over 1500 partners from both the defence and civilian sectors of the economy. The amalgamation pursues a wide range of high-profile, large-scale projects. For example, one project, valued at over 1 billion roubles, plans to create a low-orbit satellite system that will radically upgrade telephone communication. This project is to be completed by 1995.[22]

In March 1991 the Moscow-based Russian commodity exchange and one of the major defence companies made founded the Military–Industrial Commodity Exchange. In the course of a year the new exchange was joined by the two other exchanges also established by Russian military enterprises and private and semi-private companies and banks. All three exchanges serve the needs of defence enterprises by making their resources available for civilian production within these enterprises and offering them to their civilian counterparts. The Soviet Space Flights Control Centre has offered its computer equipment and

[22] *Trud*, 30 May 1991.

communication channels to the largest of the new military–industrial exchanges based in the city of Krasnogorsk, near Moscow.

On the micro-economic level, the market-oriented approach to conversion manifests itself in a rapidly spreading drive towards decentralization of the control and ownership structure of defence enterprises. When the conversion process began, liberals in the military–industrial complex suggested that factories with a high proportion of civilian production be 'collectively owned'.[23] In 1990–91 these ideas began to materialize, usually against the will of the ministerial bureaucracies. For example, in 1990 the Leningrad Kirov machine-building concern—one of the largest tank and tractor producers in the former USSR—managed to secure a lease agreement from its supervising ministry (the Ministry of Defence Production); the Frunze enterprise in Penza did the same at the end of 1989. Even more radical innovations were made at two concerns of the Ministry of Aviation Industry in the city of Ulyanovsk: shareholders replaced state ownership, and employees were given nearly half of the stocks free of charge. The balance will gradually be paid by the employees, who will then become the sole owners of their enterprise.[24]

The defeat of the 1991 communist coup and the introduction of radical market reforms in 1992 markedly accelerated the privatization drive in the military industries. With the power of industrial ministries severely curtailed (and many of them disbanded altogether), enterprises are given much more freedom to rethink their ownership and organizational structures. This is not to say that massive privatization of defence industries has already begun. In fact, the national privatization programme of the Russian Government which was unveiled in February 1992 does not provide for any large-scale privatization in the military–industrial complex until 1993. At the same time the new government does not resist the creation of commercial structures in the defence sector. In early 1992 it also gave the green light to privatization of practically all military enterprises on the territory of the Moscow region (excluding Moscow itself). The new ownership arrangements naturally open the way to full economic independence and have already helped individual defence enterprises acquire more freedom to manœuvre in the field of non-military production.

The question of conversion in the former Soviet nuclear and space industries is of particular importance for the outside world. The fate of the Soviet Army's nuclear weaponry, materials, production equipment and, in particular, nuclear industry engineers and scientists attracts considerable international attention.

Most of the data on the size and structure of the former Soviet nuclear complex have not yet been declassified. Unofficial reports cite about 10 major research and production centres scattered around Russia, Ukraine and Kazakhstan, employing 500 000–1 000 000 people. Of these, by the authoritative account of Russian Deputy Minister of Atomic Energy Victor Michailov,

[23] See, for example, Isaev, A., 'The reform and the defense industries', *Kommunist*, no. 5 (1989).
[24] *Kommersant*, no. 2 (1991).

10 000–15 000 previously worked directly in the nuclear programme and 2000–3000 possess the knowledge necessary to produce nuclear weapons.[25]

In the wake of the radical breakthroughs in disarmament negotiations of the past two years and in particular since the breakup of the USSR, military-related nuclear research and production has had to be cut. This has naturally made redundant substantial amounts of supplies, technology and thousands of employees of the nuclear complex. With conversion lagging far behind the cuts in military orders, the plight of the former Soviet nuclear industry has raised serious fears of uncontrolled nuclear proliferation, in particular the 'brain drain' of nuclear scientists to the less developed countries needing scientists for their nuclear weapon programmes.

So far, fears of an 'intellectual nuclear proliferation' have not been substantiated. On the other hand, since the autumn of 1991 producers of nuclear materials from the former USSR have substantially stepped up their export-oriented activity. Thus large amounts of uranium ore and nuclear space reactors have been offered to buyers in the United States. At the same time the Russian Chetek corporation, a joint venture of the Russian Atomic Energy Ministry and the huge Arzamas-16 research complex, is actively advertising peaceful nuclear explosions for commercial purposes.[26] Major space-related hardware such as the world's most powerful rocket, Energia, and the Mir space-orbit station are also being put up for sale at bargain prices.

V. Future issues

Given the inability of the new CIS governments to organize and finance speedy conversion of the former Soviet military–industrial complex, the trend towards its commercialization cannot but accelerate. To keep this process from becoming uncontrollable, the countries of the West should step up their efforts to provide assistance to conversion within the context of their overall aid to the successor states of the former USSR.

Lack of investment resources is the single most important obstacle to conversion of the defence industries. According to estimates by the Russian Government, in Russia alone the overall costs of conversion may reach $150 billion.[27] So far, in spite of much publicity about the threats to international security posed by the breakup of the Soviet military–industrial complex, Western efforts to aid its peaceful conversion have been few and far between. In order to provide the converting enterprises with sources of financing, Western governments could consider the creation of an International Conversion Fund. Its capital could be financed by the initial contribution of governments and financial institutions, to be further increased on a regular basis by allocations from mili-

[25] See interviews with Victor Michailov in *New York Times*, 8 Feb. 1992; and *Newsweek*, 24 Feb. 1992.
[26] *New York Times*, 7 and 19 Nov. 1991.
[27] *Kommersant*, no. 7 (1992).

tary budgets. The money would be used to finance various conversion projects in the former Soviet Union on a loan basis.

In turn, the Russian and other CIS governments should provide contributors to such a fund with information on their military industrial facilities. Although it is unrealistic to demand the total declassification of all military industrial units, it is highly desirable that information be provided on those that undergo conversion. All relevant information supplied by enterprises would be collected by a special information centre. Such measures would help in setting up direct production and commercial ties between converting enterprises and research institutions of the former Soviet Union and private business investors in the West.

Western countries should also significantly broaden the scope of their assistance in the establishment of international research and training and education centres for scientists, engineers and employees of the former Soviet military industry (both inside the former Soviet Union and abroad). Such centres, financed by the International Conversion Fund and other sources, could help to employ and retrain employees of the military–industrial complex, thus facilitating their reintegration into civilian life.

In the final analysis, the solution to the conversion dilemmas facing the former Soviet military industries can come only with the full implementation of the radical market reforms that are now under way, in which the decisive role is to be played by the converting enterprises themselves. International co-operation can lessen the pains and shorten the time of this major economic adjustment. Successful co-operation in this area cannot be started without proper understanding between governments and parliaments of the countries involved. Therefore, it is highly desirable that Western and CIS governments consider convening in the very near future a special conference to discuss the conversion agenda in detail and to work out a concrete programme of co-operation.

7. The Soviet Union and the successor republics: arms exports and the struggle with the heritage of the military–industrial complex

Herbert Wulf

I. Introduction

'There is no equivalent in Russian for a "Going Out of Business Sale," but this is what we are now witnessing in the crumbling state sector of the former Soviet Union. Desperate for hard currency, the emerging Commonwealth's new "entrepreneurs" are selling everything from formerly classified KGB files to high performance military aircraft.'[1] This sobering assessment of the gloomy situation of the state industry, and in particular the military–industrial complex of the Commonwealth of Independent States (CIS), must be contrasted with the situation of the 1970s and 1980s, when the Soviet Union was the largest arms exporter for most years of these two decades and the world's largest producer of major conventional weapon systems. Faced with military cuts and a limited scope for conversion to non-military production, it is not surprising that companies are desperately trying to prevent closures and secure their existence by trying to export arms to any country that has the money to pay. However, among other reasons, the financial constraints of potential customers and uncertainties about the reliability of suppliers are a barrier to large-scale CIS arms exports. In fact, their arms exports—in contrast to the political concerns about them—dropped dramatically in 1990 and 1991.

II. The historical context: a short history of Soviet arms trade[2]

Soviet arms supplies to countries outside the Warsaw Treaty Organization (WTO) had two main functions during the cold war period.

1. Soviet arms deliveries were seen as a foreign policy instrument in the struggle between the two major alliances. Assistance to socialist and non-

[1] Potter, W., 'Exports and experts: proliferation risks from the new Commonwealth', *Arms Control Today*, Jan./Feb. 1992, p. 32.

[2] For a detailed account of Soviet military assistance to developing countries up to the mid-1980s, see Krause, J., *Sowjetische Militärhilfepolitik gegenüber Entwicklungsländern* (Nomos Verlagsgesellschaft: Baden-Baden, 1985).

aligned countries throughout the world was supposed to help Soviet efforts to weaken Western- (mainly US-) sponsored military alliances, especially on the Soviet southern border in Asia.

2. The Soviet ideological and political intention was to strengthen its anti-colonial and anti-imperialist cause in the Third World.[3]

At the beginning of the 1950s, during the first 'hot' period of the cold war, half of all arms deliveries to Third World countries were supplied by the United States. The Soviet Union was not even the second largest supplier, as it became in later years; according to SIPRI data, it accounted for roughly 10 per cent of the trade in major conventional weapons with the Third World. At that time the United Kingdom supplied many more weapons to the Third World than the USSR did. In the second half of the 1950s Soviet supplies increased but were still only about half the amount of US arms supplies and still below British arms exports.[4]

The Western dominance in arms supplies to the Third World was primarily because of Soviet restraint. In the early period, Soviet weapon supplies and military assistance were exclusively directed at countries in the socialist camp, particularly to China and North Korea, although the volume of deliveries was small. Governments in the less developed countries and opposition groups—not only communist or communist-influenced parties—tried to convince the Soviet Government to supply them with weapons. During the Stalin period, the Soviet leadership mistrusted any kind of alliance with the elite in the decolonialized countries and did not consider the rising national freedom movements as revolutionary forces. In general, the attitude of the Soviet Union to the Third World was rather reserved; the newly independent states (such as Egypt under Nasser, India under Nehru and Indonesia under Sukarno) were considered an appendix of the capitalist system.

After the Stalin era this situation changed. The narrow, black-and-white world view of two antagonistic systems was rejected during the Khrushchev period. With reference to Lenin's thinking, the principle of the co-existence of different social systems was adopted. Although many countries were not perceived as being pro-Soviet, they were seen as anti-imperialistic and could thus be regarded as potential allies. Premier Nikita Khrushchev visited Afghanistan and India in 1955 and offered both technical aid for industrialization and military assistance. China and North Korea profited from the new foreign policy assessment of the Soviet leadership, and China's demands for the supply of arms production facilities were met.

Soviet arms supplies to non-communist Third World countries began gradually. To the distress of the US Government, in 1954 the Soviet Union sent a shipload of Czechoslovakian weapons to assist the reformist Guatemalan Presi-

[3] SIPRI, *The Arms Trade with the Third World* (Almqvist & Wiksell: Stockholm, 1971), pp. 180–214.

[4] These data (and the data given below, unless otherwise stated) are from the SIPRI data base. For details on the early period, see SIPRI, *SIPRI Yearbook of World Armaments and Disarmament 1968/69* (Almqvist & Wiksell: Stockholm, 1969), pp. 228–29.

dent Guzmán Jacobo Arbenz. The Soviet engagement and its perceived communist infiltration into the region also functioned as a pretext for the US Government to instigate the toppling of President Arbenz.[5] In 1955 the first Soviet weapons were delivered to Egypt, after the USA had rejected several Egyptian requests in order not to endanger its relations with Israel. A year later Syria received weapons from the Soviet Union and its ally Czechoslovakia.

This new foreign policy allowed the Soviet Union to counter successfully the US attempt to create a belt of forward defence countries against communist expansion. The Soviet Union was able to break several countries out of this front. In 1958, after the fall of the Iraqi monarchy, the first Soviet weapons were supplied to that country. A year later Iraq left the Baghdad Pact, which was renamed CENTO (the Central Treaty Organization).

In the early 1960s Soviet influence in Africa began to grow as well. The US Government had to realize, after the first decade of arms supplies and military assistance programmes to the Third World, that the Soviet Union's influence was not limited to its immediate vicinity. The US policy of containment of communism was not successful, especially in the Middle East and Africa. The success of the 1959 Cuban revolution proved that military assistance programmes and the supply of heavy military equipment were no ensurance against expanding communist influence.

The Soviet Government continued systematically to pursue its policy and in 1962 became for the first time the largest supplier of weapons to the Third World. Supplies at that time went primarily to Israel's Arab adversaries. In 1961 Egyptian soldiers were trained by about 1300 military advisers from the Soviet Union and its WTO allies. In Africa the USSR directed its assistance mainly at Algeria, and for a certain time also at Congo-Brazzaville, Ghana, Guinea, Mali, Somalia, Sudan and Tanzania. However, the volume of Soviet arms exports to these countries was small.

In contrast, Soviet relations with India were much more intensive. After the Indo-Chinese War in 1960–61 and the disrupted Sino-Soviet relationship, a long phase of armaments co-operation between India and the Soviet Union began. For many years India remained the only country outside the socialist group receiving licences to produce Soviet weapons such as fighters, warships, helicopters, missiles and tanks. In addition to India, the other main Asian recipients at that time were Afghanistan, Burma, Cambodia, Indonesia, North Korea, Laos and North Viet Nam.

During the 1960s the Soviet Government took an unambiguously pro-Arab position in the Middle East conflict, often at the expense of communist parties in the Arab countries. The large but low-cost arms supply programmes had created a strong impact among Arab leaders. However, the 1967 Six-Day War illustrated that the Egyptian and Syrian Air Forces were inferior to Israel's, but according to Soviet assessments this was not the result of inferior Soviet

[5] Wolpin, M. D., *Military Aid and Counterrevolution in the Third World* (DC Heath and Co: Lexington, Mass., 1972) has described the US Government reaction to reformist or left-wing leaders in the Third World and their relationship to the USSR.

Table 7.1. Average annual number of recipients of major conventional weapons supplied to the Third World by the four largest suppliers, 1971–85

Supplier	1971–75	1976–80	1981–85
USSR	16	24	24
USA	40	37	38
France	24	34	35
UK	22	21	19

Source: Brzoska, M. and Ohlson, T., SIPRI, *Arms Transfers to the Third World 1971–85* (Oxford University Press: Oxford, 1987), p. 5.

weapons but of misconceived military concepts in the recipient countries. The 1973 Arab–Israeli War and the US diplomatic offensive afterwards eventually led to the breach of relations between Egypt and the Soviet Union and the termination of Soviet arms supplies to Egypt.

During the Khrushchev era Soviet weapon exports to Third World countries increased considerably, but the number of recipients was small compared to the United States. According to SIPRI data, until 1970 only 29 Third World countries had received major conventional weapons from the Soviet Union. Compared to other major suppliers, Soviet arms were concentrated on a smaller number of recipients, as table 7.1 indicates. This fact is confirmed by US Government data as well. Until the break in relations with Egypt in 1973, more than three-quarters of all Soviet exports to the Third World went to only eight countries: Cuba, Egypt, India, Iran, Iraq, North Korea, Syria and North Viet Nam.[6]

As in the case of the United States, Soviet assistance was not given without strings attached. Both governments tried to influence the general political system and economic development pattern of the recipient countries. Such foreign policy goals of the Soviet Union were, at first, successfully implemented. Many governments in the Third World considered the Soviet Union an ally in its fight against colonialism and an alternative to the capitalist pattern of development. Many governments reacted favourably to Soviet offers of military assistance, its clear, verbal, pro-freedom movement stance and its few but spectacular development projects, such as the Assuan dam in Egypt and Bhilai steel work in India.

III. Arms exports in recent years

Lack of information

Traditionally, Soviet arms exports have been shrouded in secrecy. However, this situation has changed somewhat: information on the general level of previ-

[6] US Arms Control and Disarmament Agency, *World Military Expenditures and Arms Transfers 1963–73* (US Government Printing Office: Washington, DC, 1974).

ous exports and the main importers have been released by the authorities, often with reference to Western sources. Despite these recent increases in official information on Soviet (and now CIS) military issues, the assessment of arms exports is still based primarily on non-Soviet or non-CIS sources. Some of the information released is contradictory or unclear. SIPRI records on arms sales are still based almost exclusively on information that becomes public only after the arms have been delivered. Concrete assessments of developments are therefore complicated. There is still seldom any comprehensive information about the size and content of arms agreements. Thus, the reality of exports is often not known, and speculations and rumours are reported as if they were facts.

In an interview at the end of 1990, Soviet arms export figures were publicly mentioned by I. S. Belousov for the first time.[7] According to his information, in the past five years (probably the 1986–90 Five-Year Plan), weapons and weapon technology worth 56.7 billion roubles were exported—of which 9.7 billion roubles in 1990—and 8.5 billion roubles worth were exported free of charge. He confirmed that, aside from the WTO countries, the main recipients were (in order of size of imports) Afghanistan, Iraq, Syria, the People's Democratic Republic of Yemen, India, Viet Nam, North Korea, Algeria, Libya, Ethiopia, Mozambique, Angola and Cuba. Belousov further claimed that arms exports were drastically reduced at the beginning of the then current Five-Year Plan (1991–95): missiles by 64 per cent, tanks and armoured personnel carriers by 25–30 per cent, artillery by 48 per cent, aircraft by 53 per cent and ships by 56 per cent.

The general information given—both on the primary recipients and on the decline in weapon exports in recent years—is confirmed by the most recent arms transfer statistics.[8]

The scale of arms exports

During the 1970s and 1980s the Soviet Union and the United States were the two largest arms exporters, together accounting for roughly 70 per cent of the deliveries of major conventional weapon systems. According to SIPRI data, the USSR remained the largest exporter of major conventional weapons in the period 1987–91, despite a substantial reduction recorded for 1990 and 1991.

The USSR accounted for roughly 40 per cent of the global trade in major conventional weapons for most of the 1980s. In 1991 Soviet exports represented less than 20 per cent of the total (see table 7.2). The value of Soviet exports of

[7] Published in *Pravitelstvenny vestnik* [Government News], no. 80 (2 Jan. 1991). I. S. Belousov was then Deputy Chairman of the Council of Ministers of the USSR and Chairman of the State Committee of the Council of Ministers of the USSR on military industry issues.

[8] US Arms Control and Disaramament Agency, *World Military Expenditures and Arms Transfers 1990* (US Government Printing Office: Washington, DC, 1991); and Anthony, I., Courades Allebeck, A., Miggiano, P., Sköns, E. and Wulf, H., SIPRI, 'The trade in major conventional weapons', *SIPRI Yearbook 1992: World Armaments and Disarmament* (Oxford University Press: Oxford, 1992), chapter 8, pp. 272–359.

Table 7.2. Soviet exports of major conventional weapons, 1987–91

Figures are in US $m., at constant (1990) prices

	1987	1988	1989	1990	1991	Total
To the less developed world	13 420	10 762	10 869	6 846	3 516	**45 412**
To the industrial world	4 324	4 353	4 019	2 817	414	**15 927**
Total	**17 745**	**15 115**	**14 887**	**9 663**	**3 930**	**61 339**
World arms exports	45 870	39 317	38 228	29 004	22 114	**174 532**
USSR as % of world arms exports	*38.7*	*38.4*	*38.9*	*33.3*	*17.8*	***35.1***

Source: SIPRI data base; *SIPRI Yearbook 1992: World Armaments and Disarmament* (Oxford University Press: Oxford, 1992), table 8.1, p. 272. See also appendix 7A.

major conventional weapons in 1991 was roughly 22 per cent of the value recorded for 1987. With Soviet exports totalling over $9 billion in 1990 and less than $4 billion in 1991, the value of US arms deliveries exceeded that of Soviet arms exports, reversing the established rank order (see figure 7.1). While Soviet exports dropped substantially, the United States sustained its level of supplies and was thus able to increase its share in a shrinking market.

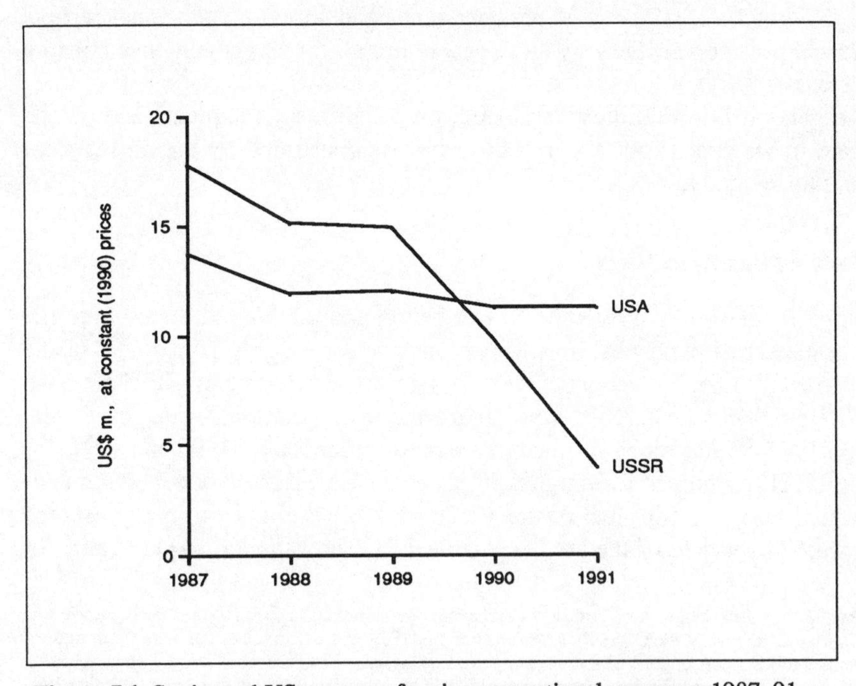

Figure 7.1. Soviet and US exports of major conventional weapons, 1987–91

Sources: SIPRI data base; *SIPRI Yearbook 1992: World Armaments and Disarmament* (Oxford University Press: Oxford, 1992), table 8.1, p. 272.

The rapid fall in Soviet exports of major conventional weapons at the beginning of the 1990s is a result of drastically reduced exports to Third World countries as well as a consequence of the virtual breakdown of the division of labour within the former WTO. According to publicly available information, exports to Iraq were completely halted to comply with the United Nations arms embargo. Reductions in exports to Afghanistan and Angola reflected US–Soviet discussions of conflicts that played a particularly important role in bilateral relations during the cold war: in 1991 agreements were reached between the USSR and the USA to halt arms supplies to both these countries. In the case of North Korea a decision was made in late 1990 to improve Soviet–South Korean relations. It is not known whether the governments of the new states emerging on the territory of the former USSR intend to uphold these decisions.

India—another important Soviet arms recipient—has reconsidered its weapon import policy both because of internal difficulties, particularly its balance-of-payments deficit, and because of the economic and political chaos in the former Soviet Union, which had an immediate impact on the Indian armed forces. Shortages of spare parts reduced the availability of several weapon systems. In 1992 these differences seem to have been resolved, but arms sales to India have continued at a lower level than during the 1980s.

During the second half of the 1980s, transfers of advanced aircraft—in particular the MiG-29 fighter and the Su-24 fighter-bomber—became more important than the traditional Soviet export, armoured vehicles.

The recipients

While the Soviet Union was the largest arms exporter—in terms of the value of major weapons supplied—it was not the supplier with the largest number of recipients. The traditional pattern mentioned above of supplies to a smaller number of countries continued during the late 1980s and early 1990s as well. As table 7A.1 (appendix 7A) shows, 58 countries—mainly from the less developed world—imported weapons from the USSR in 1980–91. Within this group of recipients, Soviet exports were heavily concentrated: the 10 leading recipients accounted for over three-quarters of total Soviet arms exports—India, Iraq, Syria, Afghanistan, Libya, Angola, and North Korea among the less developed countries, and Czechoslovakia, the GDR and Poland in the WTO. The five leading importers of Soviet weapons accounted for over half of all weapons traded by the USSR (see table 7.3).

Deliveries of major conventional weapons to former WTO member states declined to a marginal fraction of the previous volume.

In contrast to major Western arms suppliers, the Soviet Government has restricted the supply of arms production technology to many recipients. According to publicly available information, the Soviet Government has supplied production licences outside the former WTO only to China, India, Iraq and North Korea.[9]

[9] Information from the SIPRI data base.

Table 7.3. The 10 leading recipients of Soviet arms exports, 1980–91

Figures for the value of imports are in US $m., at constant (1990) prices; shares are in percentages.

Recipient	Value of arms imports	Share of total Soviet exports of major conventional weapons	Cumulative share
India	25 415	*15.3*	*15.3*
Iraq	19 519	*11.7*	*27.0*
Syria	16 703	*10.0*	*37.0*
Czechoslovakia	14 244	*8.6*	*45.6*
Afghanistan	10 626	*6.4*	*52.0*
Libya	9 105	*5.5*	*57.5*
German, DR	7 911	*4.8*	*62.3*
Angola	7 791	*4.7*	*67.0*
Poland	7 519	*4.5*	*71.5*
Korea, North	7 417	*4.4*	*75.9*
Total of the 10 leading recipients	**126 250**	***75.9***	
Total of all Soviet arms exports	**166 436**	***100***	

Source: SIPRI data base.

Export dependence

Although arms exports have in the past been a major component of Soviet foreign trade, there is no evidence that arms sales brought significant benefits to the Soviet economy as a whole. Several of the key recipients of Soviet weapons were not in a position to pay for the imported weapons, let alone in hard currency. The scattered evidence available on terms of payments suggests that the USSR could not always recover the real cost of production of exported weapons.

Most of the largest debtors to the USSR were also major importers of Soviet weapons. Among the 10 major debtors are 5 of the 10 leading weapon importers, as table 7.4 indicates. In the above-mentioned interview,[10] Belousov confirmed the total outstanding debt of approximately 86 billion roubles and said—without giving any details—that some of this debt was due to arms supplies, particularly supplies to Third World countries, although most countries had been on time in their debt service. Although a correlation between debt and arms imports is suggested by the table, and is plausible, neither set of figures should be regarded as truly reliable because information remains limited.

At the end of 1991, however, the weekly *Moscow News* published a list of African debtors to the Soviet Union.[11] The article pointed out that some of the debt with 'less solvent socialist-oriented countries such as Angola, Mozam-

[10] See note 7.

[11] *Moscow News*, no. 50 (1991), p. 11.

Table 7.4. A comparison of the major recipients of Soviet arms and major Soviet debtors, 1980–91

Figures for debt are in m. roubles; shares are percentages.

Major Soviet debtors	Debt, 1989	Shares of Soviet arms exports, 1980–91	Rank as importer of Soviet arms
Cuba	15 490.6	*3*	12
Mongolia	9 542.7	*0*	35
Vietnam	9 132.2	2	14
India	8 907.5	*15*	1
Syria	6 742.6	*10*	3
Poland	4 955.0	5	9
Iraq	3 795.6	*12*	2
Afghanistan	3 055.0	*6*	5
Ethiopia	2 860.5	*1*	19
Algeria	2 519.3	2	15
Cumulative share of the 10 leading debtors' debt and arms exports	*78*	*56*	

Sources: SIPRI data base; *Izvestia*, 1 Mar. 1990 for statistics on debt.

bique, Ethiopia etc.' might be difficult to recover, but the Russian Government hopes to be repaid in precious products such as coffee, cocoa beans, citrus fruit, bananas and oil-bearing crops. Table 7.5 shows that nearly 90 per cent of the accumulated debt of all African countries was incurred from arms supplies or military assistance. In 3 of the 14 countries, military-related debt amounted to 100 per cent and in four additional cases to 90 per cent or more.

A realistic assessment of the extent to which the former Soviet Union was dependent on arms exports is still difficult to ascertain. To arrive at an estimate of the dependence of the Soviet military–industrial complex on arms exports during the 1980s, SIPRI has compared the number of weapon systems produced (according to US Department of Defense information) and the weapon systems exported to the Third World, the WTO countries and all other countries with the total number of weapons produced in three categories of weapon systems. All three selected categories—main battle tanks, fighter aircraft and major surface ships—were important export products of the Soviet military–industrial complex. In certain areas (such as strategic bombers, intercontinental ballistic missiles and submarine-launched ballistic missiles), no exports have been recorded at all.[12] Two periods were chosen for comparison: 1980–84 and 1985–89 (the five years just before and the five full years of the Gorbachev Government). According to the available information, of the 13 800 tanks produced in the period before Mikhail Gorbachev took office in 1985, approximately 14 per cent were exported; most of these main battle tanks were

[12] Production figures are from *Soviet Military Power 1984–1990* (US Government Printing Office: Washington, DC, 1984–90); export figures are from the SIPRI arms trade data base.

Table 7.5. Africa's debt to the USSR as of 1 January 1991

Figures for debt are in m. roubles; shares are percentages.

Debtor	Total debt to the USSR	Military-related debt	Share of military-related debt
Algeria	2 706	2 478	*92*
Angola	2 100	2 100	*100*
Congo	215	153	*71*
Egypt	1 698	1 682	*99*
Ethiopia	3 075	2 630	*86*
Guinea	259	95	*37*
Libya	1 589	1 589	*100*
Madagascar	113	25	*22*
Mali	300	118	*39*
Mozambique	891	648	*73*
Nigeria	26	26	*100*
Somalia	271	159	*59*
Tanzania	330	296	*90*
Zambia	223	221	*99*
Total for Africa	**13 937**	**12 347**	*89*

Source: *Moscow News*, no. 50 (1991), p. 11.

imported by Third World countries.[13] This percentage dropped considerably after 1985, although production was kept at about the same level until it fell in 1989.

In contrast to exports of main battle tanks, a higher percentage of Soviet fighter aircraft has been exported: 21.5 per cent of total fighter production in the pre-Gorbachev years and 31.5 per cent in 1985–89. This percentage change is the result of reduced production of fighter aircraft but constant export figures. Again, most of these fighter planes were exported to less developed countries—about twice as many as to WTO countries.

The Soviet export dependence on major surface ships was particularly high. Exports of major surface ships fell from 42.5 per cent of total production in the first half of the 1980s to 28 per cent in the second half. The less developed countries were the main customers for this category of conventional weapon systems as well.

These results correspond to the general pattern of Soviet exports. In the Gorbachev period there was not an abrupt change but rather a gradual change from the previous Soviet arms export policy. Traditionally, the Soviet Union transferred about 70 per cent of its arms exports to the less developed world. This pattern was not changed in the Gorbachev period; in 1985–91 arms exports to less developed countries amounted to 73 per cent of total Soviet arms exports.

[13] For further details and for the methodological reservations, see Wulf, H., SIPRI, 'Arms production', *SIPRI Yearbook 1991: World Armaments and Disarmament* (Oxford University Press: Oxford, 1991), pp. 303–306.

The major political changes attached to the terms *perestroika* and *glasnost* began to be translated into a new arms export pattern after 1989. The changing arms export pattern was as much a by-product of the turmoil within the USSR, thus unintentional, as it was a consequence of new policy priorities.

Foreign policy changes and the results

In the 1950s and 1960s, Third World leaders perceived Soviet foreign policy and military assistance as an alternative to their reliance on Western sources. This perception changed in later years, as many governments became dissatisfied with Soviet performance. Soviet assistance remained limited, except for arms supplies. The USSR often could not deliver the goods that revolutionary ideology required. The Soviet economic potential was simply too small to afford many Cubas. The Soviet Union remained a superpower only in military terms but could not provide the technological or economic assistance that was needed by the less developed countries. The breach in relations with Ghana, Indonesia, Mali and Somalia all signaled the disappointment of the governments of these countries with the Soviet Union.

The major failure of Soviet foreign policy in the 1970s was the forced Soviet departure from Egypt. In the 1980s it was the sad experience in Afghanistan—a war that neither the Soviet military nor the Soviet-trained and -supplied Afghan Army could win—that troubled both Soviet politicians and the military. Last but not least, the experience with the long-term Soviet friend and ally Iraq proved that arms supplies were not a basis for long-term, stable economic and political relations but rather a form of Russian roulette.

Iraq's armed forces used weapons imported from two dozen countries in addition to a limited number of weapons produced in Iraq. However, the major bulk of the weapons used in the 1990 Iraqi invasion of Kuwait had been supplied by the USSR, whose leadership could neither prevent the invasion nor enforce a withdrawal by political means. Of the major conventional weapons imported by Iraq during the period 1980–90, over 50 per cent came from the USSR, which backed the UN Security Council arms embargo after the Iraqi invasion of Kuwait.[14]

The pattern of Soviet arms exports began to change in the last few years before the collapse of the USSR in response to several factors. One important factor, of course, was the above-mentioned critical state of the economy. The Soviet Union could no longer afford to subsidize its clients. This factor contributed to restraining the generous supplies of the past. At the same time, however, the need for hard currency worked and still works in the opposite direction. The highly developed arms industry was seen as a likely hard currency earner in the situation of a deteriorating foreign trade position.

The general foreign policy change in the Gorbachev era and its concept of 'new thinking' were important. In the concept of reconciling relations with its

[14] For details on the arms supply to Iraq, see *SIPRI Yearbook 1991* (note 11), chapter 7, pp. 201–204.

superpower rival the USA, arms export control played a role—although a less prominent role than the control of nuclear weapons. During the period after 1985, when General Secretary Gorbachev came to power, a change in Soviet arms export policy became increasingly apparent. As pointed out in a comment in *Komsomolskaya Pravda*: 'Lately, the phrase "assistance to the national-liberation movements in their just struggle", which has long been part of our political vocabulary, has been gradually supplanted by another one: "the search for peaceful ways of settling regional conflicts". The USSR is becoming more cautions in choosing partners and more critical of their requests for arms supplies.'[15] As a consequence of the 'new thinking', the Soviet leadership was prepared for compromises in Afghanistan, Central America and South Africa. Gorbachev spoke in 1987 about the 'need for new ideas, new points of view and collective efforts' and with regard to South Africa, even of the possibility of 'national reconciliation'.[16]

Before the 1991 Persian Gulf War, the Soviet Government had already expressed doubts about its past arms transfer policies with regard to the Middle East. Then Foreign Minister Eduard Shevardnadze had criticized the creation 'of powerful weapons arsenals . . . in close proximity to Europe'.[17] Soviet declarations in support of arms transfer control translated into a willingness to participate in negotiations—in the UN Security Council, in the UN Register of Conventional Arms as well as in the Missile Technology Control Regime (MTCR).

In quantitative terms, however—as pointed out above—the new policy of arms export control became apparent only gradually, particularly in 1990 and 1991.

IV. What next: restraint or expansion?

Policy dilemma: economic pressures versus political motives for restraint

'Soviet foreign trade has been unstable since the mid-1980s. Steady growth continued until 1986, then turned into stagnation, followed by a slump in 1990. The slump took a sharp downturn in 1991.'[18] This assessment of the general trade development is paralleled by the export of arms. There is considerable evidence that a serious domestic debate has taken place, at least in the Russian Federation, concerning arms transfer policy. Two possibly contradictory trends are emerging—the economic necessity to export arms versus the political intention to control arms exports.

[15] *Komsomolskaya Pravda*, 12 June 1990.

[16] Quoted in Tichomirow, W., 'Revolutionäre Parolen—handfeste Geschäfte. Die Südafrikapolitik der UdSSR' [Revolutionary language—serious business: the Soviet policy on South Africa], in *Der Überblick*, vol. 28, no. 1 (1992), p. 22. Tichomirow was until Sep. 1991 project leader at the Institute for Africa Studies of the Soviet Academy of Sciences.

[17] Quoted in Sivers, A., *Conventional Arms Control: Considering New Directions*, Faraday Discussion Paper no. 13 (Council for Arms Control: London, 1989), p. 51.

[18] *Moscow News*, no. 43 (1991), p. 11.

Three potential sources for the supply of arms or the proliferation of weapon design and production knowledge are obvious.

1. Hundreds of thousands of workers, engineers and scientists are threatened by lay-offs. In the chaotic economic and political situation, specialists who have lost their jobs are looking for new employment; thus their skills have become available on the world market.

2. Large numbers of weapon stocks have become available as a result of the Treaty on Conventional Armed Forces in Europe (CFE I Treaty), requiring the former Soviet Union to reduce its military holdings drastically—especially armoured combat vehicles and tanks.

3. Factory managers are trying to prevent closure by exporting the products that they have successfully built for many years—weapons—as arms procurement for the CIS armed forces is reduced and arms production decelerated or even stopped in numerous factories.[19] Their immediate aim is to generate enough funds to keep paying wages and to prevent massive lay-offs.

The problem of the distribution of existing stockpiles must be considered in all the successor republics; the design and production of weapons are of concern primarily carried out in Russia and to a lesser extent in Ukraine and Belarus.

On the other hand, the intention to reduce the level of both arms production and arms exports has been explicitly stated in Russia, reflecting the changed international environment. Newly emerging states have claimed control over the arms and facilities located on their territory but at the time of writing, with the exception of Russia, no legislative or administrative mechanism for export control has been introduced.

While the Soviet Union had become a constructive participant in the conventional arms control process, including arms transfer control, the economic imperatives to export arms seem overwhelming. Mikhail Maley, Russian State Counselor for Defense Conversion, said: 'The thing is that the weapons market does exist and will exist. Weapons are sold and produced by highly moral Germans and by Americans concerned about human rights. Why cannot we do likewise, finding ourselves in a very grave crisis?'[20]

Russian weapon systems, particularly fighters of the highly developed aerospace sector, have been actively marketed and advertized in journals, at air shows and even by Western sales promotion companies.[21] Russian officials, emphasizing their intention to sell arms, stressed their readiness to do everything possible so that these arms will not reach zones of conflict either inside or

[19] According to a Moscow Foreign Ministry official, at the beginning of 1992 only 15 per cent of the original procurement is taking place. Interview by the author on 5 June 1992 with the official. According to other sources, the aviation factories have faced a 50 per cent reduction in a 12-month period during 1991–92. *International Defense Review*, no. 4 (1992), p. 372. In 1992 production may fall by 30 or even 50 per cent according to reports in Russia, quoted in *Daily Report–Soviet Union* (hereafter *FBIS-SOV)*, FBIS-SOV-92-083, 29 Apr. 1992, p. 31.

[20] Quoted in *Russian Aeropace and Technology*, 23 Mar. 1992, p. 5.

[21] *International Herald Tribune*, 21 June 1991, p. 2; *Financial Times*, 21 June 1991, p. 4; *The Guardian*, 21 June 1991, p. 4; *Jane's Defence Weekly*, 16 Nov. 1991.

outside the CIS.[22] Reports about arms exports or alleged sales by Russia and other republics are appearing in abundance in the press, such as: Russian MiG-29s to Malaysia;[23] SA-5 surface-to-air-missiles, T-72 tanks, Su-24 fighter-bombers and submarines to Iran;[24] aviation technology exports and 20 Su-27 fighters to China;[25] South Africa is studying the purchase of MiG-29s;[26] Turkey to purchase $300 million worth of arms from Russia;[27] armoured combat vehicle sales to the UAE;[28] and purchase of an aircraft-carrier by China or India from the Ukraine;[29] Ukraine plans to sell Tu-160 bombers;[30] and Turkmenistan is said to review the sale of 223 military aircraft abroad.[31] 'Free-for-all in Moscow's arms bazaar' was the title of an article in *The Independent,* and the paper reported among other deals that Russia is about to sell Scud missiles to South Korea.[32] The German paper *Frankfurter Allgemeine Zeitung* reported that Russia is offering weapons to any country in the world for hard currency.[33]

Despite concerted efforts to promote the export of arms and other military equipment, there is little evidence that producers are recapturing markets lost during the collapse of the Soviet Union. There are many rumours about new sales, but 'where is the beef'? At the end of 1992, at the time of writing, it seems clear that the long established co-operation between India and the Soviet Union is being continued by Russia, but arms sales to India have slumped after months of argument about deliveries and pricing.[34] The deal for technology imports by China, agreed upon in 1990, is also continuing. Iran apparently has received several submarines and the United Arab Emirates are testing armoured combat vehicles. These deals, however, are far from the quantitative level of previous exports—for the reasons mentioned above.

Export control

A preliminary new system of export controls was introduced in Russia as of 2 January 1992. Two bodies were created to execute arms export policy: a special arms export and import department to make a political evaluation of the prospective customers for arms and war *matériel*; and a directorate for monitoring the export and import of arms to be set up under the Parliamentary Committee for Conversion Affairs. The export control mechanism was revised in May 1992 and will operate pending the formal passing of a law by the Russian Par-

[22] For example, Russian Minister of Industry Aleksandr Titkin, quoted in FBIS-SOV-92-047, 10 Mar. 1992, p. 23.
[23] *Far Eastern Economic Review*, 13 Aug. 1992, p. 20.
[24] *Moscow News*, 5 Apr. 1992 (no page given).
[25] FBIS-SOV-92-158, 14 Aug. 1992, p. 9.
[26] FBIS-SOV-92-158, 14 Aug. 1992, p. 9.
[27] *Neue Zürcher Zeitung*, 1 July 1992, p. 4; and FBIS-SOV-92-157, 13 Aug. 1992, p. 11.
[28] *Jane's Defence Weekly*, 1 Aug. 1992, p. 7.
[29] FBIS-SOV-92-149, 3 Aug. 1992, p. 12.
[30] FBIS-SOV-92-150, 4. Aug. 1992, p. 47.
[31] FBIS-SOV-92-106, 2 June 1992, p. 35.
[32] *The Independent*, 10 July 1992, p. 10.
[33] *Frankfurter Allgemeine Rundschau*, 25 June 1992, p. 2.
[34] According to Itar-Tass, quoted in *Financial Times*, 8 Sep. 1992, p. 8.

liament. Two decrees passed in April 1992 regulate the control in the Russian Federation 'concerning export of nuclear material, equipment and technologies' and 'on measures concerning the establishment of export control systems'[35]. A detailed list of raw materials, equipment, technologies and scientific information used for manufacturing weapons and war materials is attached to the decrees.

Similarities to Western countries are apparent. The new features in Russia—compared to previous Soviet systems of control—are the clear definitions of responsibilities of the different ministries and participation of the parliament (the Supreme Soviet); and, regarding commercial sales, it is the fact that private companies can apply for export licences. However, given the uncertainties surrounding political and economic developments, it is clear that the implementation of policy will not be easy, and it is not clear how effective control can be maintained.

Furthermore, export control does not necessarily mean restriction. Russian President Boris Yeltsin, responding to the pleas of military producers, authorized and gave the go-ahead to the international sale of weapons in January 1992.[36]

Is the export option viable?

Parallel to the debate on a reduction of arms production and conversion of production facilities to non-military production, it has been suggested that factories could, instead of converting their facilities, try to export more weapons. The 'export option' to rescue factories of the former Soviet military–industrial complex entered the debate even before the collapse of the USSR.[37] Managers of arms production plants demanded more freedom to export arms to finance capital investment and product development.[38]

The general development of introducing free market incentives and creating self-financed factories—even in the military sector—led to such considerations of exploiting the export potential. Russian Government officials give different estimates of anticipated arms exports. Expectations of hard currency income range between $6 billion and $10 billion for 1992,[39] and 1600 aircraft that have been authorized for sale by President Yeltsin are expected to earn $9 billion.[40] Estimates of the stockpile suggest there is at least $1000 billion worth of equipment.[41]

There are several reasons why the export concept is not a realistic option for keeping the existing facilities occupied.

[35] *Rossijskije Vesti*, vol. 37, no. 83 (Apr. 1992).

[36] *Kommersant*, no. 3 1992; *The Independent*, 4 Feb. 1992, p. 1.

[37] For a discussion of this, see Cooper, J., 'Soviet exports and the conversion of the defense industry', in ed. L. Bozzo, *Exporting Conflict* (Cultura Nuova: Florence, 1991), pp. 135–42.

[38] Izyumov, A., 'Conversion: an export version', *Moscow News Weekly*, no. 16 (1990).

[39] *Moscow News*, no. 7 (16–23 Feb. 1992), p. 11.

[40] *Soldat und Technik*, no. 4 (1992), p. 223.

[41] *The Independent*, 4 Feb. 1992, p. 1.

1. Arms industries in potential importer countries—including those in Western Europe—are struggling with their own over-capacities, and it is difficult to imagine that in this situation Russian or other successor republics' weapons will be bought on a large scale by Western governments, even if they are offered at lower prices. This does not exclude sales on concessionary terms. Russia most likely remains an important major arms exporter. It is doubtful, however, whether it will reach the quantitative level that the USSR occupied during the peak period of the mid-1980s.

2. The decision to conduct all foreign trade in hard currency has major implications for arms exports not only within the former WTO but also with less developed countries which previously traded on a soft-currency basis with the USSR. The dramatic changes of recent years have undermined one rationale for buying arms from the successor states of the USSR: obtaining political support from a superpower. If traded on a commercial basis, these goods may no longer retain another comparative advantage—low cost. Many of the former Soviet customers in the Third World do not have the hard currency funds available which producer companies are looking for. In the past, Soviet weapons were sold on soft terms or not paid for at all.

3. Some of the largest importers of Soviet weapons, the former WTO allies of the USSR, are reducing their inventories as a result of the CFE Treaty and the breakdown of the WTO as an alliance, and they are closing or converting at least some of their own arms factories to civilian production. Their exports cannot be expected to return to the previous level nor can they be expected to totally disappear.

4. Finally, great numbers of surplus weapons are available in the armed forces of the newly emergent states on the territory of the former Soviet Union. Thus the governments themselves are likely to compete with the factories for customers.

Appendix 7A

Soviet arms exports, 1980–91

Prepared by Herbert Wulf

Table 7A.1. Exports of major conventional weapon systems by the former Soviet Union, 1980–91

All figures are in US$ m., at 1990 prices and exchange-rates.

Recipient	1980	1981	1982	1983	1984	1985	1986	1987	1988	1989	1990	1991	1980–91
Afghanistan	504	488	314	139	219	79	758	756	1 177	2 581	2 414	1 197	**10 626**
Algeria	557	534	777	429	242	48	30	176	176	580	–	–	**3 550**
Angola	105	217	278	734	1 068	775	1 069	1 586	1 144	74	740	–	**7 791**
Bangladesh	–	18	–	–	–	–	–	–	–	–	–	–	**18**
Burundi	2	–	–	–	–	–	–	–	–	–	–	–	**2**
Benin	–	–	–	4	–	–	–	–	–	–	–	–	**4**
Burkina Faso	–	–	–	–	10	–	–	–	–	–	–	–	**10**
Botswana	–	4	5	–	–	–	–	–	–	–	–	–	**9**
Bulgaria	495	1 275	852	608	617	840	815	697	227	53	422	347	**7 248**
Canada	–	–	–	–	–	–	–	–	–	3	–	–	**3**
Cape Verde	–	–	18	1	–	–	–	–	–	–	–	–	**19**
Central African Rep.	–	–	–	–	–	–	–	6	–	–	–	–	**6**
China	2	2	2	0.5	0.5	0.5	0.5	0.5	1	–	86	410	**505**
Congo	6	–	–	–	1	–	–	–	–	–	–	–	**7**
Cuba	80	629	809	971	374	457	415	29	5	143	190	143	**4 246**

Recipient	1980	1981	1982	1983	1984	1985	1986	1987	1988	1989	1990	1991	1980–91
Cyprus	–	–	–	1	1	–	–	–	–	–	–	–	**2**
Czechoslovakia	1 194	1 638	1 166	1 368	1 350	1 558	1 285	1 167	1 197	1 557	716	47	**14 244**
Equatorial Guinea	2	–	1	–	–	–	–	–	–	–	–	–	**4**
Ethiopia	246	59	312	122	55	205	202	77	80	41	41	–	**1 441**
Finland	195	31	2	1	9	235	171	58	63	74	37	–	**877**
Guinea Bissau	1	24	6	–	4	–	–	–	–	–	–	–	**35**
German DR	772	538	669	1 229	1 146	772	569	430	601	600	585	–	**7 911**
Guinea	2	–	17	5	11	–	78	–	–	22	–	–	**135**
Guyana	–	–	1	–	–	18	–	–	–	–	–	–	**19**
Hungary	658	769	666	248	3	604	304	405	–	36	36	–	**3 730**
India	1 427	1 633	1 298	1 799	1 004	1 402	2 981	4 125	3 276	3 549	1 277	1 644	**25 415**
Iran	–	–	–	–	–	–	17	–	–	–	607	107	**732**
Iraq	1 679	1 274	1 429	1 467	2 953	2 255	1 413	3 923	1 532	1 157	437	–	**19 519**
Jordan	–	–	–	281	–	–	293	33	14	7	–	–	**628**
Kampuchea	12	–	–	4	84	69	–	–	–	112	58	–	**339**
Korea, North	76	43	60	93	760	1 120	1 047	576	1 503	1 511	612	15	**7 417**
Kuwait	112	–	–	–	55	60	102	–	–	43	169	–	**540**
Laos	–	19	3	80	115	49	–	113	12	–	–	–	**390**
Libya	2 098	1 826	1 177	418	191	1 067	1 526	198	15	589	–	–	**9 105**
Madagascar	12	13	–	6	15	1	–	–	–	–	–	–	**46**
Mauritius	–	–	–	–	–	1	–	–	–	–	–	–	**1**
Mali	5	81	5	2	1	–	–	–	–	8	23	–	**125**
Mongolia	–	19	33	27	–	–	–	–	–	–	–	–	**79**
Morocco	–	1	–	–	–	–	–	–	–	–	–	–	**1**
Mozambique	24	23	30	41	106	255	3	–	–	–	–	–	**482**

Malawi	–	–	–	2	–	–	–	–	–	–	–	–	**2**
Nicaragua	–	30	22	28	96	156	197	260	30	57	71	–	**949**
Nigeria	30	18	–	–	–	176	–	–	–	–	–	–	**224**
Oman	1	–	–	–	–	–	–	–	–	–	–	–	**1**
Papua New Guinea	–	–	–	–	–	–	–	–	–	–	19	–	**19**
Peru	183	118	–	100	–	143	43	164	71	–	100	–	**923**
Poland	353	563	367	442	434	508	1 071	1 012	1 247	1 225	286	10	**7 519**
Romania	55	181	181	185	185	212	265	269	281	86	631	10	**2 541**
South Africa	–	–	–	–	–	–	–	–	0	–	–	–	**0**
Seychelles	2	–	10	–	2	–	10	–	–	–	–	–	**23**
Syria	1 697	1 255	2 133	2 552	2 043	1 998	1 844	1 392	1 393	395	–	–	**16 703**
Uganda	–	–	–	–	–	–	–	–	12	–	–	–	**12**
Viet Nam	758	1011	74	554	253	556	432	6	–	–	–	–	**3 643**
Yemen, North	568	121	38	1	38	154	–	–	27	–	–	–	**947**
Yemen, South	509	383	104	71	31	27	246	–	292	–	–	–	**1 663**
Yugoslavia	385	447	662	227	107	106	106	286	737	385	103	–	**3 549**
Zambia	185	199	30	1	40	–	–	–	–	–	–	–	**455**
Zimbabwe	0	–	–	–	–	–	–	–	–	–	–	–	**0**
Total	**14 995**	**15 481**	**13 551**	**14 244**	**13 625**	**15 908**	**17 293**	**17 745**	**15 115**	**14 887**	**9 663**	**3 930**	**166 436**
To industrialized countries	4 106	5 441	45 65	4 309	3 851	4 836	4 586	4 324	4 353	4 019	2 817	414	**47 622**
To developing countries	10 889	10 039	8 986	9 935	9 775	11 072	12 707	13 420	10 762	10 869	6 846	3 516	**118 814**

Source: SIPRI data base.

Part IV
Europe

8. Western Europe: facing over-capacities

Herbert Wulf

I. Introduction

Of the three major centres of arms production in the world, the United States, the former Soviet Union and Western Europe, the West European industrial base is the smallest. Three indicators illustrate the different magnitude of arms production in the United States, the technologically most advanced arms industrial base in the world, and in Western Europe.

1. In the early 1990s, the US Government spent more than twice as much on the procurement of major weapon systems than all European NATO countries together, including France.
2. Most of the procurement budget is being spent domestically, in spite of declarations in favour of international co-operation. Quantitative differences in procurement are thus reflected in company sales. The list of the 20 largest arms-producing companies in the OECD (Organization for Economic Co-operation and Development) countries and the less-developed countries in 1990 includes only 4 European but 15 US companies.
3. In arms exports, the United States ranks higher than all the West European countries combined. According to SIPRI statistics, in the period 1987–91 US exports of major conventional weapons (in value terms) were twice the volume exported by all the West European countries.[1]

Governments and industry are faced with a dual problem. First, the smaller size of the West European military market and its fragmentation will handicap industries in competition with their US competitors.[2] Second, with the dominant trend of a decrease in military budgets and the rising unit production costs for major military equipment, the present arms industrial base cannot be maintained. The capability of producing high-technology equipment will be limited in the future by the inability of governments to finance such weapon systems in large enough quantities to sustain the industrial base. As was summarized in a study for the European Community (EC) Commission: 'Only by producing for the entire European region or for the American demand as well would some of this capability be justified in the future. But European producers are still in a

[1] All these data are from *SIPRI Yearbook 1992: World Armaments and Disarmament* (Oxford University Press: Oxford, 1992).

[2] Fouquet, D., Kohnstam, M. and Noelke, M. (eds), *Dual-Use Industries in Europe*, vols I, II and Executive Summary, Study carried out by Eurostrategies for the Commission of the European Communities, DG III: Brussels, 1991, p. A-12.

phase in which they are sustaining substantial over capacity and must decide whether to continue to compete, abandon, diversify, collaborate or take other action'.[3]

II. Country characteristics

The arms industry in Europe is far from a unified industrial branch, and no multilateral institution in Europe has the authority to take procurement or industrial policy decisions. Arms-producing industries are constrained by national research and development (R&D) and procurement budgets. Even in international collaborative projects, companies of a particular country can expect to receive a share of procurement orders that correlates highly to the financial burden which that country carries. Thus, the arms industry—although it is becoming increasingly internationalized—is still largely organized on a national scale.

France

In terms of procurement expenditure and exports, France is in first place in Europe. Funds for new equipment orders have been gradually reduced since 1989. Although their exports of arms have declined considerably, French companies are unlikely to face ruin under the current French three-year military spending bill (1993–95). While the procurement budget will not shrink substantially, the way it is divided up among French companies is likely to change because of a shift in military priorities. Expenditure for French nuclear forces will be reduced and reallocated to projects for intelligence-gathering systems.[4]

The arms industry (including the large nuclear sector and later the space technology sector) has played a key role in government industrial and technological policy since the 1960s. It has become a priority sector and has been sponsored and subsidized by the military budget, in addition to generous tax allowances and export credits.[5] A cluster of over a dozen large firms—partly government-owned—form the core of the French arms industry around the Délégation Générale pour l'Armement (DGA), the government agency responsible for military R&D and production. Employment in the French arms industry has declined from 290 000 in the mid-1980s and 270 000 in 1988 to around 255 000 in 1991.[6]

[3] Fouquet, Kohnstam and Noelke (note 2), p. A-67.

[4] *Interavia Air Letter*, 14 July 1992, p. 4.

[5] Serfati, C., 'Reorientation of French companies', in M. Brzoska and P. Lock (eds), SIPRI, *Restructuring of Arms Production in Western Europe* (Oxford University Press: Oxford, 1992), pp. 95–107; Kolodziej, E. A., *The Making and Marketing of Arms: The French Experience and its Implications for the International System* (Princeton University Press: Princeton, N.J., 1987).

[6] *Avis présenté au nom de la commission de la défense nationale et des forces armées sur le projet de loi de finances pour 1991 par M. Jean-Guy Branger*, Tome VI, Défense, recherche et industrie d'armement, Assemblée Nationale, no. 2258 (9 Oct. 1991), p. 14; *Interavia Air Letter*, 10 Sep. 1991, p. 6. Almost half of these employees work in the aerospace sector.

The United Kingdom

With an estimated work-force of 400 000 at the beginning of the 1990s—down from 560 000 a decade ago—in terms of employment the United Kingdom has the largest arms industrial base in Western Europe.[7] In terms of arms procurement expenditure and exports, Britain has traditionally had the second largest industry in Western Europe. The industry has been suffering from a significant drop in orders owing to the end of the cold war. In July 1991 Tom King, the Secretary of State for Defence, presented a White Paper on 'Options for Change', projecting reductions of real military expenditure by 6 per cent annually over the next few years. It is envisaged that military expenditure will fall from an average level of 5.2 per cent of the gross domestic product (GDP) in the period 1980–84 and 4.0 per cent in 1990 to 3.4 per cent over the medium term.[8]

The conservative government under Prime Minister Margaret Thatcher initiated a denationalization programme and has been firm in its belief that arms production is best managed by private companies rather than being in government ownership. This concept led to the implementation of a policy to privatize nine arms-producing units: Rolls-Royce, British Aerospace, VSEL, Swan Hunter, Vosper Thornycroft, Yarrow Yard, Royal Ordnance, Devonport and Shorts.[9] The government has also stressed the principle of allowing foreign ownership of British arms firms but in practice does not always like the prospect.

The Federal Republic of Germany

The third largest arms-producing country in Western Europe is Germany in terms of procurement and export orders combined. Arms production in Germany has for most of post-World War II political history been a non-issue. Without much political discussion, a sizeable arms industry was built up during the 1950s and 1960s—initially on the basis of technology imports, mainly from the United States.[10] Since the 1970s German companies have been engaged in developing and producing the full range of modern major weapon systems,

[7] Secretary of Defence, *Statement on the Defence Estimates* (Her Majesty's Stationery Office: London, several issues). Some estimates are higher; usually not only the employment effects of the production of arms and military equipment are included in these figures but all supplies and services to the armed forces in addition to arms exports. According to the director-general of the British Defence Manufacturers' Association, in *Defence Industry Digest*, Mar. 1991, p. 7, arms industry employment in the UK is forecasted to fall from 618 000 in 1990 to about 495 000 in the mid-1990s.

[8] Biship, P. and Trinder, D., 'The impact of the defence review on high-technology firms in South-West England', *Defense Analysis*, vol. 8, no. 1 (1992), pp. 91–93; Southwood, P., *The UK Defence Industry at the Crosroads: An Analysis of the 'Options for Change' Study and the DFE Process* (Enterprise for Defence and Disarmament: Chippenham, 1990).

[9] Taylor, T., 'The British restructuring experience', in Brzoska and Lock (note 5), pp. 81–96.

[10] Perdelwitz, W. and Fischer, H., *Waffenschmiede Deutschland* (Gruner und Jahr: Hamburg, 1984); Brzoska, M., Guha, A. and Wellmann, C., *Das Geschäft mid dem Tod. Fakten und Hintergründe der Rüstungsindustrie* (Eichborn: Frankfurt/Main, 1982).

partly as collaborative projects with companies from other NATO countries. Compared to other industrial branches, the arms industry is small.[11]

Procurement expenditure has been cut back since 1988 and R&D expenditure since 1990. These two components of the budget are scheduled to be further reduced from roughly DM 17 billion in 1988 to DM 12 billion (about $8 billion at 1992 exchange-rates) in 1995.[12] Several long-term projects have already been cancelled, facing the arms industry with severe over-capacities. This trend is exacerbated by the inheritance of arms production, repair and service facilities in the former GDR after the unification of the country.

A concentration of arms-producing companies within Germany took place during the 1980s. Arms production is now highly concentrated in several subsidiaries of Daimler Benz. In 1991 the arms sales of Daimler Benz were over four times larger than those of the second and third largest arms producers (Siemens and Rheinmetall). From a peak of around 310 000 employees in West and 40 000 employees in East Germany during the mid-1980s, employment has fallen to below 250 000 at the beginning of the 1990s.[13]

Italy

The Italian arms industry is much smaller than the industry in France, the United Kingdom or Germany. Procurement of weapons continued to grow until 1990. This postponed the process of down-sizing production capacity and the need for rationalization that was experienced earlier in other West European countries.[14] Although arms production is largely controlled by three major industrial groups, it is fragmented in small production units with little international orientation.

Two-thirds of the top 50 Italian military firms and subsidiaries are controlled by three major industrial groups: the state-owned holdings IRI and EFIM, and the private group FIAT, accounting for four-fifths of employment in the top 50 companies. However, 'looking at individual companies and their production units the fragmentation of the industry becomes apparent, with a low average number of employees involved in military production and the activities of firms and subsidiaries scattered across many different product areas, often with little integration even within the three largest industrial groups'.[15]

Italian arms production is based mainly on US licences and collaborative projects with other European NATO countries. Even the largest companies

[11] Berger, M. *et al.*, *Produktion von Wehrgütern in der Bundesrepublik Deutschland,* IFO Studien zur Industriewirtschaft (IFO Institut: Munich, 1991).

[12] Bebermeyer, H., 'Die deutsche Rüstungsindustrie im Anpassungsprozeß', *Europäische Sicherheit,* no. 3 (1992), pp. 170–75.

[13] Schomacker, K., Wilke, P. and Wulf, H., *Alternative Produktion statt Rüstung* (Bund: Köln 1987), p. 43; Opitz, P., *Rüstungsproduktion und Rüstungsexport der DDR,* Arbeitspapiere der Berghofstiftung für Konfliktforschung, no. 45 (Berlin, 1991); Berger *et al.* (note 11).

[14] Perani, G. and Pianta, M., 'The slow restructuring of the Italian arms industry', Brzoska and Lock (note 5), pp. 140–53.

[15] Perani and Pianta (note 14), p. 147.

have difficulty in acquiring a capacity to develop entire modern and sophisticated weapon systems.

Sweden

Sweden has a technologically highly developed arms industry. Its neutrality is the basis for a policy of national independence and high self-sufficiency in arms production. Although the Swedish military budget is comparatively small in absolute terms (about 12 per cent of the British budget), Sweden's arms industry develops and produces most types of major conventional weapons and ammunition, including surface ships and conventionally powered submarines, combat aircraft, tanks and other armoured vehicles, heavy artillery, and guided missiles.[16]

The general trend of ever-increasing weapon development and unit production costs for modern weapon systems has, despite the policy of self-sufficiency, forced Sweden to increase international co-operation. Swedish arms production is characterized by increasing imports of foreign components and subsystems for Swedish-produced systems and an increasing number of joint ventures. It is expected that this trend will be strengthened if Sweden joins the EC.

A small group of six companies form the core of Sweden's arms industry. A highly skilled work-force of roughly 30 000 people is employed in the arms industry, over half of them by three major producers—the restructured Celsius and Swedish Ordnance and Saab-Scania's aircraft division. 'The cost of sustaining this industrial base is substantial in terms of money, skills and technology. The military accounts for over 20 per cent of the nation's R&D'.[17]

Other West European arms-producing countries

All other West European arms-producing countries are comparatively small and do not have the capacity for the development and production of a large variety of major weapon systems. Employment in the arms industry ranges from about 100 000 in Spain to 25 000 each in Belgium, Switzerland and Turkey, 20 000 each in Austria and the Netherlands, 14 000 in Greece, 10 000 each in Finland, Norway and Portugal, and 7000 in Denmark.

The growth strategy for the Spanish arms industry relied heavily on exports to less-developed countries. When arms exports declined, the Spanish arms industry found itself in a difficult position, along with several other late-comer arms industrial nations. Spain is increasingly engaged in international projects.

Belgian, Danish and Norwegian arms production is concentrated on licensed production of major weapon systems for their own armed forces. A similar situation prevails in the Netherlands, which in addition, has been a producer of

[16] Hagelin, B., 'Sweden's search for military technology', Brzoska and Lock (note 5), pp. 178–94.
[17] Hagelin (note 16), p. 185.

major surface ships as well. Arms production in Switzerland is located in the state-run Eidgenössische Rüstungsbetriebe arsenal and the internationally operating Oerlikon-Bührle company.

Other countries play an even more limited role. Their marginal technological capabilities, limited industrial infrastructure and small internal arms markets limit possibilities for arms production. The major activity is assembly work directed at import substitution. 'In Greece and Turkey the regional arms race and the search for increasing national independence provided the main rationale for the development of military industry.'[18]

III. Producers and industrial sectors

The most outstanding characteristic of the largest West European arms-producing companies is the high concentration of a cluster of companies from the United Kingdom, France and Germany.[19] Of all the companies with arms sales of $200 million or more in 1991, 17 were from the UK, 14 from France and 10 from Germany—accounting for 70 per cent of all the West European companies in that category and for 80 per cent of their arms sales. Companies from six other countries had annual arms sales of $200 million or more (see table 8.1). The major share of arms sales is concentrated in a very small number of companies. A dozen companies in Western Europe share among themselves more than half of all the arms sales.

Table 8.1. The major West European arms-producing companies, 1991

Figures are in US$ m., at current prices and exchange-rates.

Country/company	Industrial sector	Arms sales
United Kingdom		
British Aerospace	Ac A El Mi SA/O	7 550
GEC	El	4 280[a]
Rolls-Royce	Eng	1 680
Lucas Industries	Ac	1 120[a]
VSEL Consortium	Mv Sh	920
Westland Group	Ac	530
Smith Industries	El	490[a]
Racal Electronics	El	480[a]
Hawker Siddeley	El	480[a]
Devonport Management	Sh	470[a]
Dowty Group	Ac El	450[a]
Thorn EMI	El	450[a]

[18] Bartzokas, A., 'The developing arms industries in Greece, Portugal and Turkey', Brzoska and Lock (note 5), pp. 178–94.

[19] In 1988 there were 28 British, 25 German and 17 French companies among the top 100 West European arms producers. Companies from 8 other European countries were among the 100 largest companies of Western Europe. See Anthony, I., Courades Allebeck, A. and Wulf, H., *West European Arms Production,* A SIPRI research report (SIPRI: Stockholm, 1990), pp. 64–69.

Country/company	Industrial sector	Arms sales
Ferranti-International Signal	El	440[a]
Hunting	SA/O	420[a]
Vickers	Eng Mv SA/O	320[a]
Vosper Thornycroft	Sh	230[a]
FR Group	Oth	200
France		
Thomson S.A.	El Mi	4 800
DCN	Sh	3 710
Aérospatiale	Ac Mi	3 450
GIAT Industries	A Mv SA/O	2 000[a]
Dassault Aviation	Ac	1 870
CEA Industrie	Oth	1 750
SNECMA Groupe	Eng Oth	1 320
Matra Groupe	Mi OTH	1 180[a]
Alcatel-Alsthom	El	1 100
SAGEM Groupe	El	590
Dassault Electronique	El	490
Labinal	El	340[a]
SNPE	A SA/O	330[a]
Renault	Eng Mv	300[a]
Germany		
Daimler Benz	Ac Eng Mv El	3 920
Siemens	El	900
Rheinmetall	A SA/O	860
Diehl	A Mv El SA/O	860[a]
Bremer Vulkan	El Sh	780
Thyssen	Mv Sh	770
Mannesmann	Mv	400
Lürssen	Sh	400[a]
HDW	Sh	300
Wegmann	Mv	270[b]
Italy		
IRI	Ac Eng El Sh	2 670[a]
EFIM	Ac Mv El	1 710[a]
FIAT	Eng	1 180[a]
Aermacchi	Ac	270[a]
Sweden		
Celsius	Sh	870
Swedish Ordnance	A SA/O	710
Saab–Scania	Ac Eng El Mi	520
Nobel Industries	El Mi SA/O	300
Ericsson	El	270
Volvo	Eng	230
Spain		
INI	Ac A Mv El Sh SA/O	1 560[a]
BROROS	A Oth	230[a]

Table 8.1 *contd*

Country/company	Industrial sector	Arms sales
Switzerland		
Oerlikon-Bührle	Ac A El SA/O	1 100
Eidgenössiche Rüstungsbetriebe	Ac Eng A SA/O	700[a]
Netherlands		
De Schelde	Sh	220
Fokker	Ac	200[a]
DAF	Mv	200
Belgium		
Mercantile Beliard	Sh	250[a]

Note: A = artillery, Ac = aircraft, El = electronics, Eng = engines, Mi = missiles, MV = military vehicles, SA/O = small arms/ordnance, Sh = ships, and Oth = other.

[a] Figure is for 1990.

[b] Figure is for 1989.

Source: SIPRI data base.

Industrial sectors

The various industrial sectors receive differing shares of the procurement expenditures and specialize in arms production to different degrees. These sectors vary in size and structure. The largest shares of procurement orders go—on average—to the aerospace and the electronics sector (see table 8.2).

Table 8.2. Share of industrial sectors in military procurement expenditures, 1986–90

Figures are in percentages.

Country/period	Aerospace	Ships	Machinery/ industry	Electronics	Munitions/ weapons	Transportation
Germany (1987–90)	17.5	6.9	25.0	25.2	. .	7.5
UK (1987–88)	31.0	7.0	. .	23.0	8.0	. .
France (1986–88)	34.0	10.0	8.0	27.3	. .	. .
Belgium (1986–87)	10.0	4.0	. .	16.0	50.0	20.0

Source: Huffschmid, J. and Voß, W., *Militärische Beschaffungen–Waffenhandel–Rüstungskonversion in der EG. Eine Studie im Auftrag des Europäischen Parlamentes*, PIW-Studien, no. 7 (PIW: Bremen, 1991), p. 19.

Except for the small and highly specialized munitions and weapon sector, the aerospace sector is the most dependent on arms production (up to over two-thirds in France and one-half in the UK, Italy and Spain). Many aerospace companies do the major part of their business with the military. The more a company is specialized exclusively or to a high percentage on arms production, the more vulnerable it is in the present situation (see table 8.3).

Table 8.3. Share of military production in total production, 1986–90
Figures are percentages.

Country	Aerospace	Ships	Machinery/ industry	Electronics	Munitions/ weapons	Transpor- tation
Germany	45	20	. .	. .	. .	. .
UK	50	50	. .	39	100	3
France	69	50	. .	57	60	. .
Italy	50	25	. .	10	. .	2
Belgium	35	10	3	4	80	2
Netherlands	10	30	. .	5	100	3
Spain	50	70	. .	8	20	6

Source: Huffschmid, J. and Voß, W., *Militärische Beschaffungen–Waffenhandel–Rüstungskonversion in der EG. Eine Studie im Auftrag des Europäischen Parlamentes*, PIW-Studien, no. 7 (1991), p. 19. (These data are partly based on estimates by Huffschmid and Voß.)

This specialization—although most dominant within the groups of weapons and aerospace producers—is not limited to these two sectors: 116 West European companies and company subsidiaries recorded arms sales of $100 million or more in 1990/91 (see tables 8.4–8.8, below). Many of them, particularly subsidiaries, were highly dependent on arms sales. The policy of orienting the different sectors of the arms industrial base towards national requirements has sheltered companies from competition and fostered the trend of duplication and triplication of technical capabilities and production units. This policy has contributed to the down-sizing and rationalization that are now required to adapt to the new political and economic environment.

Aerospace and missiles

Military fighter aircraft programmes are the most costly weapon projects in Europe, and their development and production are the backbone of the aircraft industry. At the end of the 1980s this industrial sector—in both military and civilian projects—employed approximately 500 000 people. The projected number of European fighter aircraft is, however, limited. The future of the European Fighter Aircraft (EFA), co-developed in the United Kingdom, Germany, Spain and Italy, is questionable as a result of German reluctance to enter from co-development to the co-production stage. The multi-role British–German–Italian Tornado aircraft will enter the final stages of production, unless additional export orders are acquired. French Mirage aircraft—although aggressively marketed in many countries—have not been exported in recent years. The French Rafale fighter aircraft is still under development.

The first aircraft is to be delivered in 1996, if no funding or technical difficulties are encountered. 140 Swedish JAS Gripen have been ordered by the Swedish Air Force, but attempts for exports (both to Switzerland and Finland)

Table 8.4. The major West European arms producers in aerospace, 1991[a]

Figures are in US$ m., at current prices and exchange-rates.

Company (parent company)	Country	Arms sales	Total sales	Employment
British Aerospace	UK	7 550	18 687	123 200
DASA (Daimler Benz)	Germany	3 620	7 441	56 465
Aérospatiale	France	3 450	8 614	43 287
SNECMA Groupe	France	1 320	4 241	27 236
Dassault Aviation	France	1 870	2 544	11 914
Alenia (IRI)[b]	Italy	1 840	3 069	21 981
Rolls-Royce	UK	1 680	6 219	57 100
MBB (DASA)[b]	Germany	1 420	2 853	23 229
MTU (DASA)	Germany	1 180	2 148	17 052
Lucas Industries[b]	UK	1 120	4 221	54 942
Matra Défense (Matra)[b]	France	920	925	..
SNECMA (SNECMA Groupe)	France	850	2 566	13 816
CASA[b]	Spain	780	961	10 050
Agusta (EFIM)[b]	Italy	560	927	8 117
Westland Group	UK	530	827	9 060
FIAT Aviazione (FIAT)	Italy	460	841	4 666
Dornier (DASA)	Germany	350	1 431	9 527
Turbomeca (Labinal)	France	290	471	3 934
Aermacchi[b]	Italy	270	281	2 736
Saab-Scania Combitech (Saab-Scania)	Sweden	270	369	2 349
Saab Aircraft (Saab-Scania)	Sweden	260	816	6 909
Volvo Flygmotor (Volvo)	Sweden	230	612	3 700
Fokker[b]	Netherlands	200	1 758	13 176
SEP (SNECMA Groupe)	France	180	804	3 976
Hispano Suiza (SNECMA Groupe)[b]	France	150	372	2 914
FFV Aerotech (Celsius)	Sweden	140	190	1 789
Saab Missiles (Saab-Scania)	Sweden	140	145	504
Messiers Hispano Bugatti (SNECMA)[b]	France	130	433	2 900
BMW/Rolls-Royce (BMW/RR)[b]	Germany	120	124	1 000
SABCA (Dassault)[b]	Belgium	110	164	1 698
Piaggio[b]	Italy	110	213	1 929
Raufoss Konsernet	Norway	100	294	2 772

[a] Several producers in this table are engaged in more than one industrial sector; they have been placed in the sector in which they have their largest share of arms sales.

[b] Figures for this company are for 1990.

Source: SIPRI data base.

were not successful; the orders went to the United States for the purchase of US F-18 fighters.

Although the list of companies in the aerospace sector is long, there is only one national company each capable of producing fighter aircraft in the UK

(British Aerospace), France (Dassault), Germany (DASA) and Sweden (Saab-Scania). All other aerospace companies in Western Europe are engaged in other than fighter projects (missiles, helicopters, space or trainer aircraft) or produce under licence.

Equally limited is the number of companies in the missiles, helicopters and engines. In most of these subsectors of aerospace, a small number of companies operate. In helicopters there are four major companies (MBB, Aérospatiale, Agusta and Westland). Westland is under the control of the US helicopter producer Sikorsky; MBB and Aérospatiale co-operate both in missile and helicopter production and marketing. Agusta and Westland co-operate in a project on a transport helicopter.[20]

A similar situation exists in engine production. The few producers—Rolls-Royce, SNECMA, MTU, FIAT-Aviazione and Turbomeca—co-operate with each other bi- or trilaterally, partly also with US companies; thus they are both collaborators and competitors at the same time.

Many of the aerospace companies co-operate in collaborative projects or have formed joint companies to co-develop or market their systems. The level of concentration is high in the aerospace sector. Fewer than a dozen companies generate over two-thirds of the turnover. Despite this concentration, West European companies are small compared to their US competitors (except for British Aerospace, ranking as the third largest arms-producers of the world in aerospace). Anticipated domestic procurement and possible exports require further reduction in capacities and, as a consequence, further collaboration and even regrouping of companies.

Military electronics

The situation is quite different in the military electronics sector, in which a number of major multinational companies operate. Major restructuring with cross-border acquisitions and change of ownership in that sector took place in recent years.[21] Only small and specialized companies are to a high degree dependent on the military budget. One exception is Thomson-CFS, a French major military electronics producer with 77 per cent of its sales in military electronics. Other top electronics producers have much smaller fractions of their sales in military electronics. Although 30 military electronics producers and company subsidiaries with arms sales of $100 million or more operate in this sector (see table 8.5). Despite the high number of companies, this is not a competitive market. Many of these companies are subsidiaries of larger concerns and specialize in key products or sectors in which they have only limited competition from other companies.

Many of these companies feel well placed and suffer less from cuts in the budget, since modernization programmes are often highly electronics-intensive.

[20] On international co-operation, see Sköns, E., 'Western Europe: internationalization of the arms industry', chapter 9 in this volume.

[21] See note 20.

Table 8.5. The major West European arms producers in electronics, 1991[a]

Figures are in US$ m., at current prices and exchange-rates.

Company (parent company)	Country	Arms sales	Total sales	Employ-ment
Thomson-CSF (Thomson S.A.)	France	4 800	6 235	44 500
GEC[b]	UK	4 280	16 923	118 529
Alcatel-Alsthom	France	1100	28 373	213 100
Siemens	Germany	900	43 994	402 000
Telefunken System Technik (DASA)	Germany	680	1 045	9 372
SAGEM Groupe	France	590	2 081	15 076
Smith Industries[b]	UK	490	1 201	13 100
Dassault Electronique	France	490	685	3 416
Hollandse Signaalapparaten (Thomson-CSF)[b]	Netherlands	490	515	4 522
Racal Electronics[b]	UK	480	3 719	38 461
Hawker Siddeley[b]	UK	480	3 887	44 600
Dowty Group[b]	UK	450	1 372	15 022
Thorn EMI	UK	450	6 532	57 932
Ferranti-International Signal[b]	UK	440	817	10 325
Systemtechnik Nord (Bremer Vulkan)	Germany	430	570	2 441
Sextant Avionique (Thomson-CSF/Aérospatiale)[b]	France	400	1 119	9 152
Labinal[b]	France	340	1 361	18 500
Atlas Elektronik (Bremer Vulkan)	Germany	290	486	4 259
Thomson-TRT-Défense (Thomson-CSF)[b]	France	290	293	. .
Ericsson	Sweden	270	7 572	71 247
Standard Elektrik Lroenz (Alcatel–Alsthom)[c]	France	260	3 247	19 503
Siemens Plessey (Siemens)[b]	UK	250	. .	. .
SAT (SAGEM Groupe)[b]	France	200	599	. .
Inisel (INI)[b]	Spain	190	294	3 151
Elettronica[b]	Italy	160	162	1 363
Pilkington[b]	UK	160	4 744	60 300
Bosch	Germany	150	20 247	181 498
Elmer[b]	Italy	120	120	1 060
Microtechnica[b]	Italy	100	116	1 044
SFIM[b]	France	100	192	1 500

[a] Several producers in this table are engaged in more than one industrial sector; they have been placed in the sector in which they have their largest share of arms sales.

[b] Figures for this company are for 1990.

[c] Figures for this company are for 1989.

Source: SIPRI data base.

Thus, military electronics is considered by many as a niche in the otherwise shrinking arms market. Whether or not this is a realistic long-term prognosis cannot be substantiated on the basis of the budget figures and the development of electronics companies during recent years.

Table 8.6. The major West European arms producers in shipbuilding, 1991[a]
Figures are in US$ m., at current prices and exchange-rates.

Company (parent company)	Country	Arms sales	Total sales	Employ-ment
DCN	France	3 710	3 715	30 000
VSEL Consortium	UK	920	920	13 028
Celsius	Sweden	870	1 832	14 508
Bremer Vulkan	Germany	780	2 006	15 021
Devonport Management[b]	UK	470	500	7 942
EN Bazan (INI)[b]	Spain	460	530	9 613
Blohm & Voß (Thyssen)	Germany	400	816	5 758
Lürssen[b]	Germany	400	495	1 080
HDW	Germany	300	708	4 866
Mercantile Beliard[b]	Belgium	250	422	1 775
Fincantieri (IRI)[b]	Italy	240	1 633	20 065
Vosper Thornycroft[b]	UK	230	251	2 052
Thyssen Nordseewerke (Thyssen)	Germany	190	299	2 084
De Schelde	Netherlands	220	496	3 600
Santa Barbara (INI)[b]	Spain	150	157	3 925
RDM[b]	Netherlands	100	229	1 315

[a] Several producers in this table are engaged in more than one industrial sector; they have been placed in the sector in which they have their largest share of arms sales.
[b] Figures for this company are for 1990.

Source: SIPRI data base.

Shipbuilding

As a consequence of the crisis in the shipbuilding industry which was experienced in merchant shipbuilding in the 1970s and 1980s, the dependence of companies on fighting-ship construction has increased considerably. Many shipbuilders survived largely because of the domestic procurement of ships and exports. Many others had to close their yards. The work-force in the EC countries was reduced during the 1980s from around 120 000 to less than 70 000. In contrast to the aerospace sector, West European companies have dominated the world market in frigates, submarines and fast missile craft. Over half of total sales of the shipbuilding sector are sales of fighting ships. Arms sales dependence is therefore high among the major shipbuilders in Europe, ranging up to 100 per cent.

Military vehicles

The tank and armoured vehicle producers in some of the West European countries have earlier than most other industrial sectors experienced a contraction in military orders. This trend is exacerbated by both the 1990 Treaty on Conventional Armed Forces in Europe (CFE), which sets ceilings that required the re-

Table 8.7. The major West European arms producers in military vehicles, 1991[a]

Figures are in US$ m., at current prices and exchange-rates.

Company (parent company)	Country	Arms sales	Total sales	Employment
GIAT Industries[b]	France	2 000	2 020	18 000
Diehl[b]	Germany	860	1 779	15 108
Oto Melara (EFIM)[b]	Italy	780	783	2 245
Krauss-Maffei (Mannesmann)	Germany	400	853	5 004
Vickers[b]	UK	320	1 389	12 613
Renault[b]	France	300	30 048	157 378
Mercedes Benz (Daimler)	Germany	300	40 436	237 442
Wegmann[c]	Germany	270	530	4 750
FIAT IVECO (FIAT)[b]	Italy	220	4 477	22 649
DAF	Netherlands	200	2 871	14 643
MAN[b]	Germany	190	11 721	65 900
MAK System Gesellschaft (Rheinmetall)[b]	Germany	190	190	660
Mainz Industries[b]	Germany	190	190	3 300
Thyssen Henschel (Thyssen)	Germany	180	394	3 245
GKN[b]	UK	180	3 641	33 904
United Scientific Holding[b]	UK	180	204	3 100
Hägglunds Vehicle (ABB)	Sweden	150	156	700
Steyr Daimler Puch (MAN)[b]	Austria	110	1 304	8 647

[a] Several producers in this table are engaged in more than one industrial sector; they have been placed in the sector in which they have their largest share of arms sales.

[b] Figures for this company are for 1990.

[c] Figures for this company are for 1989.

Source: SIPRI data base.

duction of the number of major battle tanks (over 5000) and armoured combat vehicles (over 4000), and by a limited export potential.

Only four years ago, in 1988, five European NATO countries (France, the United Kingdom, Germany, Italy and Spain) in addition to Sweden each had the ambition to develop a new generation of battle tanks. The French GIAT company started production of the Leclerc main battle tank in 1992. Although the original intention was to procure up to 1400 tanks, this figure is likely to be reduced to 1100 or even less. The first production orders for the British Challenger 2 tank was placed in 1991, but the overall British requirements have been reduced significantly. The German Leopard III tank (also called Kampfwagen 2000), the Italian Ariete and plans for the production of a main battle tank (MBT) in Spain have been given up. The first orders for the Combat Vehicle 90, an MBT for the Swedish Army, were place in 1991. This tank is produced in Sweden by Hägglunds Vehicle and Bofors (now part of Swedish Ordnance).

With a ceiling of a maximum of 20 000 battle tanks in the European NATO countries and a lifetime of at least 15 years, annual replacement amounts on average to fewer than 1500 MBTs. The respective figure for armoured combat vehicles in the NATO alliance ceiling is 30 000. Assuming a life cycle of 10 years, the annual replacement will be 3000. The battle tank and armoured combat vehicle producers are in tough competition, which will force some to exit the market. Those that are protected by privileged treatment from their government are those that are most likely to survive. Most of the battle tank producers have announced lay-offs or have already reduced their tank production capacities. Some repair and maintenance facilities have been closed or are extremely vulnerable. MAK, a major German battle tank systems integrator, has been broken up into two parts and was sold.

The economic situation of military truck producers is different. The big automobile producers usually have a limited stake of their truck production in military sales.

Artillery, small arms and munitions

Table 8.8. The major West European arms producers of artillery, small arms and munitions, 1991[a]

Figures are in US$ m., at current prices and exchange-rates.

Company (parent company)	Country	Arms sales	Total sales	Employment
Oerlikon-Bührle	Switzerland	1 100	2 527	19 138
Rheinmetall	Germany	860	2 092	13 661
Swedish Ordnance	Sweden	710	745	6 291
Eidgenössische Rüstungsbetriebe[b]	Switzerland	700	738	4 672
Hunting[b]	UK	420	1 377	6 918
SNPE[b]	France	330	727	7 129
Contraves Italiana (Oerlikon–Bührle)	Switzerland	330	330	961
Nobel Industries	Sweden	300	4 087	22 922
BROROS[b]	Spain	230	2 266	15 229
Thomson Brandt Armements (Thomson-CSF)[b]	France	220	220	975
FR Group	UK	200	297	3 658
Dynamit Nobel (Metallgesellschaft)	Germany	190	1 067	8 500
MKEK[b]	Turkey	190	296	7 986
Norsk Forsvarsteknologi	Norway	180	203	1 867
FN Herstal (GIAT Industries)[b]	Belgium	160	165	1 516
GAMESA[b]	Spain	130	157	550
IWKA	Germany	120	1 171	8 503
SIG	Switzerland	120	1 073	7 503
Raufoss[b]	Norway	110	274	2 777
Breda Meccanica Bresciana (EFIM)[b]	Italy	110	115	692

[a] Several producers in this table are engaged in more than one industrial sector; they have been placed in the sector in which they have their largest share of arms sales.

[b] Figures for this company are for 1990.

Source: SIPRI data base.

Among the arms industrial sectors, the West European small arms and munitions producers experienced great difficulties due to reduced demand.

Many of them, as a result of regional conflicts and generous interpretations of export regulations, had secured their existence in the past. A major reshuffle took place as a result of financial difficulties during recent years. GIAT and British Aerospace (which owns Royal Ordnance) bought FN in Belgium and Heckler and Koch in Germany, respectively, after these companies were on the verge of bankruptcy. PRB in Belgium, owned by Astra in the UK, went bankrupt in 1990 and Eurometaal in the Netherlands in 1991. Small arms production in Western Europe is now dominated by GIAT and British Aerospace's Royal Ordnance. Mergers and acquisitions of companies in this sector also took place in Sweden, with the formation of Swedish Ordnance (merging parts of Bofors and the state-owned FFV). Oerlikon Bührle, with its arms production for years in the red, tried unsuccessfully to sell its Italian and German Contraves subsidiaries' artillery units. Rheinmetall, the major German artillery producer, bought a part of MAK to acquire system integration capacities.

IV. Perspectives

A greater willingness by West European governments to take the changed East–West political situation seriously, financial constraints and problems in keeping abreast of the ever-increasing cost of modern weapon systems are all forcing industries and governments to react. All sectors of the arms industry suffer from over-capacities. Only a few companies can hope to expand their arms sales in niches. National markets are too small to sustain the present industrial base. Exports, although not an alternative for the industry as a whole, are regarded by some companies as a way to compensate for lost business in the domestic market. The industry trend of becoming more international makes control more difficult.

In addition to these exogenous factors, pressure is mounting from endogenous factors as well. In most sectors of the arms industry, production has been characterized by increasingly high development costs and ever-shorter production runs. This is not a new development, but the conditions in which industry is operating have changed—particularly with regard to the availability of financial resources. Permanent modernization and increased sophistication of weapon systems have caused an escalation of development costs. Fewer weapons are required as a result of the political changes in Europe, but at the same time funds needed to develop and produce the next generation of weapons have increased.

Both industry and governments are faced with a dilemma: at the government level there is an inherent conflict between plans to cut procurement and R&D expenditure and plans to maintain a viable and competitive arms industrial base. In industry more resources are required for the next generation of weapon systems at a time when down-sizing is the order of the day.

On the one hand, the number of producers of tanks, fighter aircraft, ships and missiles is too large to be maintained by present military budgets. On the other hand, competition is diminishing, with producers leaving arms production and others forming joint companies or strategic alliances. The process of shrinkage that has begun will probably have to go much further.

There seem to be two different prospects for the West European arms industry. One is development of a more co-operative and streamlined European industrial base with intensified collaboration between companies and a reversal of national decision making. However, European-wide procurement and defence industrial policies have never been put into practice on a large scale because of the prevalence of national idiosyncrasies and powerful industrial interests. With the increasing economic pressures upon industry, severe budget constraints and the disappearance of the Soviet threat, the solution might be different during the 1990s. If a more co-ordinated procurement policy and a co-operative industrial structure are implemented, the problem of over-capacities becomes even more pressing.

Second, the failure of political co-ordination will increase economic pressures and competition. Such a trend will force companies either to collaborate with other producers, including the larger US companies, or to abandon arms production.

The first prospect comes close to the always preached but seldom practised concept of West European defence planners. The second prospect comes close to the always preached but seldom practised free market concept of economic policy makers. The most likely outcome is the usual muddling-through strategy with a compromise of both perspectives: that is, intensified co-operation where national 'play it alone' policies no longer work and protracted periods of intensified economic pressures that will not allow all the present arms producers to generate enough business to survive. Assuming the continuation of the present gradual, modest trend of arms control and the continuation of strong financial pressures, the arms industry is bound to scale down its capacities—with or without government guidance and support.

9. Western Europe: internationalization of the arms industry

Elisabeth Sköns

I. Introduction

The West European arms industry is undergoing a profound transformation as a result of the changing economic and political environment in which it operates. This process, which began in the late 1980s, has several dimensions. The increasing internationalization of the arms industry is one of these. Other dimensions include national concentration of the arms industry through mergers and acquisitions, diversification into civilian production activities, privatization of state arms-producing enterprises and redutions in output. These developments take place simultaneously, they overlap, and they are sometimes combined in the strategies of one and the same company.[1]

The escalating research and development (R&D) costs of weapon systems in combination with a stagnating or declining domestic and foreign demand for weapons are the primary economic incentives behind the expanding internationalization of arms production. Developments in technology simultaneously move in a direction which facilitates internationalization.[2]

These incentives are especially powerful for companies which develop and manufacture technologically advanced, and therefore R&D-intensive, products, and for short production runs. Both these conditions are valid for many of the major arms-producing companies in Western Europe. There are thus strong driving forces for West European arms-producing companies to seek and pursue co-operation with foreign companies.[3] For companies, internationalization is one strategy of consolidation for long-term survival in the market.

[1] For recent studies on this transformation, see Brzoska, M. and Lock, P. (eds), SIPRI, *Restructuring of Arms Production in Western Europe* (Oxford University Press: Oxford, 1992); Steinberg, J. B., *The Transformation of the European Defense Industry,* RAND Report R-4141-ACQ (RAND Corp.: Santa Monica, 1992); 'Defense technology and industrial base policies of allied nations', in US Congress, Office of Technology Assessment (OTA), *Building Future Security*, OTA-ISC-530 (US Government Printing Office: Washington, DC, June 1992), appendix A; and Renner, M., *Economic Adjustment after the Cold War: Strategies for Conversion* (United Nations Institute for Disarmament Research [UNIDIR]: Aldershot and Dartmouth, 1992).

[2] See, for example, the case studies and general analysis in Creasey, P. and May, S. (eds), *The European Armaments Market and Procurement Cooperation* (MacMillan: London, 1988).

[3] This is probably also the case for part of the arms industry in Russia and some of the former Comecon (Council for Mutual Economic Assistance) countries, while in the USA the domestic arms market is larger, and in some of the other major arms-producing countries the ambition of self-sufficiency in advanced weapon systems is not as pronounced as in Western Europe.

Companies can be expected to progressively take over from governments part of their role as initiators in the process of internationalization.

Internationalization of arms-producing activities has important arms control implications. Hitherto, efforts to restrict the trade in weapons have, with a few important exceptions, been focused on national policies, national decision making, national regulations and national implementation.[4] However, these multilateral agreements also require implementation exclusively on the national level.

With the increasing internationalization of arms production, the conditions for the control of military technology transfers will change, and this change will probably be fundamental.[5] It will require closer international co-operation in efforts to monitor and control the trade in arms (probably at both the government and the company level) and will add to the urgency to achieve an international arms transfer control regime. The first step in such a process, however, is to understand the mechanisms of arms industry internationalization.

II. Forms of internationalization

The process of industrial internationalization is multi-faceted, ranging from international trade, to cross-border restructuring of ownership, to a variety of international co-operation schemes between companies—sometimes associated with ownership changes, sometimes not. Each form has its own characteristics and implications for military technology transfers and for the possibilities to control such transfers.

For an analysis of the pattern of internationalization in the arms industry, it is helpful to have a simple classification system which captures the most important elements. The criterion of simplicity is essential because of the lack of statistics and systematic data on the arms-producing sector. The classifications shown in table 9.1 are one such system, which can provide a starting-point.[6]

[4] The exceptions are COCOM (Co-ordinating Committee on Multilateral Export Controls), the NPT (1968 Non-Proliferation Treaty), the MTCR (Missile Technology Control Regime of 1987, revised in 1992), the Australia Group (a group which has met semi-annually since 1985, in 1992 consisting of 22 countries with participation of the EC, to discuss chemicals to be subject to regulatory measures) and, more recently, the United Nations Register of Conventional Arms, the talks among the five permanent members of the Security Council (the so-called Big Five Talks, abandoned by China in Oct. 1992), and the UN arms embargoes on individual countries. The latter initiatives are described in *SIPRI Yearbook 1992: World Armaments and Disarmament* (Oxford University Press: Oxford, 1992), chapter 8, pp. 297–99.

[5] For an interesting discussion of the problems associated with democratic control of international collaboration in military R&D, see Blunden, M., 'Collaboration and competition in European weapons procurement: the issue of democratic accountability', *Defense Analysis*, vol. 5, no. 4 (1989), pp. 291–304; and, 11 years earlier, Hagelin, B., 'International cooperation in conventional weapons acquisition: a threat to armaments control?', *Bulletin of Peace Proposals*, vol. 9, no. 2 (1978), pp. 144–55.

[6] Different types of classification than by organizational mode, can be used depending on the purpose of the analysis, e.g. by dominant motive, geographic scope, ownership arrangements, etc. See, for example, Contractor, F. J. and Lorange, P. (eds), *Cooperative Strategies in International Business* (D.C. Heath and Company: Lexington, Mass./Toronto, 1988). An interesting attempt to categorize the corporate strategies which are currently used in the European arms industry is found in Latham, A. and Slack, M., *The European Armaments Market: Developments in Europe's "Excluded" Industrial Sector*, University of Manitoba Occasional Paper no. 13 (Department of Political Studies: Winnipeg, Oct. 1990).

Table 9.1. Forms of internationalization of industrial activities in the OECD area

Form	Explanation	Purpose	Sector examples
International trade	Export and import of goods and services	To expand markets and increase returns on investment	Many sectors
Foreign investment	Establishment of new production or assembly facilities abroad	To increase market access and open local production	Food products Chemicals Automobiles
International subcontracting	Transfer of production tasks to foreign firms which possess cost, technical or locational advantages	To allow specialization in core components and increase flexibility	Aerospace Construction Automobiles
International licensing	Sale of access to technology or know-how in exchange for a fee or royalties	To broaden markets and increase returns on R&D expenditures	Pharmaceuticals Semiconductors
Cross-border mergers and acquisitions	Purchase of companies abroad	To increase scale economies and open local production	Electronics Financial services
International joint ventures	Equity holdings in a venture are divided among partners who share the costs and risks of technology development, production or marketing	To increase market access and pool resources	Minerals Machine tools
International inter-firm agreements	More flexible forms of collaboration which can be designed to suit the needs of all parties. Such alliances may be formed for R&D, production, marketing and a variety of other purposes.	To share costs and risks of investments	Aerospace Electronics

Source: Adapted from OECD Industry Committee, *Globalisation of Industrial Activities: Four Case Studies* (Organization for Economic Co-operation and Development: Paris, 1992), table 1, p. 13.

This system also has the advantage of being used in studies by the Organization for Economic Co-operation and Development (OECD), which facilitates comparisons with patterns in the general industry. (The term 'general industry' is used in this chapter to denote the entire manufacturing industry, including arms production.)

This classification system is used as a point of reference for the review in this chapter of the trends in arms industry internationalization. The term 'co-operation' is used as a general concept for all these forms of internationalization except trade and mergers and acquisitions.

The economic rationales for different forms of internationalization identified on the aggregate level in the manufacturing industry are probably similar in the arms industry, since the arms industry is a sub-sector of the manufacturing industry which extends across various industrial branches. There are reasons to believe, however, that there are also different purposes for arms industry internationalization, owing to the stronger influence of government policies and regulations on arms production and possibly also because of the more uneven international spread of arms production capabilities.

Forms of international co-operation can also be categorized by their link in the value-added chain: R&D, component manufacture, assembly, marketing and distribution/customer service. Actual co-operation may cover a single link or two or more links in the chain. A large proportion of arms production activities is in the high-technology sectors, where the share of R&D costs is high and rising. High R&D costs in combination with a dwindling market are often cited by corporate managers as factors behind different measures of international co-operation. Therefore, it can be expected that sharing the risks and costs of R&D projects is an important purpose of several forms of arms production internationalization.

International trade in weapons is excluded from the analysis in this chapter since the point of departure is that internationalization of the arms industry is increasingly turning from trade to other forms.[7]

Foreign investment is a common form of internationalization in those sectors of the general industry which derive advantages from market proximity, while it is a negligible form in sectors which have high fixed costs and scale economies, such as aerospace and steel.[8] Thus, further study may show that it is not a common form of internationalization in the arms industry.

Sub-contracting can in general be regarded as a substitute for foreign investment which has been made feasible by technological developments in the direction of a more fragmented production process, facilitating the spin-off of production tasks to the country receiving the contract. In the arms industry, foreign sub-contracting is often part of an offset arrangement made to compensate a country placing a large military order abroad for the employment, technology and other losses incurred.

Licensing, that is, the sale of access to technology or know-how, has in the general industry provided an alternative to foreign investment in sectors where R&D investments are protected by patents. It is not an international strategy in sectors with strong links between research and marketing, such as the automobile and telecommunications sectors. In the weapon sector, licensing has for decades been an important substitute for, or complement to, the trade in finished goods, often at the request of the buyer country. Licensed production

[7] The trade in major conventional weapon systems is regularly treated in a chapter in the *SIPRI Yearbooks: World Armaments and Disarmament* (Oxford University Press: Oxford, annually).

[8] The trends in the general industry, as summarized in this section, are the conclusions of the OECD in OECD Industry Committee, *Globalisation of Industrial Activities: Four Case Studies* (OECD: Paris, 1992), pp. 12–14.

of a weapon from a foreign design covers very different degrees of technology transfer. It can entail little more than indigenous assembly of a weapon system from basically imported kits and limited indigenous production. At the other end of the spectrum it can include purchase of a licence to be used as a starting-point for further development work, often in co-operation with the licenser (supplier of the licence).[9]

Mergers and acquisitions is a form of internationalization which is becoming increasingly more common in the general industry. This rising trend is part of a strategy to gain access to and to supply foreign markets from local production rather than by exports. In the arms industry it is a form which has recently gathered momentum, on the national level as well as internationally within Europe and in transatlantic relations. In a period of shrinking markets and subsequent production cuts, mergers and acquisitions are a natural corporate strategy for staying in the market and possibly maintaining market shares.

Joint ventures will probably be a popular choice for internationalization by multinational firms in the 1990s, since joint ventures allow more rapid market access than exports and incur lower costs than foreign investment. The number of joint ventures with shared R&D as a specific company objective has gradually increased in recent decades. In the West European arms industry there are a number of international joint ventures, some of which are quite large. In view of the very high R&D costs of advanced weapon systems and the stagnation in the quantities of systems which are ordered, it is generally believed that the objectives which play a major role in this trend include sharing of research efforts and spreading of risks. Other important objectives are sharing of fixed costs and capturing of economies of scale.

Inter-firm agreements[10] have become a major focus of internationalization strategies in the early 1990s, even among firms which are competitors. Such agreements are most common in the aerospace, electronics and automobile sectors. International R&D agreements are a direct outcome of the spiralling costs of R&D and are prevalent in high-technology sectors. These technology alliances range from collaboration in a single stage of the R&D process to complete product development and joint marketing.

Teaming arrangements for the purpose of contract bidding is a common form of international company co-operation in the arms industry owing to the large size of many arms procurement orders and the difficulty to gain entry to foreign arms markets. The transformation of these teams into actual international co-operation projects depends among other things on the extent to which governments allow competition to overrule national security or national employment considerations, but even the bidding process itself may be considered part of a process of internationalization.

[9] For a list of licensing agreements, see the annual register in the *SIPRI Yearbooks* (note 7).

[10] Global arrangements with foreign firms to obtain technology, inputs or entry for production and marketing with a regional or global scope have variously been called strategic or corporate alliances, strategic or international coalitions, and global strategic partnerships.

III. General trends in the internationalization of industrial activities

Internationalization of industrial activities in the OECD countries during the post-World War II era is for analytical purposes commonly divided into three broad stages.[11]

The first stage, the so-called *golden age of trade* of the 1950s and 1960s, was characterized by an expansion in international trade, as protectionist barriers to trade were dismantled through multilateral agreements within the framework of the General Agreement on Tariffs and Trade (GATT).

The second phase of industrial internationalization was dominated by a surge in *foreign direct investment,* primarily in the form of international mergers and acquisitions, which became an important form of industrial expansion during the 1970s.

The third and current phase started in the 1980s as a result of the increasing role of technology as a factor of industrial competitiveness and the ensuing higher R&D costs. This phase is characterized by new patterns of global corporate industrial linkages through a range of external alliances or *international inter-firm agreements*. In addition to the traditional forms of international trade and foreign investment, firms are establishing intricate international networks of research, production and information. This 'technology networking' has become so prominent that it has been assigned a new term, 'techno-globalism'.

Internationalization of R&D is an important part of the third phase of techno-globalization, since the sharing of R&D costs and risks is a dynamic factor behind the rising trend in cross-border corporate co-operation.[12] The dramatic rise in international inter-firm agreements with an impact on technology transfers in the OECD industry since the early 1980s is shown in table 9.2. The form of co-operation which has increased most rapidly is joint R&D agreements without the formal structure of a joint venture.

Internationalization of the arms industry

In the arms-producing sectors of industry, the internationalization process has been much more limited and irregular than in the general industry. This is a consequence of the predominance of national security and other political considerations over economic considerations in the arms-producing sector as well as an effect of the tendency in arms procurement to give priority to national

[11] See, for example, OECD 1992 (note 8); and Ostry, S., 'Beyond the border: the new international policy arena', *Strategic Industries in a Global Economy,* OECD International Futures Programme (OECD: Paris, 1991).

[12] For surveys of current trends in R&D internationalization, see, for example, Brainard, R., 'Internationalising R&D' and Aubert, J-E., 'What evolution for science and technology policies?', *OECD Observer,* no. 174 (Feb./Mar. 1992), pp. 7–10 and 4–6, respectively; Drilhon, G., 'Science and technology sans frontières?', *OECD Observer,* no. 175 (Apr./May 1992), pp. 20–23; and Hagedoorn, J., 'Organizational modes of inter-firm co-operation and technology transfer', *Technovation,* vol. 10, no. 1 (1990), pp. 17–30.

Table 9.2. Trends in inter-firm agreements on technology transfers in the OECD industry, by form of co-operation, pre-1972 to 1988

Figures are for 4-year periods, the absolute number and share (percentage, in italics) of total agreements recorded.

Form of co-operation[a]	Pre-1972	1973–76	1977–80	1981–84	1985–88	Total
Joint ventures and research corporations	83	64	112	254	345	**858**
	53	*42*	*23*	*21*	*18*	***22***
Joint R&D[b]	14	22	65	255	653	**1 009**
	9	*14*	*13*	*21*	*34*	***26***
Technology exchange agreements[c]	6	4	33	152	165	**360**
	4	*3*	*7*	*12*	*9*	***9***
Direct investment	27	29	168	170	237	**631**
	17	*19*	*34*	*14*	*12*	***16***
Customer-supplier relationships[d]	5	19	47	133	265	**469**
	3	*12*	*9*	*11*	*14*	***12***
One-directional technology flows[e]	21	15	71	259	271	**637**
	14	*10*	*14*	*21*	*14*	***16***
Total	**156**	**153**	**496**	**1 223**	**1 936**	**3 964**
	4	*4*	*13*	*31*	*49*	*100*

[a] Forms of co-operation are listed according to the degree of organizational interdependence, starting with the highest degree.
[b] Research pacts, joint development agreements, etc.
[c] Technology sharing and cross-licensing.
[d] R&D contracts, etc.
[e] Licensing and second-sourcing.

Source: Adapted from Hagedoorn, J., 'Organizational modes of inter-firm co-operation and technology transfer', *Technovation*, vol. 10, no. 1 (1990), table 2, p. 20.

projects and national employment. The design, production, marketing and sale of weapon systems take place under conditions that are very different from those prevailing in the general industry. Under the combined forces of political and military detente and economic pressures, however, the arms industry is now becoming increasingly more subject to the same economic conditions as the general industry.

The extent and pattern of arms industry internationalization have, with few exceptions, not been studied in depth. This is a difficult task, mainly for two reasons: there is no definition of the arms industry, either in industrial statistics or elsewhere,[13] and there are very few statistics available on the arms industry, since it is so intertwined with other activities on the company level, especially at the sectoral level.

[13] An interesting starting-point to devise 'workable taxonomies, or classification schemes, of the technologies and products found in the military sector' is made in Walker, W., Graham, M. and Harbor, B., 'From components to integrated systems', eds P. Gummett and J. Reppy, *The Relations Between Defence and Civil Technologies* (NATO and Kluwer: Dordrecht, Boston and London, 1988), pp. 17–37.

The ambition of this chapter is therefore to make an exploratory overview based on available data, studies and other types of information and to develop some tentative hypotheses about the level and pattern of internationalization in the West European arms industry.

From the available literature on traditional forms of arms industry co-operation[14] and on the more recent trends of internationalization,[15] it is possible to make the following very general description of the trends in and forms of arms industry internationalization.

International trade, licensed production and international sub-contracting are traditional forms of internationalization in the arms industry. During the 1960s, 1970s and most of the 1980s, international co-operation in weapon systems was as a rule dependent on government-to-government agreements, which have been associated with several types of difficulties (such as differences in required systems, problems in agreeing on work shares and high costs). Successful co-operation in the arms industry was therefore the exception rather than the rule and has almost exclusively been confined to the NATO countries. International restructuring of ownership (such as mergers, acquisitions and joint ventures) has previously been unusual in the arms industry.[16]

This limited, more gradual and irregular process of internationalization in the arms industry has since the late 1980s been replaced by a more dynamic process taking different forms than previously. The new trend can be sub-divided into two major developments. A restructuring of ownership in the form of international take-overs is taking place, while major arms-producing companies are entering into international co-operation arrangements or alliances, sometimes associated with ownership changes.

IV. Internationalization by arms industry sector

Aerospace

National restructuring of the West European arms industry is rather advanced in the field of aerospace.[17] The process of national concentration, although un-

[14] The history of NATO arms co-operation has been the subject of numerous studies, e.g., Draper, A. G., *European Defence Equipment Collaboration* (RUSI: Macmillan, 1990); Covington, T. G., Brendley, K. W. and Chenoweth, M. E., *A Review of European Arms Collaboration and Prospects for its Expansion under the Independent European Progam Group*, RAND Note N-2638 (RAND Corp.: Santa Monica, Calif., July 1987); and Hagelin, B., 'Multinational weapon projects and the international arms trade', SIPRI, *World Armaments and Disarmament: SIPRI Yearbook 1984* (Taylor & Francis: London, 1984), pp. 151–63; and Hagelin, B., 'Towards a West European defence community?', *Cooperation and Conflict*, no. 4 (1975), pp. 217–35.

[15] The more important surveys include Brzoska and Lock (note 1); Latham and Slack (note 6); OTA Report (note 1); Steinberg (note 1); Walker, W. and Gummett, P., 'Britain and the European armaments market', *International Affairs*, vol. 65, no. 3 (1989), pp. 419–42.

[16] In the non-NATO OECD area, international armaments co-operation has been limited and has to a large extent taken the form of single-source licensing agreements for primarily US designs. This is especially true for Japan.

[17] The aircraft, aero-engine and missile sectors are so intertwined on the company level that it is difficult to treat them separately, at least in terms of ownership restructuring. They are therefore combined under the heading 'aerospace'.

even, appears to be reaching its limits, especially in the aircraft and helicopter sectors.[18] In the missile sector, national consolidation is not as advanced yet, perhaps because of company specialization on different types of missiles and the limited costs involved in production of advanced combat aircraft and helicopters.

International acquisitions

While the process of ownership concentration is also taking place on the international level in the commercial aerospace industry, this is not happening to the same extent in the military aerospace sector (see table 9.3). The main exception is electronics (see the sub-section on military electronics below).

Developments in commercial aerospace are inter-related with military aerospace, since aircraft technology is largely dual-use technology. Thus, most

Table 9.3. International take-overs in the military aerospace sector, 1988 to September 1992

Buyer company	Country (head office)	Purchased company	Country (head) office)	Year of take-over	Comments
Bombardier	Canada	Short Brothers	UK	1989	. .
Matra	France	Fairchild Space, Fairchild Communications and Electronics, Fairchild Control Systems	USA	1989	. .
SNECMA	France	FN Moteurs	Belgium	1989	. .
SABCA	Belgium	Dassault Belgique Aviation	Belgium	1990	Seller: Dassault, France
Lucas Aerospace	UK	Tracor Aviation	USA	1991	. .
Aérospatiale Alcatel Alenia	France France Italy	Space Systems Loral	USA	1991	49% purchased
DASA	FRG	Fokker	Netherl.	1992	51% purchased

Sources: *SIPRI Yearbook 1990: World Armaments and Disarmament* (Oxford University Press: Oxford, 1990); *SIPRI Yearbook 1991*; and the SIPRI arms production archive.

[18] For documentation of national concentration in recent years, see Brzoska and Lock (note 1); for concentration of the French, German and British aerospace industries in the period 1950–80, see Hagelin, B., *Faelles Vestlig Våbenproduktion* [Joint Western Arms Production] (SNU: Copenhagen, 1986), figures 1–3.

military aerospace companies are also producers of commercial aircraft. Furthermore, trends observed in commercial aerospace can be expected to develop in military aerospace under more competitive conditions. According to industry representatives, the number of major producers of regional (medium-distance) passenger transport aircraft in the world has to be cut from nine to about three in a relatively short period, one of which would be European.[19] The industry is therefore intensively engaged in international restructuring. Aérospatiale and Alenia, which are already partners in development of the ATR (Avions de Transport Regional) family of regional aircraft, tried in 1991 to take control of the Canadian de Havilland aircraft company. However, this acquisition, which would have given the new company a 50 per cent share of the world market and a 67 per cent share of the European Community (EC) market for commuter aircraft, was blocked by the EC in October 1991.[20] This was the first time the EC used its then one-year-old anti-trust laws to veto a major business transaction.[21] De Havilland was instead purchased by Bombardier,[22] another Canadian aerospace and defence group, which recently also acquired Shorts (UK) and Canadair and thereby achieved a strong market position.

A European group is being formed around Deutsche Aerospace (DASA) and Fokker and may in the future also include Aérospatiale and Alenia. DASA won approval in 1992 to acquire a controlling share (51 per cent) of Fokker. In combination with the consortium for a regional airliner, formed in 1991 between Aérospatiale, Alenia and DASA, this may become the origin of a wider European commercial aircraft alliance. British Aerospace (BAe), the major European company in the field outside this alliance, decided in late 1992 to sell half of its regional transport activities to the recently formed Taiwan Aerospace.[23]

In military aerospace, a major international take-over deal in the USA failed in 1992 on national security grounds. This was the attempt by the French missile and electronics group Thomson-CSF, in co-operation with the US firms Carlyle and Northrop, to acquire the missile division of the US LTV company. This attempt was curtailed through intervention by the US Department of Defense.

As a result of national merger and acquisition activities, the European aerospace industry today consists of a relatively small number of major producers which are more or less national oligopols (see table 8.4, chapter 8 in this volume). This is true on the commercial side as well as in the arms production part of the industry.[24]

[19] *The Guardian*, 4 Sep. 1992.

[20] See also Courades Allebeck, A., 'The European Community: from the EC to the European Union', chapter 10 in this volume.

[21] *Wall Street Journal*, 3 Oct. 1991.

[22] Together with the Government of Ontario.

[23] *The Guardian*, 24 Sep. 1992.

[24] See Reppy, J., 'The United States: unmanaged change in the defence industry', chapter 3 in this volume.

The major West European producers of *military fixed-wing aircraft* are Aérospatiale, BAe, DASA, Dassault Aviation and Saab Military Aircraft. Medium-sized producers include Alenia and CASA. The main producers of *military helicopters* are Aérospatiale, Agusta, MBB (a DASA subsidiary) and Westland. In *aero-engines* the major producer in Europe is Rolls-Royce. Other significant producers in Europe include BMW/Rolls-Royce, MTU (a DASA subsidiary), SNECMA and Turbomeca (a Labinal subsidiary). In *missiles,* there are five main European producers: Aérospatiale, BAe, DASA, Matra and Thomson-CSF. In *avionics, guidance systems and other types of aerospace electronics*, the major producers are the British General Electric Company (GEC) and Thomson-CSF.

These companies form the core of the military aerospace sector in Europe. Around and between these leading producers, there is a close-knit structure of international co-operation which is currently being expanded. If the entire range of these cross-border links could be charted, the emerging pattern would probably show a high degree of internationalization. Inherent in current international co-operative structures is also a tendency towards stronger international concentration in the European aerospace sector, a tendency which may also be true for the rest of the OECD area.

International co-operation: history

The aerospace industry has a relatively long history of international co-operation in weapon systems. These co-operation projects have as a rule been government-initiated and government-sponsored and a consequence of the requirement for modern equipment in at least two and often more countries in NATO Europe.

Licensed production from foreign designs has been a common alternative to trade. This is true for both licensed production in Europe of foreign, mostly US, weapon systems and more recently for sales of European licences. Since the end of World War II, a usual practice in licensed production has been that a number of European NATO countries have agreed to produce a US system, with a labour division structure where firms in smaller countries have contributed sub-assemblies and some of the larger countries have contributed final production lines. Major examples of traditional European licensed production of US aircraft include the F-104 Starfighter (involving most of the NATO countries and Japan) and the F-4 Phantom fighter aircraft produced in the UK, initiated in the 1950s and 1960s, respectively. The most important examples of the same arrangement in missile production are the AIM-9 Sidewinder air-to-air missile (AAM) and the MIM-23 Hawk surface-to-air missile (SAM), starting in the 1950s and 1960s, respectively. FR Germany and most of the smaller European NATO countries participated in both of these missile programmes, while France took part only in the Hawk programme; the United Kingdom did not participate in any of them.

In licensed production the pattern has been one of mainly one-directional sales of US designs to Europe. The main exceptions are the AV-8B Harrier II fighter and, more recently, the T-45 Goshawk trainer, both cases involving original designs by BAe further developed in co-operation with the McDonnell Douglas US aerospace company.

Other forms of international co-operation have existed in parallel. *Sub-contracting* is one such form. Although not sufficiently documented to establish volumes and trends, it is clear that international sub-contracting is extensive both within Europe and between European companies and firms outside of Europe.[25]

The co-production arrangement for F-16 fighter aircraft, beginning in the 1970s, represents an intermediary form of international co-operation, combining licensed production with a structure for sales and sub-contracting. In this arrangement US aircraft were produced in two European countries (Belgium and the Netherlands) for all European partner countries (which initially included Denmark and Norway and later also Greece and Turkey).[26] Non-producing European partners were guaranteed a work share in sub-assemblies for the entire European and US production as an offset arrangement.

Full-scale collaboration projects, including *R&D co-operation* and agreements on industrial work shares,[27] have been relatively common in the fields of aircraft and missiles. The aircraft projects have, with few exceptions, been exclusively European. The most well-known of these are the Jaguar and Alpha Jet fighter aircraft programmes of the 1970s and the subsequent Tornado multi-role combat aircraft programme now approaching its end.

The major co-operative missile projects include the Sea Sparrow, Sky Flash and Improved Hawk of the 1970s and the RAM of the 1980s. These were projects for European procurement in co-operation with US companies. The major purely intra-European projects are the Milan and HOT anti-tank missiles (ATMs) and Roland SAMs, all three developed and produced by the Euromissile consortium, formed in 1962 and today made up of Aérospatiale and MBB; the Kormoran ASM developed and produced by the same two companies; and the Anglo-French (BAe–Matra) Martel ASM project initiated in the 1960s.[28]

[25] US DoD statistics show that the value of major US sub-contracts to other NATO countries (including Canada) was about $1 billion per year in the mid-1980s; Webb, S., *NATO and 1992: Defense Acquisition and Free Markets,* Rand Report R-3758-FF (RAND Corp.: Santa Monica, Calif., 1989), p. 23.

[26] Turkey now produces its own F-16 aircraft; see Günlük-Senesen, G., 'Turkey: the arms industry modernization programme', chapter 13 in this volume.

[27] 'Co-development' programmes in NATO terminology.

[28] For a comprehensive listing of international co-operation projects in the missile industry, see Wilén, C., *Internationellt samarbete inom robotindustrin* [International Co-operation in the Missile Industry], FOA Report A 10035-1.3 (FOA 1, Huvudavd. för försvarsanalys, Sundbyberg: Sep. 1992). The main report is in Swedish, but a summary in English is forthcoming.

Table 9.4. International joint ventures in the military aerospace sector, formed 1988 to September 1992

Company (shares)	Country	Year	Merged/new company	Purpose
CGE (50%) GEC (50%)	France UK	1989	GEC Alsthom	Power systems company
Aérospatiale (33%) Alenia (33%) Thomson-CSF (33%)	France Italy France	1989	Eurosam	Development of new Family of Anti-air Missile system (FAMS) for naval and ground-based air defence
BMW Rolls-Royce	FRG UK	1990	BMW/Rolls-Royce	Development of next-generation aero-engines
Aérospatiale (60%) MBB (40%)	France FRG	1991	Eurocopter	Production, marketing and sales of helicopters
Alenia BAe CASA Inisel MBB	Italy UK Spain Spain FRG	1991	Euroteam	Development of automatic test systems for military and civil applications, incl. for EFA
MTU Pratt & Whitney	FRG USA	1991	CSC	Development of next generation aero-engines
Hughes Electronics (40%) Inisel (60%)	USA Spain	1991	Gyconsa	Development of anti-tank missiles
GEC-Marconi Matra Defense	UK France	1991	Matra Marconi Space	Space systems
Elettronica Syseca	Italy France	1992	Eisys	Development of software for missiles
Alenia Honeywell	Italy USA	1992	Space Controls Alenia Honeywell	Construction of space equipment
Aérospatiale Alenia DASA Dassault Aviation	France Italy FRG France	1992	Euro-Hermespace	Development of Hermes

Sources: SIPRI Yearbook 1990: World Armaments and Disarmament (Oxford University Press: Oxford, 1990); *SIPRI Yearbook 1991*; and the SIPRI arms production archive.

International co-operation: the current pattern

While international take-overs of military aerospace activities have been only moderately common in the past five years (as shown in table 9.3 above), other internationalization activities have been more intensive. A number of joint ventures have been formed in all sectors of military aerospace during this period (see table 9.4). Other forms of co-operative arrangements also appear to be common in military aerospace. The major cases entered during the past two years are presented in table 9.5.

Table 9.5. Selected international co-operation arrangements in the military aerospace sector, 1991 to September 1992

Co-operating companies	Countries	Year	Weapon category/ type of co-operation	Purpose of co-operation
BAe General Dynamics	UK USA	1991	All types/ Co-operation alliance	On 'defence and technology projects'
Aérospatiale Alenia BAe CASA DASA	France Italy UK Spain FRG	1991	Aircraft/ Consortium: 'Euroflag'	Development of military cargo aircraft (FLA, Future Large Aircraft) to replace C-130 and C-160
OGMA SABCA SONACA TAI	Portugal Belgium Belgium Turkey	1992	Aircraft/ Consortium	Joined Euroflag
Aérospatiale Alenia BAe CASA DASA Fokker	France Italy UK Spain FRG Neth.	1992	Aircraft/ Planned joint venture: 'Europatrol'	Development of maritime patrol aircraft (MPA)
BAe Dassault Aviation	UK France	1992	Aircraft/ Joint R&D	Joint approach to studies of new technologies for successor fighter to Rafale and EFA
BAe Beech	UK USA	1992	Aircraft/Teaming arrangement	Collaboration on the US JPATS competition
CATIC Eurocopter Singapore Aerospace	China France/ FRG Singapore	1991	Helicopters/ Joint development	Joint development of light helicopter for civil and military applications: 'P120L'

Table 9.5 *contd*

Co-operating companies	Countries	Year	Weapon category/ type of co-operation	Purpose of co-operation
IBM Westland	USA UK	1991	Helicopters/ Teaming arrangement	UK MoD prime contract for the EH-101 Merlin
BAe Eurocopter	UK France–FRG	1992	Helicopters/ Teaming arrangement	To bid for UK MoD contract for new attack helicopters
McDonnell Douglas Westland	USA UK	1992	Helicopters/ Teaming arrangement	To bid for UK MoD contract for new attack helicopters
MTU Pratt & Whitney	FRG USA	1991	Aero-engines	Preferred partnership in future engine programs
Rolls-Royce Westinghouse	UK USA	1992	Aero-engines/ Strategic alliance	15-year co-operation agreement
Alenia Matra Defense	Italy France	1991	Missiles/Co-operation agreement	To co-ordinate their AAM activities, particularly R&D
MBB Raytheon	FRG USA	1991	Missiles/ Consortium	Production of Patriot SAM systems
Short Brothers Thomson-CSF	UK France	1991	Missiles/R&D	Joint development of new generation of very short-range air defence (VSHORAD) missile systems
BAe Dynamics Raytheon	UK USA	1991	Missiles/ Teaming arrangement	To bid for UK MoD contract for medium-range SAM
Eurosam GEC-Marconi	France/ Italy UK	1991	Missiles/ Teaming arrangement	To bid for UK MoD contract for medium-range SAM
Hughes Norsk Forsvars-teknologi	USA Norway	1991	Missiles/ Teaming arrangement	To bid for UK MoD contract for medium-range SAM
BAe Defence SAAB Missiles	UK Sweden	1992	Missiles/R&D	Joint development of medium-range missile
BAe Dynamics Hughes	UK USA	1992	Missiles/Teaming arrangement; sub-contracting	Sub-contracting work on short-range AAM for RAF

Co-operating companies	Countries	Year	Weapon category/ type of co-operation	Purpose of co-operation
GEC-Marconi Matra	UK France	1992	Missiles/Teaming arrangement	To bid for UK MoD contract for short-range AAMs

Note: For similar lists of international co-operation in the arms industry and related sectors in 1990–91, see Hébert, J-P., 'L'européanisation de l'industrie d'armement 1990–91' in *Memento défense-désarmement 1992: L'Europe et la sécurité internationale* (GRIP: Brussels, Apr.–July 1992), pp. 235–46.

Source: The SIPRI arms production archive.

In *combat aircraft*, the major co-operative structure in Europe has for two decades been Panavia Aircraft, the consortium formed in 1969 to manage the design, development and production of the Tornado combat aircraft for the FRG, Italy and the UK. The Tornado programme is commonly described as one of the largest European industrial ventures undertaken.[29] The prime contractors and owners of Panavia Aircraft are Alenia, BAe and DASA. The list of subcontractors and component suppliers is very long and includes most companies in the West European aerospace industry. The Tornado programme is nearing completion unless additional export orders are secured.

Another large fighter aircraft consortium, the Eurofighter Jagdflugzeug company, was formed in 1986 by Alenia, BAe, CASA and MBB (now DASA) to manage the European Fighter Aircraft (EFA) programme for the air forces of the FRG, Italy, Spain and the UK. With the announcement by the German Defence Minister in July 1992 that Germany would withdraw from this project, its future content and company constellation were uncertain in the autumn of 1992, although the British Ministry of Defence declared its determination to continue the project as planned. France and Sweden are pursuing their own fighter aircraft programmes: the Rafale by Dassault Aviation and the JAS 39 Gripen by Saab Military Aircraft.

While European governments may provide the European aerospace industry with three parallel fighter programmes from now into the early 21st century, it is generally believed that the European market cannot support more than one fighter aircraft programme for the next generation of aircraft. The negotiations in 1992 between BAe and Dassault Aviation on joint studies of new technologies for a successor fighter reflect this belief.[30]

The financial and technological need for international co-operation in major aircraft programmes has also been confirmed in recent European decisions relating to the development of a new *military transport aircraft* and for a new *maritime patrol aircraft*. Most European airframe producers are joined in the

[29] See, for example, *Jane's All the World's Aircraft 1992–93* (Butler & Tanner: Frome and London, 1992), p. 130.

[30] 'Dassault et BAe développent un projet d'avion de combat', *La Tribune*, 19 Feb. 1992, p. 11.

Euroflag consortium,[31] formed in June 1991 'to manage, develop, manufacture and support' the European Future Large Aircraft (FLA)[32] to replace the C-130 and C-160 Transall military transports in the early part of the next century. In 1992 it was also reported that an agreement had been signed by the six major European aircraft producers to co-operate on a future maritime patrol aircraft, the MPA, in a new industrial grouping called Europatrol.[33]

In military aircraft there are few examples of successful co-operative European projects with *non-European* governments or companies. The exceptions include the joint development and production project of Italy (Alenia/Aermacchi) and Brazil (Embraer) for the AMX small fighter/ reconnaissance aircraft, which has been going on since 1980; and German–US company co-operation on the Egrett intelligence-gathering aircraft and on the X-31 fighter technology demonstrator. In the current competition for a US Air Force trainer, the JPATS (Joint Primary Air Training System), European and US companies are teaming together—for example, BAe with the US Beech aircraft company. Still, it is unlikely that extensive collaboration projects, such as the EFA, could be formed with US companies.[34]

In *military helicopters* there are two major co-operative structures in Europe. In addition to the EHI Industries, there is the Eurocopter International venture formed in December 1991 by Aérospatiale and MBB with the transfer of their helicopter divisions to national subsidiaries of a joint holding company. Another important co-operative project, although not a joint venture, is the government co-operation programme for the NH-90 (NATO helicopter for the 1990s) which involves Eurocopter, Agusta and Fokker; it was started in 1985, and the first deliveries are planned for 1998.

Co-operation with *non-European* partners in military helicopter production is not uncommon: in the competition for the British Ministry of Defence (MoD) contract for EH-101 Merlin naval helicopters, the winning team was formed by IBM and Westland. BAe and Eurocopter joined in 1992 to bid for a British MoD contract for new attack helicopters. Another team in the same competition was formed by GEC-Marconi and Bell Helicopters.[35]

In the field of *aero-engines* it is difficult to make a distinction between military and non-military work. Three producers dominate the global market, of which only one, Rolls-Royce, is European, the other two being the US companies General Electric and Pratt & Whitney.

[31] Participating companies are: Aérospatiale, Alenia, BAe, CASA, DASA and, since 1992, SABCA and SONACA (Belgium), OGMA (Portugal) and TAI (Turkey); Jane's (note 29), pp. 123 and 709.

[32] Original work started in 1982 under the acronym FIMA (Future International Military Airlifter).

[33] *Jane's Defence Weekly*, 12 Sep. 1992, p. 17.

[34] This was recently refuted by a top manager of MDC, who, citing the Harrier example, insisted that the prospects for corporate co-operation with US companies were 'very bright'; 'McDonnell Douglas ready to build new Harrier', *The Guardian*, 9 Sep. 1992.

[35] West European companies show strong interest in co-operating with Russian helicopter producers, mainly on civil versions. One example is the advanced negotiations between Eurocopter and Kamov to develop jointly a new light helicopter. *Interavia Air Letter*, 31 Jan. 1992, p. 1; and *La Tribune*, 30 Oct. 1992, p. 12.

Aero-engines is a sector in which both 'the investment in R&D and the risks are immense', to quote the chairman of Daimler Benz.[36] Therefore, it is almost impossible for a single company alone to develop and build a new aircraft engine. Inter-firm co-operation is therefore regarded as imperative. Most major producers of aero-engines seem to be interlinked. Rolls-Royce formed a joint venture with BMW in 1990, mainly in order to spread the high R&D costs for the next generation of aircraft engines. The smaller aero-engine producers MTU and SNECMA have for the same reason become partners with Pratt & Whitney and General Electric, respectively. Four of the European aero-engine manufacturers are also combined in the Eurojet consortium, formed for the development of a power plant for the Eurofighter.

Co-operation between any of the three big companies has so far been limited to minor projects, but they may soon have to enter into more close co-operative structures. Rolls-Royce, however, is trying to resist this development through expansion in the commercial jet engine market and diversification into industrial power generation activities.

The recent trend of internationalization in the *missile* sector appears to be characterized less by ownership restructuring and more by co-operative inter-firm arrangements. The attempt to form a joint venture, Eurodynamics, from the missile-producing activities of BAe and Thomson-CSF would have created a dominant position for this venture on the European market. However, these plans to combine Thomson-CSF's expertise in guidance systems with BAe's weapon platforms were abandoned in early 1991 because of the failure to agree on the financial terms. After Thomson's failure also to acquire LTV's missile division, Thomson instead reached an agreement with Short Brothers of the UK (a subsidiary of Bombardier of Canada) to develop a new generation of surface-to-air missiles.[37]

Licensing is still a common form of internationalization in aerospace. A major new programme in Western Europe is that for licensed production of the US Stinger shoulder-launched anti-aircraft missile by the Stinger Project Group, a West European consortium, consisting of Germany, Greece, Italy, the Netherlands and Turkey (production planned to begin in 1992) and Switzerland (in 1993).[38]

The co-operation network of the West European missile industry is rapidly expanding. Missile manufacturers join in different groupings for each major MoD contract. In the competition for a British MoD contract for a short-range AAM, GEC-Marconi teamed with Matra while BAe teamed with Hughes. In the competition for a British MoD contract on a medium-range SAM, BAe is teaming with Raytheon, while GEC-Marconi is teaming with Eurosam, and Hughes teamed with Norwegian NFT. Among the teams receiving contracts for

[36] 'German industrial goliath seeks synergy in diversity', *Signal*, May 1992, p. 60.
[37] As successors to Shorts' Starstreak system; *Financial Times*, 2 Sep. 1992.
[38] *World Missile Forecast*, Forecast Associates Inc. (Newtown, Conn.), Apr. 1991.

a US Army SAM system to replace the Hawk missile, there are at least four teams with European participation.[39]

Military electronics

There are two major producers of military electronics in Europe, Thomson-CSF and GEC-Marconi. These two are far ahead of their European competitors in total sales of military electronics. Other significant companies in this sector include Alcatel-Alsthom, Dassault Electronique, SAGEM Groupe and Sextant Avionique in France; Bremer Vulkan (Systemtechnik Nord and Atlas Elektronik), DASA (Telefunken System Technik), and Siemens in Germany; Hollandse Signaalapparaten in the Netherlands (although owned by French Thomson-CSF); Elettronica and IRI (Finmeccanica) in Italy; Bofors in Sweden; and Dowty Group, Ferranti-International, Hawker Siddeley, Racal Electronics, Siemens-Plessey, Smiths Industries and Thorn EMI in the UK (see table 8.5, chapter 8 in this volume).

International acquisitions

In the military electronics industry there have been many international acquisitions since 1988 (table 9.6). Thomson-CSF is the most active company in this respect.

The pattern and strategies behind the national and international acquisition activities taking place in the West European military electronics industry since the mid-1980s have been thoroughly described and analysed in a recent study.[40]

The companies acquired have been of several different types:

Firstly, there are companies with a predominant focus on subsystems in electronics, such as Ferranti and Plessey. In addition, there are companies which manufacture parts and smaller subsystems used for e.g. radar, communication and guidance systems. Among these are for example TRT, parts of the former Plessey and Pilkington Optronics. Several of these can be defined as strategic sub-contractors, which manufacture special-developed products to many systems companies. Thirdly, there are also companies which manufacture whole systems, e.g. Link Miles, HSA and Ferranti.[41]

[39] DASA in co-operation with Raytheon, E-Systems and Loral; BAe and Dornier together with Martin Marietta, Booz-Allen and others; BAe with Martin Marietta and Booz-Allen; Plessey and Norsk Forsvarsteknologi with Hughes, GE and others; *Jane's Defence Weekly*, 18 July 1992, p. 9.

[40] Sandström, M., *Strukturförändringar inom den europeiska försvarselektronikindustrin* [Structural Changes in the European Defence Electronics Industry], FOA-Rapport A 10036-1.3 (FOA 1, Huvudavd. för försvarsanalys, Sundbyberg, Sep. 1992). A summary in English is forthcoming. A business forecast on the military electronics industry is summarized in Buckles, G. A., 'Protracted program pruning fosters manufacturing glut', *Signal*, June 1992, pp. 63–66.

[41] Sandström (note 40), pp. 106–107 (author's translation).

Table 9.6. International take-overs in the military electronics sector, 1988 to September 1992

Buyer company	Country (head office)	Purchased company	Country (head office)	Year of take-over	Comments
Thomson-CSF	France	Ocean Defence Corp.	USA	1988	
Alcatel	France	Defence and other divisions of ACEC	Belgium	1989	
Thomson-CSF	France	NV Philips MBLE Defence	Neth.	1990	Increase of share from 40% to 100%
Thomson-CSF	France	Hollandse Signaal	Neth.	1990	
Thomson-CSF	France	Link-Miles	UK	1990	
Thomson-CSF	France	Ferranti-Intl sonar division	UK	1990	
Thomson-TRT Défense	France	Pilkington Optronics	UK	1991	50% share
Thomson-CSF	France	MEL Communica-tions	UK	1991	Seller: Thorn-EMI
Thomson-CSF	France	Kyat	Spain	1991	Merged with Syseca
Thomson Sintra ASM	France	SAES	Spain	1992	49% share; seller: Inisel/Bazan
Siemens	FRG	Plessey Radar and Defense Systems	UK	1989	
Siemens	FRG	Cardion Electronics	USA	1991	Seller: Ferranti-Int.
ELSAG	Italy	Bailey Controls	USA	1989	
Finmeccanica	Italy	Ferranti Italiana	UK	1990	
Finmeccanica	Italy	FIAR	Italy	1990	Seller: Ericsson, Sweden
Nobel Industries	Sweden	Philips Elektronik-industrier	Neth.	1989	
Plessey	UK	Leigh Instruments	Canada	1988	
Dowty	UK	Palmer Chenard Industries	USA	1989	. .
Thorn EMI	UK	MEL	UK	1990	Seller: Philips, Neth.
Dowty Group	UK	Resdel Engineering	USA	1990	
General Motors	USA	Rediffusion Simulation	UK	1988	
BEI Electronics	USA	4 divisions of Systron Donner	UK	1990	

Sources: SIPRI Yearbook 1990: World Armaments and Disarmament (Oxford University Press: Oxford, 1990); *SIPRI Yearbook 1991*; and the SIPRI arms production archive.

The identified purposes of these activities are of three major types: (*a*) to obtain additional technical competence, mainly through vertical acquisitions to ensure the supply of subsystems and components for the acquiring company's production of larger systems; (*b*) to attain synergy effects for future production in the sense of becoming more effective on a larger market, mainly through horizontal acquisitions to enable future rationalization of production and to take over the customers of the acquired companies; and (*c*) to integrate high-technology activities with the aim of transferring these to civil production. With the exception of Thomson-CSF, the companies in this sector were found to show an intensified need to exploit dual-use technologies in the manufacture of electronic systems.

These purposes are combined in two major types of strategy, according to the study by Sandström. One strategy is to use acquisitions to create a company structure of technical width and depth within most system areas in order to become an actor of sufficient size to achieve an international position in electronics. The main representatives of this strategy are Thomson-CSF and GEC-Marconi. The other strategy is to acquire activities needed to reduce dependence of subcontractors. This strategy is represented mainly by DASA, Bremer Vulkan, and IRI/Finmeccanica.[42]

The pattern emerging from the study is summarized as follows:

> Among the studied companies, it is mainly Thomson-CSF, which has been international in its acquisitions. It is probably Thomson-CSF, which has created itself the most influential position on the market through its acquisitions in both the United Kingdom and in the Benelux countries. If these are combined with its co-operation activities in Italy, it is only in Germany where this company is not established in national industry . . . The current West European market is also characterized by the fact that the dominating companies have shown a preparedness to develop co-operation contracts with each other. This co-operation has not always resulted in production, but has in many cases been limited to development or research co-operation.[43]

International co-operation

International joint ventures and other forms of international co-operation are common in the military electronics sector (see tables 9.7 and 9.8). There are many examples of international co-operation agreements in the military electronics industry. Teaming arrangements are common in international competitions, both between West European companies and with US electronics companies.

Transatlantic company co-operation is not unusual in military electronics. Especially Thomson-CSF has a wide network in the USA, including subsidiaries, joint ventures and teaming arrangements with several important US firms for US DoD contracts.[44]

[42] Sandström (note 40), pp. 61–63.

[43] Sandström (note 40), p. 107 (author's translation).

[44] For a description of this network, see 'Corporate group reorganizes for world technology race', *Signal*, Apr. 1992, pp. 47–48.

Table 9.7. International joint ventures in the military electronics sector, formed 1988 to September 1992

Companies (shares)	Countries	Year	Merged/new company	Purpose
CESELSA SD-Scicon	Spain UK	1990	Auronautical Systems Designers	. .
Allied Ordnance Thomson-CSF	Singapore France	1990	Defence Electronics of Singapore	. .
Ferranti-International sonar division Thomson-CSF	UK France	1990	Ferranti-Thomson Sonar Systems	. .
GEC-Marconi (50%) Thomson-CSF (50%)	UK France	1991	GTAR (GEC-Thomson Airborne Radar)	Development and marketing of radar systems for fighter aircraft
IRI Thomson-CSF	Italy France	1991	SGS-Thomson Microelectronic	. .

Sources: SIPRI Yearbook 1990: World Armaments and Disarmament (Oxford University Press: Oxford, 1990); *SIPRI Yearbook 1991*; and the SIPRI arms production archive.

Thomson-CSF is among the arms-producing companies which have done most to internationalize and it has done so as part of a well-directed strategy. As is seen from tables 9.6–9.8, Thomson has exploited a broad range of internationalization methods: it has made a number of foreign acquisitions during the last five-year period; it has formed joint ventures with electronics companies in many countries; and it has entered several different forms of co-operation arrangements with other companies in the field.

GEC-Marconi, the arms-producing part of British General Electric, expanded on the national market by some British acquisitions in the late 1980s, was reorganized in 1991 to reduce operating costs, and has during the past two years been active in creating international alliances.[45]

Although GEC-Marconi and Thomson-CSF are the two major competitors in a shrinking market, there are indications that they may join forces in the future. One example is their joint venture, GTAR, formed in 1991, for development of aircraft radar systems. Another example is their teaming arrangement the same year for one of the largest military communications projects in Europe. The

[45] 'GEC boosted by joint ventures', *Financial Times*, 2 July 1992, p. 21; 'GEC at the crossroads', *Financial Times*, 7 July 1992, p. 15.

Table 9.8. Selected international co-operation arrangements in the military electronics sector, 1991 to September 1992

Companies	Countries	Year of agreement	Type of co-operation	Purpose
BASE (BAe Systems and Equipment)	UK	1991	Marketing and sales	To co-operate in marketing and sales of air defence gunsights
Bofors Aero-tronics	Sweden			
GEC-Marconi	UK	1991	Teaming arrangement	To bid for British MoD contract on Bowman battlefield communications system (lost)
Thomson-CSF	France			
Officine Galileo/ EFIM	Italy	1991	Co-operation agreement	Electro-optical applications in defence, space and environment
SAGEM/SAT	France			
GEC-Marconi	UK	1991	Co-operation agreement	Development of missile guidance systems
Alenia	Italy			
BAe	UK	1992	Sub-contracting	BAe principal sub-contractor to ITT for the British MoD contract on Bowman battlefield communications system
ITT Defense	USA			
Boeing's defence and space division	USA	1992	Exploratory study	To explore opportunitites in aerospace and defence markets, incl. anti-submarine warfare and flight-control component technology
GEC-Marconi	UK			

Source: The SIPRI arms production archive.

competition was for a contract for the British Ministry of Defence for the development of the next-generation battlefield communications system, the Bowman, reported to have a value of £1–2 billion.[46] Another team in the same competition was British–US, in which BAe was the principal subcontractor to the US company ITT.

Tanks and other military vehicles

The industrial structure in tank and other military vehicle production appears to show a lower degree of national concentration than the aerospace and electronics sectors. It is, however, difficult to state this with any satisfactory degree of

[46] *Financial Times* and *International Herald Tribune*, 24 Oct. 1991.

Table 9.9. International take-overs in small arms, ordnance and ammunition, 1988 to September 1992

Buyer company	Country (head office)	Purchased company	Country (head) office)	Year of take-over	Comments
Thomson-Brandt Armaments	France	Forges de Zeebrugge	Belgium	1988	. .
GIAT Industries	France	Munitions division of PRB	Belgium	1990	From Astra (UK) after PRB went bankrupt
GIAT Industries	France	Technologies Belcan (launch and testing centre)	Canada	1990	. .
GIAT Industries	France	FN Herstal (90%)	Belgium	1991	. .
SNPE	France	Sipe Nobel	Italy	1992	. .
Diehl	FRG	BGT (Bodenseewerk Gerätetechnik, 80%)	USA	1989	Perkins Elmer Group, USA
Astra Holdings	UK	BMARC	Switzerl.	1989	Seller: Oerlikon-Bührle
Astra Holdings	UK	PRB	Belgium	1990	. .
Royal Ordnance	UK	Heckler&Koch	FRG	1991	. .
Royal Ordnance	UK	Muiden Chemie	Neth.	1991	. .

Sources: SIPRI Yearbook 1990: World Armaments and Disarmament (Oxford University Press: Oxford, 1990); *SIPRI Yearbook 1991*; and the SIPRI arms production archive.

certainty without a more detailed study of company structures and product areas.

Most manufacturers of military vehicles are also producers of other types of land systems, such as artillery, small arms and ordnance. Military shipbuilding and military vehicle production are often two parallel activities, at least at the parent company level, owing to their historical roots in the steel industry.

The major producers of heavy tanks in Europe are GIAT in France; Krauss-Maffei, MAK Systems and Thyssen Henschel in Germany; Oto Melara in Italy; and Vickers in the UK. Several of the other European military vehicle-producing companies (for a list of these, see table 8.7, chapter 8 in this volume) make subsystems for tanks or other armoured vehicles.

National concentration in military vehicle production is perhaps most advanced in France, with GIAT as a fully vertically integrated producer of three major components of military vehicles, that is, the platform, the armaments and the ammunition. In Germany, some national restructuring has been made but there are still three major manufacturers of tanks. In Italy, a large part of the military vehicle industry is affected by the reorganization of the state-owned

Table 9.10. International take-overs in the military vehicle sector, 1988 to September 1992

Buyer company	Country (head office)	Purchased company	Country (head) ofice)	Year of take-over
MAN	FRG	Steyr Daimler Puch	Austria	1990
FIAT	Italy	ENASA	Spain	1990

Sources: SIPRI Yearbook 1990: World Armaments and Disarmament (Oxford University Press: Oxford, 1990); *SIPRI Yearbook 1991*; and the SIPRI arms production archive.

industry which took place in late 1992. The likely outcome is a stronger drive towards concentration and internationalization. In the UK, there are no indications of concentration in the vehicle industry.

International acquisitions

Several of the companies producing military vehicles have been involved in rather extensive international take-over activities, but this has been mainly in the sectors of small arms, ordnance and ammunition (see table 9.9). In the military vehicle industry, the intensity of international acquisitions is low. During the five-year period since 1988, there have been only two major take-overs in this sector: MAN's (FRG) take-over of Steyr Daimler Puch of Austria and FIAT's (Italy) take-over of ENASA of Spain, both in 1990 (see table 9.10). In both cases the company realignment was not primarily motivated by arms production considerations but was directed at the production of trucks. Thus, in contrast to the development in military aerospace and electronics, it seems as if the military vehicle sector is not yet ready for increased internationalization, at least not in the form of international concentration.

International co-operation

Internationalization in the military vehicle sector has traditionally been the result of, and taken place in the framework of, government-to-government agreements, mainly among NATO members. One example is Italian licence-production of the German Leopard-1 main battle tank, starting in the 1970s.

Although institutional and organizational arrangements have been set up by NATO to promote international co-operation from early stages in the development process, the result in terms of internationalization of industry has been rather meagre. A recent example is the NATO co-operation effort, begun in 1989, to ensure some commonality between the tanks fielded in different NATO countries in the early next century. This aim did not withstand its first

Table 9.11. International co-operation arrangements in the military vehicle sector, 1991 to September 1992

Co-operating companies	Countries	Year	Type of co-operation	Purpose of co-operation
GIAT GKN	France UK	1992	R&D 2-year	Joint development of medium-weight and light wheeled and tracked vehicles
GIAT Krauss-Maffei	France FRG	1992	R&D 2-year	Joint development of medium-weight armoured vehicle for future German and French programmes

Source: The SIPRI arms production archive.

test, which was the competition for a British Army order for a new battle tank, won by Vickers in 1991.[47]

Although few, there are examples of company-initiated international co-operation also in this sector, which may indicate an expanded structure of international inter-firm links in the future (table 9.11). The major example is the Euro-LAV joint venture, formed in 1992 by MaK Systems (Rheinmetall) and Panhard (Peugeot) for the development of a new type of light armoured vehicle for the German Army.[48] MaK also has had an agreement since 1989 with Oto Melara on joint development of an armoured vehicle, AV-90, to replace the M-113 series.[49]

GIAT and GKN have co-operated since 1991 on a study for the development of medium-weight and light armoured vehicles. In June 1992 this resulted in an agreement for joint R&D. An agreement was at the same time made with the German vehicle producer Krauss-Maffei.[50]

GIAT has an active internationalization strategy in general, including the military vehicle sector. It includes co-operation with Rheinmetall and British Royal Ordnance on a 140-mm gun for the next generation of MBTs, and since 1990 with Vickers for development of technologies to be used in the next generation of tanks.

[47] Order for 130 Challenger-2 MBTs at a contract value of £520 million after a 3-year, hard competition with an upgraded version of the German Leopard-2, the French Leclerc and the US M-1-A2 Abrams. This order was described by Krauss-Maffei as a blow to international collaboration in tanks, while Vickers' view was that this order puts Vickers in a stronger position to discuss future collaboration in tanks with other Western manufacturers.

[48] 'Panhard plans for new markets', *Jane's Defence Weekly*, 8 Aug. 1992; and *Military Technology*, no. 8 (1992).

[49] This project is described in *Jane's Armour and Artillery 1991–92* (Butler & Tanner: Frome and London, 1991).

[50] See, for example, '3 European firms to cooperate on family of combat vehicles', *Defense News*, 29 June–5 July 1992, p. 36. These agreements are based on British, French and German requirements for new families of armoured vehicles for battlefield support. Each country has a requirement for up to 1000 vehicles. The VBM concept calls for a basic vehicle, in at least 2 weight categories, to be developed for the 3 countries.

In 1992 GIAT made a major reorganization of its operations into five relatively independent companies. One explicit objective was to prepare 'for the remodelling of a European arms industry by sectors and by products', since this would enable share exchanges at this level, sector by sector.[51]

Examples of West European company R&D co-operation with US companies are Thyssen's co-operation with General Dynamics in development of the Fox (Fuchs) NBCRS vehicles and Vickers' co-operation agreement since 1985 with US FMC to develop a battle tank for export, still only in the prototype stage.

Shipbuilding

The shipbuilding industry has suffered from over-capacities for at least two decades as a result of the collapse of the market in merchant ships in the mid-1970s. Thus, the rationale for co-operation has existed for a long time in this industry. With the sharp drop in orders also for combat ships since the mid-1980s, the economic need to co-operate has become urgent.

International take-overs

International take-overs of shipyards are rare events. No West European shipyard building ships for military purposes has been involved in international ownership restructuring during the period 1988–92.[52] However, the shipbuilding industry includes not only shipyards but also the manufacture of naval electronics, weapons and sensors, which account for the larger and an increasing share of the total value of warships. The major producers in this sector overlap to a great extent with those of the military electronics industry, described above, in which international restructuring of ownership has been extensive in recent years.

International co-operation

Joint ventures are not common in West European military shipbuilding. Only one example of a new formation has been found during the past five-year period, and this is only a marketing company (table 9.12).

Although the basic assumption in shipbuilding traditionally has been that the domestic shipbuilding industry should be supported, there have been several efforts of international co-operation in military shipbuilding over the past decade.

Apart from licence-production, which is common in the construction of warships, these collaborative projects have been of two major types. A more limited type of co-operation, in which a purchase by one country of a class of

[51] Chiquet, chairman and director of GIAT, interview in *Defense and Armement Internationale*, no. 110 (May/June 1992), p. 25.

[52] Some related take-overs have taken place, however, such as the acquisition in 1990 by Bremer Vulkan and Fincantieri of the diesel-engine activities of Swiss Gebrüder Sülzer and the purchase by Kvaerner (Norway) of the merchant shipbuilders Neptun and Wernov from Treuhand (FRG).

Table 9.12. International joint ventures in the military shipbuilding sector, 1988 to September 1992

Companies (shares)	Countries	Year	Merged/new company	Purpose
Chantiers de l'Atlantique (50%)	France	1992	Eurocorvette	Marketing and sales of the jointly designed BRECA family of ships
Bremer Vulkan (50%)	FRG			

Sources: SIPRI Yearbook 1990: World Armaments and Disarmament (Oxford University Press: Oxford, 1990); *SIPRI Yearbook 1991*; and the SIPRI arms production archive.

ships designed in another country, is linked to various degrees of manufacture, and in some cases, limited co-development work, by the purchasing country. Examples include Norway's purchase of six German Ula Class submarines, ordered in 1982, which involved an industrial offset programme under which Norway supplied pressure hull sections, while final assembly took place at a German shipyard; Australia's purchase of Swedish Kockums Type 471 submarines, ordered in 1987, in which the major part of production takes place in Australia; and the FRG's procurement of Dutch Kortenaer Class frigates, approved in 1976, including joint Dutch–German development work to modify this design to German Navy requirements.[53]

The other major type of West European shipbuilding collaboration is joint international development projects initiated by governments in order to reduce costs and sometimes also to achieve interoperability, starting from the requirement/design stage. Examples are the Tripartite Minehunter co-operation programme of Belgium, France and the Netherlands; and the seven-nation effort to develop jointly and produce a NATO frigate for the 1990s, the NFR-90, which collapsed in 1989. The Tripartite Minehunter, produced from 1977 to 1991, is the best example of a successful international co-operation programme in shipbuilding. It was a co-development programme, but with separate national production of the hulls, and with a division of labour in the supply of some naval equipment: the electrical installations by Belgium; the minehunting gear and some electronics by France; and the propulsion system by the Netherlands. The NFR-90 was a large-scale NATO programme to procure a common frigate. After having spent about $85 million on the project definition phase, it came to an end in late 1989 when the UK left the programme, followed first by France and Italy and then by the FRG, the Netherlands and Spain, leaving only Canada and the USA, which meant the termination of the project. Bi- or trilateral co-operative successor projects are

[53] The first of the 6 resulting Bremen Class was completed after 8 years, in 1984, after the FRG had redefined its requirements several times.

Table 9.13. International co-operation arrangements in the military shipbuilding sector, 1991 to September 1992

Co-operating companies	Countries	Year	Type of co-operation	Purpose of co-operation
DCN Bazan	France Spain	1991	R&D	Develop and market a new class of submarines; primary design responsibility with DCN
Rolls-Royce Westing-house Electric	UK USA	1992	Sub-contracting	US Navy ($165 m.) contract for a new propulsion system for surface ships

Source: The SIPRI arms production archive.

being discussed.[54] The latest agreed international co-operation venture in warship-building is the agreement between the huge French state-owned shipbuilder DCN[55] (Direction des Constructions Navales) and the Spanish Bazan to jointly develop a diesel-powered submarine called the Scorpene (table 9.13).[56]

One form of internationalization which appears to be increasing in the construction of warships is sub-contracting. Electronics, weapons and engines are increasingly being bought from companies in other than the shipbuilding countries. A few standard radars of each type, missiles, torpedoes, guns and engines are fitted on the majority of the ships built today.

Summary and the need for further research

Internationalization in the West European arms industry appears to have increased in intensity and also shifted forms. In the words of a former corporate manager in this industry: 'The world will have to get used to forms of company organizations, which exhibit no traditional structural features, but which nevertheless function very well.'[57]

The intensive international acquisitions taking place in the commercial aerospace sector are affecting the internationalization of military aerospace, because these companies are also engaged in arms-producing activities. However, international acquisitions are also common in military aerospace. The sensitivity of cross-border take-overs in the military sectors is demonstrated by

[54] Notably a joint Anglo-French frigate project.

[55] DCN began a major reorganization in late 1992 to facilitate participation in international co-operation programmes, and possibly with a view to transform its industrial branch into a structure similar to that of GIAT; see 'Charting a new course', *Jane's Defence Weekly*, 17 Oct. 1992, pp. 30–35.

[56] 'DCN/Bazan venture', *Jane's Defence Weekly*, 19 Sep. 1992.

[57] Henri Marti, former chairman and director of Aérospatiale, quoted in *Military Technology*, no. 6 (1992).

the Thomson–LTV case. International joint ventures and other types of international co-operation are, however, becoming increasingly common in the military aerospace sector.

Electronics is the part of the arms industry in which internationalization appears to be most advanced, in terms of acquisitions, joint ventures and other types of international co-operation. Here the pattern for commercial electronics seems to be paralleled in the military sector, largely by the same companies.

In tanks and other military vehicles the company structure is still dispersed, both nationally and, even more so, internationally. In smaller and medium-sized vehicles, recent agreements may indicate an approaching change in trend. The production of tanks is, as the production of fighter aircraft, a nationally protected but also very costly activity. Therefore, international co-operation is imminent, although probably also in the future in a form which preserves national competence and employment, such as the management consortium model used in international fighter aircraft projects. Limitations are obviously put on the number of tanks to be allowed by the 1990 Treaty on Conventional Armed Forces in Europe (CFE).

The construction of warships consists of two rather different sectors: the hull builders and the manufacturers of naval weapons, electronics and other equipment. The latter sector overlaps to a great extent with the military electronics industry, in which internationalization is advanced. In hull construction, overcapacities will have to be reduced. International concentration may become one vehicle for these cuts.

The purposes of internationalization in the arms industry must for several reasons differ somewhat from those common in the general industry. The main reasons are the significance of high R&D costs in the arms industry, the foreign policy objectives served by the industry, and technology security considerations circumventing it. A tentative classification is suggested in table 9.14, by adjusting table 9.3 to the specific conditions of the arms industry.

The pattern of acquisitions in the electronics industry reflects an orientation of most of these companies towards exploiting dual-use technologies. It is sometimes argued that the use of dual-use goods and industries is a route which governments will have to choose for all weapon systems with high R&D costs and decreasing product lines.[58] It appears that companies are already implementing this on their own and on an international scale.

The pattern and purposes of internationalization in the arms industry have not been studied in the same detailed manner as the industry in general. Some of the hypotheses which deserve further study are the following.

1. The process of arms industry internationalization has entered a more dynamic phase with new common forms of company internationalization.

[58] For a summary of this view, see Rupp, R., 'Dual use industries', *Nato's Sixteen Nations*, no. 2 (1992), pp. 26–29.

Table 9.14. Forms of internationalization in the arms industry

Form	Purposes	Sector example
International trade	To serve government foreign policy; To expand markets and increase returns on investments	All sectors of conventional weapons; limitations in missiles by the MTCR
Foreign investment	To increase market access and open local production	(Minority shares not reviewed here)
International sub contracting	To reduce costs; To satisfy offset arrangements	All sectors of conventional weapons
International licensing	To provide alternative to trade at the insistence of customers (customer leverage); To broaden markets and increase returns from R&D expenditures	Aerospace Electronics Vehicles Shipbuilding
Cross-border mergers and acquisitions	To increase scale economies and open local production	Electronics
International joint ventures	To share high costs and risks, esp. in R&D; To increase market access	Aerospace Electronics
International inter-firm agreements, including: co-production management consortia teaming arrangements	To share high costs and risks, esp. in R&D; To increase market access; To produce common equipment; To bid for large contracts	Aerospace Electronics

2. Arms industry internationalization is not a homogeneous process throughout the arms industry: in some sectors internationalization is more advanced than in others.

3. Sectors with lower unit production costs will be forced to a lesser extent to internationalize.

4. When smaller companies get into trouble, bigger competitors will buy them to consolidate the market.

10. The European Community: from the EC to the European Union

Agnès Courades Allebeck

I. Introduction

Government–arms industry relations cover the entire range of all the various forms of interaction between governments and industry, itself a complex amalgamation of various industrial sectors.[1] This chapter provides a detailed discussion of one manifestation of the governmental approach to a specific industry, by presenting the European Community (EC) and its policies towards the arms industry. The work of the EC in this area is an interesting subject of study today because, while the EC traditionally did not have any responsibilities in this field, it has become progressively more involved through the 1986 Single European Act, the implementation of the internal market and the transformation of the EC into a European Union.

The arms industry environment in Western Europe is characterized by protected national markets, with a wide range of public policies concerning ownership, market structure, financing and arms procurement, as well as other policies with sometimes unintended consequences for the industry. While a common feature has long been that the arms industry should benefit from national protectionism, East–West détente as well as financial and economic constraints have brought governments to reconsider this special treatment. The profound changes that have occurred in the arms market—concentration combined with internationalization[2]—now confront different traditions of industrial policies, from French interventionism to German and British liberalism. France, whose close governmental control over the arms industry was formerly expressed in ownership and in government-funded research and development (R&D), investment and promotion of exports, maintains its policy by the creation of new governmental instruments to aid the necessary ongoing restructuring of the armament sector. In contrast, the British and the German approaches have been based on other concepts, such as competition and privatization.[3] This variety of governmental approaches to coping with a restructuring industry is the main factor determining the European Community's approach to the arms industry. It

[1] An interesting analysis and comparison of government–industry relations can be found in Wilks, S. and Wright, M. (eds), *Comparative Government–Industry Relations, Western Europe, the United States, and Japan* (Clarendon Press: Oxford, 1987).

[2] See Sköns, E., 'Western Europe: internationalization of the arms industry' chapter 9 in this volume.

[3] For country case studies, see various chapters in Brzoska, M. and Lock, P. (eds), SIPRI, *Restructuring of Arms Production in Western Europe* (Oxford University Press: Oxford, 1992).

is also the main obstacle to the definition of a comprehensive EC policy towards this industry.

From the creation of the informal EUROGROUP in 1968 to the more recent and successful attempts of the Independent European Programme Group (IEPG)[4]—bypassing efforts of other bodies such as the NATO Conference of National Armaments Directors (CNAD) and even the Western European Union (WEU)—co-ordination in the fields of arms production and procurement is an aim that has been pursued by many international institutions.[5]

The Maastricht Treaty on European Union—signed in February 1992 and at the time of writing expected to be ratified by the parliaments of the EC states and to enter into force in 1993—will establish a European Union and create new links between the European Community and the WEU.[6] The Treaty specifies that the WEU will become an integral part of the European Union and will 'elaborate and implement decisions and actions of the Union which have defence implications' (for excerpts from the Maastricht Treaty, see appendix 10A). In practice, following a gradual process of successive phases, the WEU will serve as the defence component of the European Union and reinforce the European pillar of the Atlantic Alliance. Following the first stage, which will be limited to strengthening the links between the European Union, the WEU and NATO, two dates will be important: 1996 and 1998. In 1996 the Council of Ministers plans to present a report on the Common Foreign and Security Policy (CFSP) to the European Council (composed of the EC heads of state or government and the President of the EC Commission), which shall formulate the next step to be taken. In 1998 the WEU aims to complete the planned revision of the 1948 Brussels Treaty of collaboration and collective self-defence among Western European states (modified in the Protocols of 1954), which will present an occasion to evaluate the experience of EC–WEU collaboration.

The Maastricht Treaty will have implications not only for the future of the WEU: the impetus given to the CFSP and not least to the future role of the WEU might have serious consequences for the agenda of the IEPG. In fact, although the IEPG is considered by many to be the most serious forum for European co-ordination of the production and procurement of military equipment, its relevance has been questioned since the signing of the Maastricht Treaty. Since responsibilities for collaboration in arms procurement are likely to be transferred to the WEU, as proposed by France and Germany in February 1992, it seems that the days of the IEPG might be numbered. If the activities of the

[4] For a thorough study of the IEPG, see Delhauteur, D., 'La coopération européenne dans le domaine des équipements militaires: la relance du GEIP', *Dossier "notes et documents"* (Institut Européen de recherche et d'information sur la paix et la sécurité (GRIP): Brussels), no. 159 (July 1991).

[5] For a review of the trans-Atlantic and international bodies which affect the European defence market, see Drown, J. D., Drown, C. and Campbell, K. (eds), *A Single European Arms Industry? European Defence Industries in the 1990s* (Brassey's: London, 1990).

[6] Article J.4 of the Treaty on European Union.

IEPG are transferred to the WEU, this would provide another argument for countries such as Norway and Turkey to join the WEU.[7]

One of the proposals which the WEU will examine is the creation by 1994 of a European Armaments Agency. According to French Defence Minister Pierre Joxe, the agency could take over some of the activities of the IEPG, such as the EUCLID (European Co-operative Long-term Initiative for Defence) programme.[8] European industrialists such as Henri Martre, Managing Director and Chairman of Aérospatiale, have also welcomed the creation of a European arms procurement agency aiming at promoting intra-European rather than trans-Atlantic trade.[9]

Regardless of the progress that has been made towards the achievement of a European Union, the European Community already has a significant influence on the military and military-related industry in the powers it is granted by its present constitutional and legal basis. Its main area of competence is competition policy, where the EC has supranational powers. To a lesser extent, the EC also affects the armament sector through its programmes of aid to industry and its prerogatives in trade.

The major obstacle to full EC involvement in the armament sector is Article 223 of the 1958 Treaty of Rome, which keeps the production of and trade in weapons within national competence and permits governments to derogate from the EC rules on national security grounds. Article 223 states: '1(*b*) Any Member State may take such measures as it considers necessary for the protection of the essential interests of its security which are connected with the production of or trade in arms, munitions and war material; such measures shall not adversely affect the conditions of competition in the common market regarding products which are not intended for specifically military purposes'.[10] (For excerpts from the Treaty of Rome, see appendix 10A.)

Governments have interpreted this article as a general exemption from all common market discipline, where in legal terms it is basically a derogation clause to be invoked only on particular occasions. In the past, all armament-related issues, however linked to EC policies, were left to the discretion of governments. The attitude of the EC Commission progressively changed from a broad interpretation of Article 223 towards a stricter approach. Following the adoption of the Single European Act in 1986, the Commission expressed its interpretation of Article 223 of the Treaty of Rome as not allowing derogation for so-called 'dual-use products'—that is, products with both a military and a civilian use or application.

The Single European Act was the first stage in the long process of deepening EC integration, with the ultimate aim of a European Union. It introduced new domains of co-operation among EC states. The European Political Co-operation

[7] *Defense News*, 9 Mar. 1992, p. 3.
[8] *Atlantic News*, 16 May 1992, p. 2.
[9] *Air & Cosmos*, 16 Dec. 1991, p. 6.
[10] European Communities, *Treaties Establishing the European Communities* (Commission of the European Communities, Office for Official Publications: Luxembourg, 1973), pp. 329–30.

(EPC)—an intergovernmental forum where EC member states discuss and co-ordinate actions which fall outside EC competence—was institutionalized in the Single European Act. EC countries committed themselves in Article 30.6 of the Single European Act to co-operate in the sphere of foreign policy: '(b) The High Contracting Parties are determined to maintain the technological and industrial conditions necessary for their security. They shall work to that end both at the national level and, where appropriate, within a framework of the competent institutions and bodies'.[11] (See appendix 10A for excerpts from the Single European Act.)

Article 30.6 of the Single European Act registers only statements of intent. Article 223 of the Treaty of Rome was not amended by the Single European Act but remained legally valid and binding on governments and EC institutions. In the framework of the Intergovernmental Conference on Political Union, the EC Commission proposed in 1991 a common policy for defence research and production and the suppression of Article 223.[12] Following the opposition of a majority of governments which did not wish to lose their prerogatives in this field—not least the UK—Article 223 was not amended in the Maastricht Treaty either.

The EC Commission now refers to the limitation of the scope of the derogation clause of Article 223 to purely military production and trade and claims its prerogatives on dual-use products. Dual-use products are quite significant: it has been estimated that transfers of such goods within the Community may account for around 5 per cent of intra-EC trade.[13] Once the EC Commission has exercised its new approach for a few years, Article 223 might lose its significance, especially since the difference between military and civil goods is becoming more difficult to establish.

II. EC competition policy

Competition policy has been one of the traditional fields of competence of the European Community since its creation. Provisions of the Treaty of Rome include the 'institution of a system ensuring that competition in the common market is not distorted'.[14] This system is based on common rules for ensuring a competitive environment for all companies within the Common Market. The EC Commission plays a central role in the implementation of EC competition policy, with powers of investigation and decision against non-competitive

[11] Article 30.6, Title III of the Single European Act, reproduced in *Treaties Establishing the European Communities, Treaties Amending these Treaties and Documents Concerning the Accession* (Commission of the European Communities, Office for Official Publications: Luxembourg, 1987), p. 1049.

[12] Article Y31, *First Contributions of the Commission to Intergovernmental Conference on 'Political Union'*, Working Document SEC(91)500, Brussels, 30 Mar. 1991, pp. 23 and 59. The implications of Article 223 are examined in section II of this chapter.

[13] *Financial Times*, 17 Feb. 1992, p.6; *Le Monde*, 24 Feb. 1992, p. 8.

[14] Article 3 (f), Part One—Principles, and Articles 85–94, Rules on competition, Part Three—Policy of the Community, Treaty establishing the European Economic Community, in *Treaties Establishing the European Communities* (note 10).

practices by both companies and governments. In cases of dispute between the Commission and a company or a member state, the decision lies with the EC Court of Justice, which ensures the interpretation and application of EC law.

EC competition policy is limited in scope to maintain competition as far as the Common Market is concerned. Competition within a member state remains under the control of national authorities; but, in the case of disagreement between a national and an EC interest, the latter would prevail if the case were brought before the EC Court of Justice, unless EC law states the case of superior national sovereignty as stipulated in Treaty of Rome Article 223.

Control over companies

The legal basis of EC competition policy with regard to companies is contained in Articles 85–87 and 89 of the Treaty of Rome.[15] These articles apply in principle to all sectors of the economy.[16] According to Article 223, the production of dual-use items of the armament sector falls under the competition rules. These rules permit the monitoring of cartel agreements and enable action to be taken to prevent the abuse of dominant market positions.

The Treaty of Rome only permits merger control to take place retroactively; that is, previously control could only take place after a merger agreement was signed, but now companies must notify the EC Commission prior to an agreement (the 'principle of prior notification'). In December 1989 EC competition law was completed by a merger control regulation, which entered into force in September 1990 and conferred on the EC prior control of concentrations between companies.[17] The new regulation applies to mergers having a Community dimension, which are defined by three criteria: (*a*) a threshold of a least 5 billion European Currency Units (ECUs), or about $3.6 billion, for the aggregate world-wide turnover of all firms concerned, (*b*) a threshold of at least 250 million ECUs (about $180 million) for the aggregate Community-wide turnover of each of at least two of the firms concerned, and (*c*) Community control does not apply to mergers whose impact is mainly national, that is, if each of the

[15] The Treaty of Rome as well as other EC law refers to companies under the term 'undertaking'.

[16] Except the coal and steel sectors, for which similar articles of the ECSC (European Coal and Steel Community) Treaty apply. Separate competition arrangements apply to the agriculture and transport sectors.

[17] The merger control system is set up by Council Regulation (EEC) no. 4064/89 of 21 Dec. 1989 and supplemented by the implementing and interpretative provisions adopted by the Commission on 25 July 1990. All texts can be found in 'Community merger control law', *Bulletin of the European Communities*, Supplement 2/90 (Commission of the European Communities, Office for Official Publications: Luxembourg, 1990). All mergers falling within the scope of the EC merger control are treated according to the concept of 'dominant position' referred to above. The official policy is that the creation or strengthening of a dominant position is declared incompatible with the Common Market if effective competition is impeded to a significant extent, whether within the Common Market as a whole or in a substantial part thereof: conversely, a merger which does not impede effective competition is declared compatible with the Common Market. Various aspects are taken into consideration at the assessment level. These include the structure of the markets concerned, actual and potential competition (from inside and outside the EC), the market position of the companies concerned, the scope for choice on the part of third parties, barriers to entry, the interests of consumers and technical and economic progress.

firms concerned achieves two-thirds of its turnover within the same member state.

This new merger control system has become necessary because of the increasing transnational concentration in Western Europe, following plans for full implementation of the internal market by January 1993. National regulations have become inadequate to prevent concentrations posing problems from a competition point of view at the EC level. The system is based on the so-called principle of exclusivity, by which national laws do not apply to cases falling under the EC merger control system. Derogation from the principle of exclusivity may be requested by member states, which may invoke legitimate interests, such as public security, for referral to a national authority. These legitimate interests may be recognized by the EC Commission. However, a concentration which has been prohibited by the Commission may not be authorized by a member state. The exclusive competence given to the EC Commission by this merger control regulation does not apply in cases, as defined in Article 223 of the Treaty of Rome, which are connected with the production of or trade in arms, munitions and war material. At the same time, the Commission has clearly stated that the restriction set by this article concerning products not intended for specifically military purposes should be complied with.[18] Therefore, concentration in the dual-use sector legally falls under the merger control system.

For many years the EC Commission, responsible for implementing competition policy, had shown great respect for national prerogatives in defence-related sectors and interpreted Article 223 in a broad sense, unless it was clear that the cases had to do with products strictly for civilian use. There were few exceptional cases. Controversies were never taken to court, so there has been no EC Court of Justice judgement involving Article 223 and competition. Competition policy became the field where the Commission progressively stated its determination strictly to interpret and apply Article 223. The 1989 merger control regulation increased the influence of the Commission in the defence and arms-related sector, especially since it dealt with assets and the composition of companies. Asked by a member of the European Parliament to give his interpretation of the legal basis of the Treaty of Rome with regard to mergers in the military industrial sector, Sir Leon Brittan, referring to Article 223, stated:

> . . . This provision clearly relates to State measures. It can only be invoked by member states and not by private or public undertakings. For this reason, only mergers which are imposed or encouraged by the government of a member state can benefit from the exemption under Article 223 (1) (b) of the Treaty.
>
> However, State intervention must be justified in the sense that the measures in relation to the merger are necessary for reasons of national security which are connected with the production of/or trade in arms, munitions and war materials. This means that Article 223 can only justify measures in relation to the merger which concern such

[18] Annex, Notes on Council Regulation (EEC) 4064/89, in 'Community merger control law' (note 17), p. 25.

production or trade. It should further be noted that although Article 223 of the EEC Treaty refers to the opinion of the member states, it does not create a possibility of unlimited unilateral derogation from the Treaty, since Article 225 of the EEC Treaty allows judicial control over the exercise of the derogation in Article 223.

Where and in so far as the merger extends to industrial or commercial activities of a civil nature, it is subject to the full application of Community law in general and of its competition rules in particular. The Commission has then to ensure that the conditions of competition will not be adversely affected in the markets of those products which are not intended for specifically military purposes.

A similar approach is taken by Council Regulation (EEC) No. 4064/89 of 21 December 1989 on the control of concentrations between undertakings. Article 21 (3) reserves to Member States the power to take appropriate measures to protect legitimate interests other than those pursued by the Regulation—including public security—provided that such measures are compatible with the general principles and other provisions of the Community law. A Member State can therefore prohibit a merger which does not raise problems as to its compatibility with the common market under the competition rules, in order to safeguard legitimate national interests of public security, of which defence matters are an important aspect.[19]

This interpretation of Article 223 must not conceal the fact that the EC Commission's intentions are much more far-reaching and, as mentioned above, the Commission has proposed the suppression of Article 223.[20]

Since the enactment of the merger control regulation, a few cases have arisen in which a decision of the EC Commission applying to production of dual-use products had a direct impact on the activities of arms-producing companies. In February 1991 the Commission gave formal clearance to the creation of the Eurocopter firm by a merger of the helicopter sections of the French company Aérospatiale and the German company MBB.[21] In 1991, 58 per cent of Eurocopter's turnover was in the military sector.[22] However, the Commission only considered the competition in the civil sector when it approved the merger. In late 1991, under the same regulation, the Commission also approved a joint venture of the French electronics group Thomson-CSF with Pilkington PLC of the UK, through the purchase of 50 per cent of Pilkington Optronics, a subsidiary.[23] Optronics are mainly used in the defence sector. In January 1992 the Commission approved the projected merger of two Swedish companies—Saab-Scania Combitech and Ericsson Radar Electronics—whose activities cover military and civil products.[24]

[19] Answer given by Sir Leon Brittan on behalf of the Commission of the European Communities, on May 16, 1991, to Mr James Ford, *Official Journal of the European Communities*, no. C 130/2-3 (21 May 1991), written question no. 1088/89.

[20] Perissich, R. (Director General, Commission of the European Communities, Directorate-General for the Internal Market and Industrial Affairs), 'The defense industry in Europe: competition, cooperation and rationalisation', International Herald Tribune Conference on The Future of European Security—Political, Strategic and Industrial Aspects, Rome, 3 May 1991, p. 10.

[21] Data base of the EC Commission, IP/91/161, 26 Feb. 1991.

[22] *Air & Cosmos*, 20 Jan 1992, p. 19.

[23] *Flight International*, 27 Nov. 1991, p. 14; *The Week in Europe* (Commission of the European Communities), no. 38/91 (1991), p. 2.

[24] *Defense News*, 13 Jan. 1992, p. 25.

The case which has attracted most attention to the powers given to the Commission by the merger control regulation is the veto of the venture purchase of Boeing's de Havilland subsidiary by the Franco-Italian joint venture Avions de Transport Régional (ATR), owned by Aérospatiale and Alenia. Here again, the Commission only assessed the civilian market. The merger was considered anti-competitive by the Commission since it would have given ATR and de Havilland a world-wide share of about 50 per cent of the overall commuter aircraft market and of about 65 per cent in the Community.[25] This veto was the first refusal of the Commission, which had examined more than 50 cases of mergers and acquisitions.[26]

Discontent was strongly expressed in Italy and particularly France, where the Commission was criticized by members of the government for having an exclusively legal approach to competition problems, without consideration of industrial realities. Henri Martre, Managing Director and Chairman of Aérospatiale, even accused Sir Leon Brittan, the EC Commissioner for competition, of having served the interests of the other major European competitors, Fokker and British Aerospace.[27] Sir Leon defended the decision, saying that industry does not become strong through being monopolistic, and referred to Germany, which has both Europe's strongest anti-trust policy and its most powerful industry.[28] No solution could be found to permit the merger, and the Canadian company Bombardier and the Province of Ontario finally acquired de Havilland, with shares of 51 per cent and 49 per cent, respectively.[29]

The de Havilland controversy had two important consequences. First, the powers of the EC Commissioner for competition were reassessed and somewhat limited. Until then, competition investigations were directed by the Commissioner for competition and the Chairman of the Commission alone; other members of the Commission were involved only at the last stage of the decision-making process.[30] Martin Bangemann, Commissioner in charge of industrial policy who opposed the de Havilland decision, asked for a change in the internal decision-making process of the Commission entailing the involvement of other concerned commissioners during the investigation period.[31] His request was approved, and since February 1992 the projects presented for decision on mergers and acquisitions to the collective group of Commissioners have

[25] Commission Decision of 2 Oct. 1991 declaring the incompatibility with the Common Market of a concentration (case no. IV/M.053—Aérospatiale-Alenia/de Havilland), *Official Journal of the European Communities*, no. L 334/42-61 (5 Dec. 1991).

[26] Brittan, Sir Leon, 'Transcender les intérêts nationaux', *Le Monde*, 11 Oct. 1991, p. 2.

[27] *La Tribune de l'Expansion*, 9 Oct. 1991, p. 32.

[28] Brittan (note 26); interview with Sir Leon Brittan in *International Herald Tribune*, 14 Oct. 1991, p. 2.

[29] *La Tribune de l'Expansion*, 24 Jan. 1992, p. 1.

[30] The EC Commission takes decisions by a majority of its 17 members, all having equal votes, and once a decision is taken the principle of collective responsibility applies. In the De Havilland case, the margin was 9 to 4, with 4 abstentions; *International Herald Tribune*, 3 Oct. 1991, p. 9. President of the Commission Jacques Delors abstains when he is opposed to a decision and the question goes against French interests, he says; *La Tribune de l'Expansion*, 7 Oct. 1991, p. 44.

[31] *Le Monde*, 18 Oct. 1991, p. 28.

included the considerations of other Commissioners.[32] Second, the de Havilland decision resulted in the re-emergence of the debate on EC industrial policy, on which, as explained in section III below, the EC governments are far from having a common approach.

Control over governments

EC competition policy provides for control over state aid to firms and rules concerning state monopolies of a commercial character and public firms.

State aid

The legal basis for competition policy with regard to member states is contained in Treaty of Rome Article 37 on monopolies, Article 90 on public enterprises, and Articles 92–94 on state aid to companies.[33]

According to these articles, state aid which distorts trade within the Common Market is prohibited. The Treaty requires all aid to be notified to the Commission at the proposal stage, in order to allow it to determine whether the aid is compatible with the Common Market. The Commission keeps under constant review all systems of aid existing in member states. Since the early 1980s it has noted an enormous increase in state aid and adopted a stricter policy. In assessing state aid, the Commission claims to have become more vigilant than in the past and to approve aid measures only if they both 'promote recognized Community objectives and do not constitute barriers to the completion of the internal market'.[34] Public enterprises and enterprises to which member states grant special or exclusive rights are subject to the rules of competition in so far as the application of such rules does not obstruct the performance of the particular tasks assigned to them. According to the Commission, competition rules should apply in the same manner, regardless of the ownership structures of firms, since a public enterprise in a member state can be a competitor to a private company of another member state. In July 1991 the Commission adopted a new regulation which forces governments to show more transparency concerning their financial relations with public enterprises.[35]

As explained above, Treaty of Rome Article 223 is meant to permit governments to take the special measures which they consider essential to the interests of their security and which are connected with the production of or trade in arms, munitions and war material. This escape clause of Article 223 can be invoked to permit aid for such production that otherwise would be considered as incompatible with the Common Market. However, the restriction that these measures should not distort competition in products not intended for specifi-

[32] *Le Monde*, 7 Jan. 1992, p. 15.
[33] Treaty establishing the European Economic Community (note 10).
[34] *XXth Report on Competition Policy* (Commission of the European Communities, Office for Official Publications: Luxembourg, 1991), p. 125.
[35] Brittan, Sir Leon, 'Entreprise publique et concurrence', *Le Monde*, 3 July 1991, p. 22; *Le Monde*, 26 July 1991, p. 20.

cally military purposes applies as well, and the Commission has recently stressed the fact that it would be particularly vigilant on state aid granted for production of dual-use items.[36] In implementing its state aid policy, the Commission sometimes takes a decision which directly affects arms-producing companies and therefore indirectly their military production. In 1991 the French Government had to suspend its plan to inject nearly $325 million into the state-run electronics group Thomson S.A. The French Government argued that the funds concerned capital increase and not aid. The Commission stated in clarification that capital injections by governments into their state-owned companies do not necessarily constitute illegal state aid, but that the determining factor was whether a private company would have made the same investment under similar circumstances.[37] In another case, the Commission went as far as deciding that British Aerospace should repay £44.4 million (about $25.5 million) so-called 'sweeteners' received from the British Government when it acquired the Rover automobile group in 1988. The EC Court of Justice overturned the decision, but only on procedural grounds and after a new process had been instituted by the Commission. The case over whether these funds constituted illegal state aid is still in court.[38]

Public procurement

The Treaty of Rome does not address the question of public procurement. The first EC legislation on public procurement dates from 1971, amended in 1977, 1988 and 1989, dealing with public works and public supplies.[39] The so-called Supply Directive applies only to procurement by EC ministries of defence of those products not intended for specifically military purposes, in accordance with Article 223. While the definition of the products concerned is becoming more and more blurred, dual-use products have been taking a growing share of defence procurement. Accordingly, since the Supply Directive was updated in 1988 and in accordance with its provisions, the number of notices published in the *Official Journal of the European Communities* relative to tenders by ministries of defence has been increasing.

In 1990 the Commission made the total opening of public procurement procedures one of the main priorities of the process towards full implementation of the internal market. The first step towards fighting national purchasing favouritism in traditionally protected sectors was the adoption of a regulation on public procurement in the so-called 'excluded sectors': energy, water, transport and telecommunications.[40] Two proposed directives related to public procurement of services were still, at the time of writing, at the stage of proposal to

[36] *Air & Cosmos*, 11 Nov. 1991, p. 6.

[37] *International Herald Tribune*, 21 June 1991, p. 9; *Financial Times*, 20 June 1991, p. 1.

[38] *The Week in Europe* (Commission of the European Communities), 12 Mar. 1992, p. 1.

[39] Directive 71/305/EEC, *Official Journal of the European Communities*, no. L/185 (1971), last amended by Directive 89/440/EEC, *Official Journal of the European Communities*, no. L/210; and Directive 77/62/EEC, *Official Journal of the European Communities*, no. L/13 (1977), last amended by Directive 88/295/EEC, *Official Journal of the European Communities*, no. L/127 (1988).

[40] Directive 90/531/EEC, *Official Journal of the European Communities*, no. L/297, 1990.

the Council.[41] The existing legislation provides for common procurement rules for contracts defining advertising and participation conditions and objective criteria for the selection and award of contracts—either the lowest price or the best tender.[42]

Although the ministries of defence of member states are included in the list of purchasing entities that fall under existing directives, they exclude from their scope of application security-related supplies. One of the proposed directives will apply to defence services, except for those provided on the grounds of national security by Article 223 of the Treaty of Rome.[43] The same Directive also excludes from its scope most R&D financing by member states.[44] Despite these limitations, Martin Bangemann, Commissioner responsible for the internal market and industrial affairs, has expressed the Commission's interest in applying these new EC regulations on public procurement to the defence industry.[45] However, it is clear that arms procurement will be the exception to the Community public procurement rules at least for the foreseeable future, even though member states are said to be in favour of greater competition in the dual-use product sector.[46]

Another body, the IEPG, has already laid the ground for the opening of the European armament market. The Action Plan on a Stepwise Development of a European Armament Market, based on the Vredeling Report presented to IEPG Ministers in November 1988, listed specific measures aiming at the creation of a European Armament Market open to competition, the inclusion of less-developed defence industries (LDDI) in arms co-operation, the ensurance of some kind of *juste retour* and co-operation in research and technology. The main handicaps of the IEPG are its limited administrative resources and the voluntary character of collaboration within its framework. In comparison to the legal prerogatives of the EC Commission, which will encompass the economic aspects of security, according to the Maastricht Treaty, the IEPG appears powerless. In this sense, the EC seems to be a more appropriate forum for implementing the deregulation of the armament market if governments are serious about their intention to rationalize this sector. The principle of *juste retour*, which has been criticized as incompatible with the fair competition that the IEPG is supposed to bring to the defence sector, could better be adopted at the EC level, where it could be conceived in a broader economic sense and applied taking into consideration other sectors of industry. In the context of the European Union, in February 1992 France and Germany proposed that the responsi-

[41] 'Public procurement: opening public services contracts', *Background Report* (Commission of the European Communities), 6 Apr. 1992.

[42] Note 41.

[43] Note 41; and Article 4 of Council Directive 92/.../EEC (the final number will be assigned when it has been adopted), relating to the co-ordination of procedures for the award of public service contracts, according to the text adopted for common position by the Council on 25 Feb. 1992.

[44] Council Directive 92/.../EEC (note 43), p. 3. This directive on services has gathered a common position at the Council and is at the moment of writing at the second reading stage by the European Parliament. It should be adopted before the end of 1992 without major amendments.

[45] *Defense News*, 30 Mar. 1992, p.10.

[46] *The Week in Europe* (Commission of the European Communities), 12 Mar. 1992, p. 2.

bilities for collaboration in arms procurement could be transferred to the WEU.[47] If the WEU were to take over this task, this would imply, however, that collaboration would remain an intergovernmental voluntary exercise of the same kind as the IEPG.

The opening of the armament market is not only a declared wish of Europeans. In April 1990, perhaps partially worried by what it considered to be a protectionistic attempt by the EC Commission to include some defence-related products under the EC common custom tariffs scheme, William Taft, US ambassador to NATO, launched a proposal to create a new code of conduct aiming at a more open trans-Atlantic armament market, also referred to as the 'Defence GATT'. By April 1992 the NATO Conference of National Armaments Directors (CNAD) was still confronting fundamental disagreement on the concept of this code of conduct, although they agreed to present such a regime by October of that year.[48] Some observers have described the CNAD project as having 'ambitions openly to override, under a new framework, the IEPG initiatives'.[49]

The difficulties encountered by these various forums while trying to open up armament procurement to competition illustrate the fact that most governments are inclined to adopt a protectionist attitude; this is why arms procurement will probably be the last area in which EC governments will give up their sovereignty.

III. Towards an industrial policy for the armament sector: the obstacles ahead

By mid-1992 the arms industry had received little direct financial support from the EC—only that channelled through the PERIFRA funds described below in this section. However, indirect support to most arms-producing companies of the European Community has occurred through the various EC R&D programmes, such as the Brite-Euram programme for the aeronautical industry. Obstacles for the proponents of an active EC policy towards the armament sector were of two kinds: first, the obvious implications of Article 223 and second, the preponderance of competition policy over industrial policy in the basic philosophy of the European Community. In the light of the development towards a European Union with the aim of establishing a Common Foreign and Security Policy, the arms industry has suddenly become more relevant to the work of the European Community. Furthermore, economic uncertainty has favoured more interventionist policies on the part of some governments.

[47] *Defense News*, 6 Mar. 1992, p. 3.
[48] *Defense News*, 20 Apr. 1992, p. 3.
[49] Delhauteur (note 4), p. 36 (in French).

Interventionists versus liberals

According to the spirit of the Treaty of Rome, based on economic liberalism, a free market and competition were intended to provide the best environment for EC industry. Competition policy is therefore the field in which the EC Commission is most powerful. However, given the current crisis that affects the armament sector as well as other parts of industry, interventionists are gaining ground at EC headquarters as well as in member states. The President of the EC Commission, Jacques Delors, has given priority to a new approach to industrial policy[50] believing that the EC should help industry to modernize in order to be able to face competition from the USA and Japan.[51] Indeed, interventionists argue against Sir Leon Brittan, Competition Commissioner, that real competition problems exist outside rather than inside the Community. Martin Bangemann, Internal Market and Industrial Affairs Commissioner, recently presented many proposals that could be characterized as interventionist. Among others, in a White Paper on the automobile industry he advocated EC financial support and R&D funds. He also presented an initiative to promote restructuring in the aerospace industry which calls for closer relations between governments and manufacturers, R&D efforts and the creation of a funding system to help companies with problems associated with dollar exchange-rate fluctuations.[52] In its five-year budgetary proposal for 1993–97, Delors requested that EC R&D funds increase from approximately $3.4 billion to $6 billion between 1992 and 1997.[53] The Commission planned for part of this increase to go to specific sectors such as the automobile and aeronautics sectors. However, the EC Council of Ministers rejected the idea on the grounds that such sectors were better supported by the private sector.

The Maastricht Treaty will add a new chapter to the Treaty of Rome. For the first time in EC legislation—in the new Title XIII, Article 130 added to the Treaty of Rome by the Maastricht Treaty—the need for co-ordinated support to industry is acknowledged.[54] The content of Title XIII is vague. The text invites other EC institutions to initiate or offer advice on action to be taken, but it restricts such action to mere co-operation; and for common practical measures to be taken, unanimity at the EC Council of Ministers is required, which ensures that no member state can be forced into an action. This can be interpreted as a conservative provision, since in the Single European Act, and even more so in the Maastricht Treaty, a growing number of measures concerned with the internal market and other economic matters require only a simple majority vote. Another indication that industrial policy is kept at a lower

[50] For the first attempt of the Commission to define the concept of Community Industrial Policy, see *Industrial Policy in an Open and Competitive Environment, Guidelines for a Community Approach*, Commision of the European Communities, Communication of the Commision to the Council and to the European Parliament, COM (90) 556 final, 16 Nov. 1990.

[51] *The Economist*, 2 May 1992, p. 88.

[52] *Interavia Air Letter*, 6 May 1992, p. 6; *Air & Cosmos*, 27 Apr. 1992, p. 14.

[53] *Tribune de l'Expansion*, 7 Apr. 1992, p. 10.

[54] See the text of Title XIII in appendix 10A.

level of ambition is the fact that the text twice makes clear that it should not undermine EC competition rules.

The possibility of establishing a substantial programme of aid for conversion of the armament industry to civilian production was discussed by EC Ministers of Industry, meeting informally in Lisbon in March 1992.[55] Differences between member states emerged on this occasion. France, which is traditionally more interventionist in its approach to industrial problems, favoured granting direct financial aid to defence industries in need of restructuring. Germany and the United Kingdom, as could be expected, were opposed to it, arguing, as did the Competition Commissioner, that the main defence industries are located in the rich and technologically developed regions of the Community. Apart from their fundamental disagreement on a governmental approach to industrial adjustment, the reason behind the governments' refusal to co-ordinate funds at the EC level in favour of the military sector might be their fear that the EC Commission will progressively gain more control over state aid.

The limits of the PERIFRA programme of conversion

The European Parliament has long advocated EC aid for conversion of defence industries to civilian production. It was not until September 1991 that such an EC fund was created under a programme called PERIFRA (*Régions périphériques et actions fragiles,* or special action for peripheral regions and destabilized activities). This special programme, which is not exclusively oriented towards the problems of the armament sector, was established in order to counteract effects in certain regions of 'the exceptional events of 1990'.[56] The following developments were considered: disarmament, trade concessions granted to Central European countries in the framework of the PHARE (Pologne–Hongrie: action pour la reconversion économique, or Assistance for economic restructuring in the countries of Central and Eastern Europe) programme,[57] the integration of the former GDR in the EC and the 1991 Persian Gulf War. The PERIFRA programme was awarded 40 million ECUs (about $29 million) to be allocated over a period of three years. PERIFRA was of symbolic significance, and its limited resources are insufficient to permit the implementation of vast industrial conversion in the necessary areas. The idea behind the programme was to stimulate and test pilot projects for conversion to civilian production. The PERIFRA funds co-financed up to 50 per cent (the other 50 per cent is to be paid by the government) of the costs of construction and renovation works, equipment purchasing, professional education and other support actions to enterprises in need of conversion. The remaining financing

[55] There was no press release because of the informal character of the meeting. *Atlantic News*, 25 Mar. 1992, p. 3; *Air & Cosmos*, 30 Mar. 1992, p. 7.

[56] 'Perifra', *Info Technique*, Information file published by the Directorate-General for Regional Policies (DG XVI), T-523, 1992.

[57] The EC PHARE programme was established in 1989 and extended to other Central and East European countries in 1990.

was the responsibility of governments. The member states, with the exception of Luxembourg, presented a total of 124 projects. Of the 51 which were chosen after examination by the EC Commission, 20 per cent of the funds went to Italy, 19 per cent to Germany, 17 per cent to Spain, 16 per cent to France, 10 per cent to Greece, 6 per cent to Ireland, 5 per cent to the UK, 3 per cent to Portugal, 2 per cent each to the Netherlands and Denmark, and less than 1 per cent to Belgium.[58] As much as 52 per cent of the funds were linked to defence-related events. Although many of these defence-related projects concerned support to areas which were economically depressed after the closure of military bases, some of them were directly concerned with arms-producing factories in difficulty, as for example: the shipyards in Bremen and arms-producing companies in Nordrhein-Westfalen (FRG); the Brest-located radar division of Thomson-CSF and the Tarbes-located factories of GIAT Industries (France); the Merseyside shipyard, the VSEL Cumbria shipyards, the British Aerospace plants in Lancashire and the Cowal Peninsula Enterprise Centre in Strathclyde (UK); and the Lisbon factories of INDEP (Portugal). No evaluation has yet been made of the PERIFRA programme, but most of the beneficiary projects were small in scale and period of implementation (not exceeding 18 months) and had a demonstration character for similar conversion projects in other parts of the European Community.

Following the insistence of the European Parliament, it was decided in December 1991 to renew funds for a second programme to be managed on the same terms. PERIFRA 2 was awarded 50 million ECUs (about $36.4 million)—still only a symbolic figure—and projects were granted under two of the four themes for which PERIFRA was originally created: namely, to help sensitive regions affected by the new trade concessions with countries benefiting from the PHARE programme, and regions facing major structural problems both because of arms industry conversion to civil production and because of closures of military installations.[59]

At the European Parliament, the Socialist and Christian Democrat Party groups allied with a smaller Union of the Left have been asking for the creation within the EC budget of a permanent fund to be called RECARM (conversion of armament industries) similar to EC funds devoted to the naval sector (RENAVAL), the coal mining sector (RECHAR), and the iron and steel sector (RESIDER). These programmes for industrial conversion are funded by so-called Structural Funds and fall under the EC regional policy schemes and their strict criteria of applicability.

The Structural Funds restrict assistance to five priority objectives:[60] (*a*) areas whose development is lagging behind,[61] (*b*) industrial areas with conversion

[58] Note of information from Commissioner Bruce Millan to Enrique Baron Crespo, Chairman of the European Parliament, 2 Aug. 1992, SG(91) D/15298, annexe 1, p. 2.

[59] Note 56.

[60] 'EC regional policy and the outlook for 1992', *Target 92*, Supplement, no. 1 (1992), p. 1.

[61] These are regions with a gross domestic product (GDP) per inhabitant of less than 75% of the Community average. The regions involved are the entire national territory of Greece, Ireland and Portugal; large parts of Spain; southern Italy; Corsica: the French overseas departments and Northern Ireland.

problems,[62] (*c*) combating long-term unemployment, (*d*) the occupational integration of young people, and (*e*) agriculture and rural areas. These five priority objectives result in practice in the division of the EC territory into different areas, some eligible for all funds, some for certain funds and others not eligible for any funds. The Commission's view has been that Europe's arms industry is largely located in flourishing areas[63] and that under these conditions it is not the task of the EC, as a Commissioner put it, 'to substitute itself for the signals of the market'.[64] Two conditions which were posed by the EC Commission on the first PERIFRA funds were that they had to be allocated in priority to regions already recognized for their economic problems by the EC regional funds and that EC rules on state aid should be respected. For the PERIFRA 2 exercise, however, it is stated that any part of the Community can be taken into consideration, whether or not it is eligible for assistance under the Structural Funds criteria.[65] The Commission's policy, based on Article 223, has been not to finance any direct aid to military production. The PERIFRA programme is meant to co-finance projects for conversion from military to civil activities. For example, a project aimed at switching from inefficient military production to more efficient military production would not be eligible.

In the so-called Delors II package, the Commission-proposed financial plan for 1993–97, which plans an increase of yearly resources from 66.6 billion ECUs in 1992 to around 87.5 billion ECUs in 1997, the budgetary item which benefits from the largest increase is the Structural Funds—from 18.6 billion ECUs in 1992 to 29.3 billion ECUs in 1997.[66] The Commission is considering proposing to member states a new approach to regional policy with the aim of being able to anticipate industrial adjustments before they have reached a critical point. In order to do so, it envisages less formalism and more flexibility in its criteria, especially the geographical criteria. If such a new approach were implemented, assistance for conversion in the arms industry could take place, in many cases where today such actions would not fall under regional policy criteria.

IV. The internal market and its implications for EC internal trade in defence-related goods

In 1986 the EC governments agreed to transform the Common Market created by the Treaty of Rome in 1958 into a genuine internal market comparable to a large national market. Article 8a, which was inserted into the Treaty of Rome by the Single European Act, defines the internal market as an area without

[62] This group consists of regions seriously affected by industrial decline, which are defined according to such socio-economic indicators as, for example, a higher rate of unemployment than the EC average. These regions are situated mainly in France, Germany, Spain and the UK.

[63] On the question of the location of arms industries, opinions differ between the Commission and some European Parliament (EP) members, as could be noted during the EP debates in Sep. 1991 in Strasbourg.

[64] *Financial Times*, 12 Sep. 1991, p. 3.

[65] Note 56.

[66] *Tribune pour l'Europe*, Feb. 1992, p. 2.

internal frontiers in which the free movement of goods, persons, services and capital is ensured.[67] As far as defence-related products are concerned, the abolition of frontiers between member states will have direct implications for government control over the transfer of such goods. With two exceptions,[68] most EC member states apply export controls to other member states for a range of defence-related products, according to their respective arms trade regulations.[69] Since Article 223 of the Treaty of Rome allows member states to derogate from EC law as far as production or trade in arms, munitions and war material is concerned, whenever governments consider it necessary for the essential interests of their security they will still be able to maintain export control of those products to other member states in January 1993, when the internal market is to be fully implemented. However, the scope of Article 223 is clearly restricted to goods intended for 'specifically military purposes'. As required by the Treaty, a list of products to which the derogation clause would apply was drawn by the EC Council of Ministers in 1958. This list, which was based on the 1958 COCOM (Co-ordinating Committee on Multilateral Export Controls) Munitions List,[70] has never been amended, which implies that the whole range of new technologies introduced in the armament market since that date do not legally fall under Article 223. However, in case of disagreement between the Commission and member states, it would be easy for the latter to update the list so that it covered all modern systems as well. Updating the list should be a priority now that member states have decided in the Maastricht Treaty not to cancel Article 223. The EC Commission has on several occasions in recent years officially declared that it considers that, with the exception of the goods in this list, all other goods, including the COCOM Industrial List of dual-use goods, fall under the provisions of the Treaty of Rome.[71]

The gradual disappearance of customs controls at intra-EC borders by January 1993, as required by the internal market, is seen as a dangerous incentive for illegal dealers to export goods with sensitive technologies, produced in the EC, to non-EC countries across the borders of EC countries with less efficient controls.

What could be defined as a purely internal technical problem takes on an external and political dimension. According to the EC Commission, although EC national export control regimes vary in their nature and efficacy, and in addition to the fact that there is no official co-operation between national

[67] *Single European Act,* Article 13 (Council of the European Communities, Office for Official Publications of the European Communities: Brussels, 1986), p. 18. The Act entered into force in July 1987.

[68] The exceptions are trade within the Benelux countries and exports from Ireland to the United Kingdom.

[69] Some member states, however, apply simplified export licensing procedures to each other.

[70] The COCOM regime has drawn up three lists of goods on which to apply control: a Munitions List, an Atomic Energy List and an Industrial List. For a description of these lists, see Anthony, I., 'The Co-ordinating Committee on Multilateral Export Controls', in ed. I. Anthony, SIPRI, *Arms Export Regulations* (Oxford University Press: Oxford, 1991), chapter 25.

[71] Lennon, P. J., 'Export controls and the European Community', Directorate-General, Internal Market and Industrial Affairs, Commission of the European Communities, Paper presented at the 6th Orgalime Seminar, Brussels, 22 Nov. 1991, p. 3.

administrative or political authorities outside known international frameworks such as COCOM, all member states have common legal grounds enabling the eventual harmonization of regulations. The Commission therefore proposed that, since the Treaty of Rome applies to dual-use goods, their trade within the internal market should not be submitted to control.[72] The condition *sine qua non* of the elimination of such controls would then be that all member states should establish efficient controls on exports of those goods to third countries, based on common norms. The Commission proposes that such a harmonization of regulations for dual-use products should apply to all international control of dual-use goods and technologies (by COCOM, the Missile Technology Control Regime [MTCR], and the Australia Group) as well as all national controls by member states. EC governments have already started informally to discuss the possible establishment of a common list of products to be controlled, but the Commission wishes to establish more formal procedures in order to speed up the process before the January 1993 deadline for the full implementation of the internal market.[73] In practical terms, the EC Commission proposes that several measures should be taken, some of a legal nature, in order to ensure the implementation of unified, strict control at EC borders. These include the following.

1. A common list of dual-use goods and technologies to be submitted to control. Such a list would be adopted as EC legislation in order to ensure its uniform implementation and direct applicability in all member states. The June 1992 Lisbon European Council meeting agreed on a common list of nuclear goods and nuclear-related dual-use goods to be controlled by member states.[74]
2. A common list of destinations which should be subject to control (although it has not been decided whether the list should be a list of proscribed or of 'friendly' countries).
3. Common criteria for issuing export licences.
4. A mechanism or forum to co-ordinate administrative and political export licensing procedures.
5. Clear procedures and a computer network for administrative co-operation between customs and licensing authorities from different member states. This should permit, for example, improvement in communication between the customs services of one member state and the licensing offices of another country. Two systems are already in service between most EC countries[75] and could be further developed for this purpose. One system permits the exchange of confidential information in order to combat fraud, and it could be used to keep track of companies which have already been involved in illegal deals. Another system, the TARIC (Tarif intégré de la Communauté) system for custom tariff pur-

[72] *Contrôles à l'exportation de biens et technologies à double usage et achèvement du marché intérieur*, Communication de la Commission au Conseil et au Parlement Européen, Commission des Communautés Européennes, SEC(92) 85 final, Brussels, 31 Jan. 1992.

[73] Note 71.

[74] *Atlantic News*, 30 June 1992, p. 2.

[75] It still has to be developed in Greece, Ireland, Portugal and Spain. Among these countries, one would certainly find those most in need of perfecting their arms export controls.

poses, could be used for recording information on products and technologies and their status regarding all their destinations.

The EC Ministers of Industry, meeting informally in Lisbon in March 1992 to discuss defence industries, expressed differing opinions on the EC Commission's proposal. Whereas Germany, which backed the Commission's proposal, was in favour of regulating these exports at the EC level, the UK expressed its concern that all security-related exports should remain under the sole jurisdiction of the member states.[76] On this last point, the Commission could very well ask for the interpretation of Article 223 by the EC Court of Justice. In its attempt to harmonize EC dual-use export controls, the Commission presented a so-called 'communication', which is not even a proposal for legislation and appears to strive for compromise. The European Parliament and the Council of Ministers were to give their opinions on the communication in December 1992. It is far from certain that the proposal will be approved in its totality; however, there are good chances that member states will agree on a minimum level of co-operation and even harmonization of procedures, which could result at least in a common, simplified licence system for intra-EC trade. If this were partially to solve the internal market aspect, it would not, however, address the problem of external controls.

As far as industrialists are concerned, they are generally positive to the prospect of the elimination of controls inside the Community, which would ensure them easier access for their dual-use products in the internal market. Edzard Reuter, the Chief Executive Officer of Daimler Benz, the largest arms producer in Germany, wrote in 1991 to German Chancellor Helmut Kohl and to European Commission President Jacques Delors requesting an EC arms export regime which would cover all armament products.[77] Another industrialist, Jürgen Schrempp, Chairman of Deutsche Aerospace, a subsidiary of Daimler Benz, expressed the same wish,[78] which was again reiterated by German companies in early 1992, following the strengthening of German arms export legislation.[79] German industrialists are the most enthusiastic supporters of the establishment of an EC arms export control regime, as their national legislation is among the most strict in the Community and they believe such a harmonization at the European level would result in looser regulations. Furthermore, some companies have even been lobbying, since the signing of the European Economic Area Agreement in October 1991 (to enter into force in January 1993) between the EC and the European Free Trade Association (EFTA) countries, for a regime to be extended to EFTA countries, in the eventuality that the Commission proposal would result in an EC open market for dual-use goods.[80]

[76] *Atlantic News*, 25 Mar. 1992, p. 3.
[77] *Atlantic News*, 27 Mar. 1991, p. 5.
[78] *Le Monde*, 20 June 1991, p. 34.
[79] *Air & Cosmos*, 3 Feb. 1992, p. 6.
[80] *Financial Times*, 17 Feb. 1992, p. 6.

V. The future EC arms transfer regime

After many years of resistance to such a project, in November 1991 the EC countries, with the support of Japan, initiated UN General Assembly Resolution 46/36, which has resulted in the establishment of 'a universal and non-discriminatory Register of Conventional Arms'.[81] This was a sign that EC governments finally envisaged a common approach in the field of arms exports control. No comprehensive legal mechanism for controlling arms exports exists at the EC level. Such a regime is required,[82] not least because the growing internationalization of arms production can result in situations such as that which arises when the same product is granted a licence for export to certain countries by one national authority, while it is refused by another. In practice, export is channelled through the country with the weakest arms export regulation.

Only one EC piece of legislation exists which to some extent addresses a problem of proliferation: Council Regulation (EEC) no. 428/89 concerning the export of certain chemical products, adopted in February 1989. This Regulation prohibits the export of listed products which could be used for the development or production of chemical weapons or when there is a risk of their being delivered directly or indirectly to belligerent countries or to areas of serious international tension.[83] The Regulation, whose only achievement is the establishment of a common list of products, is very limited in scope—all member states apply arms export controls to more chemical products than the eight listed—and still relies for its implementation on member states' respective export control mechanisms.

Arms embargoes

When arms embargoes have been enacted in the European Political Co-operation forum, where EC member states co-ordinate some aspects of their foreign policy on a purely voluntary basis,[84] they have been implemented through their national exports controls. In June 1992 EPC arms embargoes were in force against China, Iraq, Libya, Myanmar, South Africa, Syria and Yugoslavia, and discussions were under way on the possibility of establishing an arms embargo on the Horn of Africa—Sudan, Ethiopia and Somalia.[85] The Maastricht Treaty provides EC sanctions with a legal basis with the new Article 228a, which

[81] See Anthony, I. *et al.*, 'The trade in major conventional weapons', *SIPRI Yearbook 1992: World Armaments and Disarmament* (Oxford University Press: Oxford, 1992), appendix 8A, pp. 305–306.

[82] See for example *Regulating Arms Exports, a Programme for the European Community* (Saferworld Foundation: Bristol, Sep. 1991).

[83] Council Regulation (EEC) no. 428/89, of 20 Feb. 1989, concerning the export of certain chemical products, *Official Journal of the European Communities*, no. L50/1 (1989).

[84] EC member states' informal discussions on political aspects of security started in the framework of the EPC in the early 1970s. Formally recognized in 1981, the EPC was institutionalized in 1987 in the Single European Act.

[85] Telephone conversation with Mr Ole Neustrup from the EPC secretariat on 21 May 1992.

states that the Council of Ministers may, acting by qualified majority, take urgent measures to interrupt or reduce economic relations with third countries, in accordance with a decision taken under the CFSP.[86] This new provision should facilitate the implementation and co-ordination of embargoes and other economic sanctions. The main weakness of past EC embargoes was that they reflected reactions to specific events rather than efforts to prevent undesirable situations caused by unreasonable arms transfers.

EC arms export policy in the framework of the CFSP

In its Declaration of December 1991 on areas which could be the subject of joint action, the European Council mentioned nuclear non-proliferation issues as well as the economic aspects of security, in particular control of the transfer of military technology to third countries and control of arms exports, as issues for joint action under the CFSP. However, preparatory work for the Maastricht Treaty has shown that the various member states and EC institutions envisaged different forms of co-operation. The basic disagreement revolved around the legal nature of decisions to be taken under foreign and security policy.

A few governments, the EC Commission and the European Parliament were in favour of making foreign and security policy a field of EC competence. Other governments, concerned about their sovereignty although interested in some kind of co-ordination of this policy, wanted it to be a purely intergovernmental exercise of the type elaborated in the framework of the EPC. The Maastricht Treaty resulted in a compromise between the two approaches.

A common foreign and security policy covering all its aspects is established in the Maastricht Treaty. The objectives of the CFSP are to be pursued by establishing systematic co-operation between member states and by gradually implementing joint action in the areas in which the member states have important interests in common.[87] The European Council defines the principles of and general guidelines for common foreign and security policy.[88] On this basis, the Council of Ministers,[89] after deciding that a matter should be the subject of joint action, lays down the specific scope and objectives of the action and if necessary decides on its duration, means, procedures and conditions for implementation.[90]

The full association of the EC Commission with the work carried out in the common foreign and security policy field, and to a lesser extent the advisory role of the European Parliament, as stated in the Maastricht Treaty provisions,

[86] Council of the European Communities, *Treaty on European Union* (European Communities, Office for Official Publications: Luxembourg, 1992), Part Five, p. 89, amending Article 228a.

[87] Treaty on European Union, as finalized by the Working Party of Legal/Linguistic Experts, Conferences of the Representatives of the Governments of the Member States—Political Union—Economic and Monetary Union (CONF-UP-UEM 2002/92), UP-UEM/en 68, Internal document of the Commission of European Community, SEC(92)250, Brussels, 3 Feb. 1992.

[88] *Treaty on European Union* (note 86), Article J.3.

[89] The Council of Ministers, unlike the European Council, is the institution for the formal participation of the governments in the EC decision-making process; it does not comprise the EC Commission.

[90] *Treaty on European Union* (note 86), Article J.3.

are a sign of compromise in the direction of the federalist approach, which wanted to prevent the CFSP from being pursued in a purely intergovernmental framework. However, the Maastricht Treaty provisions do not go as far as the federalists wanted because of the fact that, contrary to most Community policies, CFSP implementation is not the responsibility of the Commission but of the Council of Ministers. It is in fact the task of the Council to ensure the 'unity, consistency and effectiveness of action by the Union'.[91] The same Presidency represents the Union in international organizations and conferences.[92] EC governments have given the Commission a more modest role in the CFSP than it has played in other fields, where its growing interference resulted from its quasi-executive prerogatives, given by the Treaty of Rome, as an initiating, supervisory and implementing body.

One can say that two main steps were taken by EC countries in the Maastricht Treaty in comparison to the past EPC framework. First, the new concept of joint action, which entails the definition of objectives and means that will commit member states.[93] This should in principle enable the European Union to achieve better performance in the field of foreign policy than the EC countries have been able to do. EC countries have not been very successful in their attempt to react in time and find efficient solutions to external events such as the Gulf War or the Yugoslavia crisis. Second, the full association of the Commission with the CFSP, through its initiative power as well as its obligation to inform the European Parliament on the development of the CFSP, is one step in the direction of a more EC-integrated CFSP.[94]

One criticism that can be made of the CFSP provisions is the lack of democratic control over decisions taken in this framework. The European Parliament is merely a consultative body, with little legal means to counter joint actions which it disapproves. The Council of Ministers does not have to comply with the Parliament's opinion. The European Parliament has no dismissive power over the members of government sitting on the Council or the Commission. The only way it can block a joint action is to use its budgetary power, through blocking it at the financial stage by refusing the adoption of the EC budget. However, as the Treaty provides for the possibility of financing joint actions with member states' contributions, a European Parliament blockade could be easily overturned.[95]

In practice and as far as arms exports are concerned, the consequences of the lack of democratic control at the European Union level are that arms export decisions which were in the past subject to parliamentary scrutiny in a few member states, as for example Germany, will no longer be subject to parliamentary control, especially since decisions taken at the European Union level will commit member states. This should be an issue of concern for national

[91] *Treaty on European Union* (note 86), Article J.8.
[92] *Treaty on European Union* (note 86), Article J.5.
[93] *Treaty on European Union* (note 86), Article J.3.
[94] *Treaty on European Union* (note 86), Articles J.7–8.
[95] *Treaty on European Union* (note 86), Article J.11.

parliaments ratifying the Maastricht Treaty. Furthermore, the future definition of the arms exports policy of the Union, which will result in compromise-seeking negotiations, might well have the insidious side-effect of loosening the criteria upon which arms exports licences are granted in the countries that have stricter regulations.

These criteria were presented as a basis for a possible future common approach to arms exports within the framework of the European Union:

– respect for the international commitments of the member States of the Community, in particular the sanctions decreed by the Security Council of the United Nations and those decreed by the Community, agreements on non-proliferation and other subjects, as well other international obligations;
– the respect of human rights in the country of final destination;
– the internal situation in the country of final destination, as a function of the existence of tensions or internal armed conflicts;
– the preservation of regional peace, security and stability;
– the national security of the member States and of territories whose external relations are the responsibility of a member State, as well as that of friendly and allied countries;
– the behaviour of the buyer country with regard to the international community, as regards in particular its attitude to terrorism, the nature of its alliances, and respect for international law;
– the existence of a risk that the equipment will be diverted within the buyer country or re-exported under undesirable conditions.[96]

Following the signature of the Maastricht Treaty in February 1992, the June 1992 Lisbon European Council meeting confirmed these criteria and added a new criterion: 'The compatibility of the arms exports with the technical and economic capacity of the recipient country, taking into account the desirability that States should achieve their legitimate needs of security and defence with the least diversion for armaments of human and economic resources'.[97]

The Maastricht Treaty has for the first time in the history of the European Community laid the legal basis for a comprehensive EC arms transfer regime. Some exports of armaments and sensitive technologies have been subject to decisions at the EC level in the past, on an *ad hoc* basis, when member states have agreed on occasions to co-ordinate arms embargoes in the framework of the EPC. With ratification of the Maastricht Treaty, arms export control will become part of the CFSP, itself to become an established policy of a more integrative character than the intergovernmental EPC forum.

[96] *Europe*, special edn, no. 5524 (30 June 1991), Annex VII, p. 16.
[97] *Atlantic News*, 30 June 1992, p. 2.

Appendix 10A

Excerpts from treaties relating to security policy, arms production and arms exports of the European Community

Prepared by Agnès Courades Allebeck

I. Treaty Establishing the European Economic Community, Treaty of Rome

PART SIX: General and Final Provisions

Article 223

1. The provisions of this Treaty shall not preclude the application of the following rules:

(*a*) No Member State shall be obliged to supply information the disclosure of which it considers contrary to the essential interests of its security;

(*b*) Any Member State may take such measures as it considers necessary for the protection of the essential interests of its security which are connected with the production of or trade in arms, munitions and war material; such measures shall not adversely affect the conditions of competition in the common market regarding products which are not intended for specifically military purposes.

2. During the first year after the entry into force of this Treaty, the Council shall, acting unanimously, draw up a list of products to which the provisions of paragraph 1 (b) shall apply.

3. The Council may, acting unanimously on a proposal from the Commission, make changes in this list.

Source: European Communities, *Treaties Establishing the European Communities* (European Communities, Office for Official Publications: Luxembourg, 1973), pp. 329–30.

LIST OF PRODUCTS REFERRED TO IN ARTICLE 223 OF THE TREATY OF ROME

Order 255/58 of 12 April 1958

The provisions of Article 223 paragraph 1 b) of the Treaty of Rome are applicable to the arms, munition and war material specified below, including nuclear arms:

1. *Portable and automatic firearms*, such as rifles, carbines, revolvers, pistols, sub-machine guns and machine guns, except for hunting weapons, pistols and other low calibre weapons of the calibre less than 7 mm.

2. *Artillery, and smoke, gas and flame-throwing weapons* such as

(a) cannon, howitzers, mortars, artillery, anti-tank guns, rocket launchers, flame-throwers, recoilless guns
(b) military smoke and gas guns

3. *Ammunition for the weapons at 1 and 2 above*

4. *Bombs, torpedoes, rockets and guided missiles*:
(a) bombs, torpedoes, grenades, including smoke grenades, smoke bombs, rockets, mines, guided missiles, underwater grenades, incendiary bombs
(b) military apparatus and components specially designed for the handling, assembly, dismantling, firing or detection of the articles at (a) above

5. *Military fire control equipment*:
(a) firing computers and guidance systems in infra-red and other night guidance devices
(b) telemeters, position indicators, altimeters
(c) electronic tracking components, gyroscopic, optical and acoustic
(d) bomb sights and gun sights, periscopes for the equipment specified in this list

6. *Tanks and specialist fighting vehicles*:
(a) tanks
(b) military type vehicles, armed or armoured, including amphibious vehicles
(c) armoured cars
(d) half-tracked military vehicles
(e) military vehicles with tank bodies
(f) trailers specially designed for the transportation of the ammunition specified at paragraphs 3 and 4

7. *Toxic or radioactive agents*:
(a) toxic, biological or chemical agents and radioactive agents adapted for destructive use in war against persons, animals or crops
(b) military apparatus for the propagation, detection and identification of substances at paragraph (a) above
(c) counter-measures material related to paragraph (a) above

8. *Powders, explosives and liquid or solid propellants*:
(a) powders and liquid or solid propellants specially designed and constructed for use with the material at paragraphs 3, 4 and 7 above
(b) military explosives
(c) incendiary and freezing agents for military use

9. *Warships and their specialist equipment*:
(a) warships of all kinds
(b) equipment specially designed for laying, detecting and sweeping mines
(c) underwater cables

10. *Aircraft and equipment for military use*

11. *Military electronic equipment*

12. *Camera equipment specially designed for military use*

13. *Other equipment and material*
(a) parachutes and parachute fabric
(b) water purification plant specially designed for military use
(c) military command relay electrical equipment

14. *Specialised parts and items of material included in this list insofar as they are of a military nature*

15. *Machines, equipment and items exclusively designed for the study, manufacture, testing and control of arms, munitions and apparatus of an exclusively military nature included in this list*

Source: IEPG, *Towards a Stronger Europe*, A Report by an Independent Study Team Established by Defence Ministers of Nations of the Independent European Programme Group to Make Proposals to Improve the Competitiveness of Europe's Defence Equipment Industry ('Vredeling Report'), (IEPG: Brussels, 1987).

II. Single European Act

Title III: Treaty Provisions on European Cooperation in the Sphere of Foreign Policy

Article 30, section 6

a) The High Contracting Parties consider that closer cooperation on questions of European security would contribute in an essential way to the development of a European identity in external policy matters. They are ready to coordinate their positions more closely on the political and economic aspects of security.

b) The High Contracting Parties are determined to maintain the technological and industrial conditions necessary for their security. They shall work to that end both at national level and, where appropriate, within the framework of the competent institutions and bodies.

c) Nothing in this Title shall impede closer cooperation in the field of security between certain of the High Contracting Parties within the framework of the Western European Union or the Atlantic Alliance.

Source: Council of the European Communities, *Single European Act and Final Act* (European Communities, Office for Official Publications: Luxembourg, 1986), p. 39.

III. Treaty on European Union, Treaty of Maastricht

TITLE II: Provisions amending the Treaty establishing the European Economic Community with a view to establishing the European Community:

Title XIII Industry

Article 130

1. The Community and the Member States shall ensure that the conditions necessary for the competitiveness of the Community 's industry exist.

For that purpose, in accordance with a system of open and competitive markets, their action shall be aimed at:

– speeding up the adjustment of industry to structural changes;

– encouraging an environment favourable to initiative and to the development of undertakings throughout the Community, particularly small and medium-sized undertakings;

– encouraging an environment favourable to cooperation between undertakings;

– fostering better exploitation of the industrial potential of policies of innovation, research and technological development.

2. The Member States shall consult each other in liaison with the Commission and, where necessary, shall coordinate their action. The Commission may take any useful initiative to promote such coordination.

3. The Community shall contribute to the achievement of the objectives set out in paragraph 1through the policies and activities it pursues under other provisions of this Treaty. The Council, acting unanimously on a proposal from the Commission, after consulting the European Parliament and the Economic and Social Committee, may decide on specific measures in support of action taken in the Member States to achieve the objectives set out in paragraph 1.

This Title shall not provide a basis for the introduction by the Community of any measures which could lead to a distortion of competition.

Part Five. Institutions of the Community

Article 228a

Where it is provided, in a common position or in a joint action adopted according to the provisions of the Treaty on European Union relating to the common foreign and security policy, for an action by the Community to interrupt or to reduce, in part or completely, economic relations with one or more third countries, the Council shall take the necessary urgent measures. The Council shall act by a qualified majority on a proposal from the Commission'.

TITLE V: Provisions on a Common Foreign and Security Policy

Article J

A common foreign and security policy is hereby established which shall be governed by the following provisions.

Article J.1

1. The Union and its Member States shall define and implement a common foreign and security policy, governed by the provisions of this Title and covering all areas of foreign and security policy.

2. The objectives of the common foreign and security policy shall be:

– to safeguard the common values, fundamental interests and independence of the Union;
– to strengthen the security of the Union and its Member States in all ways;
– to preserve peace and strengthen international security, in accordance with the principles of the United Nations Charter as well as the principles of the Helsinki Final Act and the objectives of the Paris Charter;
– to promote international co-operation;
– to develop and consolidate democracy and the rule of law, and respect for human rights and fundamental freedoms.

3. The Union shall pursue these objectives:
– by establishing systematic cooperation between Member States in the conduct of policy, in accordance with Article J.2;
– by gradually implementing, in accordance with Article J.3, joint action in the areas in which the Members States have important interests in common.

4. The Member States shall support the Union's external and security policy actively and unreservedly in a spirit of loyalty and mutual solidarity. They shall refrain from

any action which is contrary to the interests of the Union or likely to impair its effectiveness as a cohesive force in international relations. Tne Council shall ensure that these principles are complied with.

Article J.2

1. Member States shall inform and consult one another within the Council on any matter of foreign and security policy of general interest in order to ensure that their combined influence is exerted as effectively as possible by means of concerted and convergent action.

2. Whenever it deems it necessary, the Council shall define a common position.

Member States shall ensure that their national policies conform to the common positions.

3. Member States shall coordinate their action in international organizations and at international conferences. They shall uphold the common positions in such fora.

In international organizations and at international conferences where not all the Member States participate, those which do take part shall uphold the common positions.

Article J.3

The procedure for adopting joint action in matter covered by the foreign and security policy shall be the following:

1. The Council shall decide, on the basis of general guidelines from the European Council, that a matter should be the subject of joint action.

Whenever the Council decides on the principle of joint action, it shall lay down the specific scope, the Union's general and specific objectives in carrying out such action, if necessary its duration, and the means, procedures and conditions for its implementation.

2. The Council shall, when adopting the joint action and at any state during its development, define those matters on which decisions are to be taken by a qualifited majority.

Where the Council is required to act by a qualified majority pursuant to the preceding subparagraph, the votes of its members shall be weighted in accordance with Article 148(2) of the Treaty establishing the European Community, and for their adoption, acts of the Council shall require at least fifty-four votes in favour, cast by at least eight members.

3. If there is a change in circumstances having a substantial effect on a question subject to joint action, the Council shall review the principles and objectives of that action and take the necessary decisions. As long as the Council has not acted, the joint action shall stand.

4. Joint actions shall commit the Member States in the positions they adopt and in the conduct of their activity.

5. Whenever there is any plan to adopt a national position or take national action pursuant to a joint action, information shall be provided in time to allow, if necessary, for prior consultations within the Council. The obligation to provide prior information shall not apply to measures which are merely a national transposition of Council decisions.

6. In cases of imperative need arising from changes in the situation and failing a Council decision, Member States may take the necessary measures as a matter of

urgency having regard to the general objectives of the joint action. Tne Member State concerned shall inform the Council immediately of any such measures.

7. Should there be y major difficulties in implementing a joint action, a Member State shall refer them to the Council which shall discuss them and seek appropriate solutions. Such solutions shall not run counter to the objectives of the joint action or impair its effectiveness.

Article J.4

1. The common foreign and security policy shall include all questions related to the security of the Union, including the eventual framing of a common defence policy, which might in time lead to a common defence.

2. The Union requests the Western European Union (WEU), which is an integral part of the development of the Union, to elaborate and implement decisions and actions of the Union which have defence implications. The Council shall, in agreement with the institutions of the WEU, adopt the necessary practical arrangements.

3. Issues having defence implications dealt with under this Article shall not be subject to the procedures set out in Article J.3.

4. The policy of the Union in accordance with this Article shall not prejudice the specific character of the security and defence policy of certain Member States and shall respect the obligations of certain Member States under the North Atlantic Treaty and be compatible with the common security and defence policy established within that framework.

5. The provisions of this Article shall not prevent the development of closer cooperation between two or more Member States on a bilateral level, in the framework of the WEU and the Atlantic Alliance, provided such cooperation does not run counter to or impede that provided for in this Title.

6. With a view to furthering the objective of this Treaty, and having in view the date of 1998 in the context of Article XII of the Brussels Treaty, the provisions of this Article may be revised as provided for in Article N (2) on the basis of a report to be presented in 1996 by the Council to the European Council, which shall include an evaluation of the progress made and the experience gained until then.

Article J.5

1. The Presidency shall represent the Union in matters coming within the common foreign and security policy.

2. The Presidency shall be responsible for the implementation of common measures; in that capacity it shall in principle express the position of the Union in international organizations and international conferences.

3. In the tasks referred to in paragraphs 1 and 2, the Presidency shall be assisted if need be by the previous and next Member States to hold the Presidency. The Commission shall be fully associated in these tasks.

4. Without prejudice to Article J.2(3) and Article J.3(4), Member States represented in international organizations or international conferences where not all the Member States participate shall keep the latter informed of any matter of common interest.

Member States which are also members of the United Nations Security Council will concert and keep the other Member States fully informed. Member States which are permanent members of the Security Council will, in the execution of their functions, ensure the defence of the positions and the interests of the Union, without prejudice to their responsibilities under the provisions of the United Nations Charter.

Article J.6

The diplomatic and consular missions of the Member States and the Commission Delegations in third countries and international conferences, and their representations to international organizations, shall, co-operate in ensuring that the common positions and common measures adopted by the Council are complied with and implemented.

7. They shall step up cooperation by exchanging information, carrying out joint assessments and contributing to the implementation of the provisions referred to in Article 8c of the Treaty established the European Community.

Article J.7

The presidency shall consult the European Parliament on the main aspects and the basic choices of the common foreign and security policy and shall ensure that the views of the European Parliament are duly taken into consideration. Tne European Parliament shall be kept regularly informed by the Presidency and the Commission of the development of the Union's foreign and security policy.

The European Parliament may ask questions of the Council or make recommendations to it. It shall hold an annual debate on progress in implementing the common foreign and security policy.

Article J.8

1. The European Council shall define the principles of and general guidelines for the common foreign and security policy.

2. The Council shall take the decisions necessary for defining and implementing the common and foreign and security policy on the basis of the general guidelines adopted by the European Council. It shall ensure the unity, consistency and effectiveness of action by the Union.

The Council shall act unanimously, except for procedural questions and in the case referred to in Article J.3(2).

3. Any Member State or the Commission may refer to the Council any question relating to the common foreign and security policy and may submit proposals to the Council.

4. In cases requiring a rapid decision, the Presidency, of its own motion, or at the request of the Commission or a Member State, shall convene an extraordinary Council meeting within forty-eight hours or, in an emergency, within a shorter period.

5. Without prejudice to Article 151 of the Treaty establishing the European Community, a Political Committee consisting of Political Directors shall monitor the international situation in the areas covered by common foreign and security policy and contribute to the definition of policies by delivering opinions to the Council at the request of the Council or on its own initiative. It shall also monitor the implementation of agreed policies, without prejudice to the responsibility of the Presidency and the Commission.

Article J.9

The Commission shall be fully associated with the work carried out in the common foreign and security policy field.

Article J.10

On the occasion of any review of the security provisions under Article J.4, the Conference which is convened to that effect shall also examine whether any. other amendments need to be made to provisions relating to the common foreign and security policy.

Article J.11

1. The provisions referred to in Articles 137, 138 to 142, 146, 147, 150 to 153, 157 to 163 and 217 of the Treaty establishing the European Community shall apply to the provisions relating to the areas referred to in this Title.

2. Administrative expenditure which the provisions relating to the areas referred to in this Title tail for the institutions shall be charged to the budget of the European Conununities.

The Council may also:

– either decide unanimously that operating expenditure to which the implementation of those provisions gives rise is to be charged to the budget of the European Communities; in that event, the budgetary procedure laid down in the Treaty establishing the European Community shall be applicable;

– or determine that such expenditure shall be charged to the Member States, where appropriate in accordance with a scale to be decided.

Source: Council of the European Communities, *Treaty on European Union* (European Communities, Office for Official Publications: Luxembourg, 1992), pp. 52–53, 89, 123–29.

IV. Declaration by the European Council on areas which could be the subject of joint action

The European Council hereby declares that the implementation of joint action, in accordance with the procedures laid down in Article C of those provisions of the Treaty on European Union relating to the common foreign and security policy, is proposed as from the entry into force of the Treaty in areas connected with security, in particular:

– the CSCE process;

– the policy of disarmament and arms control in Europe, including confidence-building measures;

– nuclear non-proliferation issues;

– the economic aspects of security, in particular control of the transfer of military technology to third countries and control of arms exports.

The European Council invites the Ministers for Foreign Affairs to begin preparatory work with a view to defining the necessary basic elements for a policy of the Union by the date of entry into force of the Treaty.

The European Council furthermore invites the Council to prepare a report to the European Council in Lisbon on the likely development of the common foreign and security policy with a view to identifying areas open to joint action vis-à-vis particular countries or groups of countries.

Source: Annex to the Treaty on European Union, '*Europe*', Documents No. 1750/1751, 13 Dec. 1991, p. 29.

11. Poland: declining industry in a period of difficult economic transformation

Maciej Perczynski and Pawel Wieczorek

I. The historical background

The traditions of contemporary arms production in Poland are rooted in the inter-war period. After regaining independence in 1918, Poland commenced building its arms industry in an industrial infrastructure void. In the 1930s tremendous efforts were made by the state and society to establish the foundations of modern production of armaments and military equipment. At that time several technologically advanced factories were erected in the Central Industrial Region of Poland, in the middle and south-eastern part of the country. The most important plants—those in Starachowice, Pionki, Swidnik, Mielec, Rzeszow and Stalowa Wola—produced basic military equipment for the infantry and artillery as well as for mechanized infantry and air forces (fighter and bomber aircraft).

Polish arms production has never been on a very large scale, but the restoration of the pre-war potential after it was destroyed in World War II and its further modernization remain the backbone of the Polish arms industry. After World War II the dynamics of military construction fluctuated owing to changes in the international situation and the nature of its Warsaw Treaty Organization (WTO) obligations. Investment for military purposes peaked in 1951–55. At that time a wide range of new kinds of combat vehicle and weapon entered production: the T-34 battle tank, the MiG-15 fighter aircraft, some new anti-aircraft guns and radars. In the first half of the 1950s a specific 'division of labour' in arms production among the WTO member countries crystalized and determined the future structure of the Polish arms industry. However, starting in 1957 and for several years thereafter (at least until the early 1960s), investments in the military sector were considerably curtailed. This was primarily because of changes in internal policy (resulting from the so-called 1956 'Polish October'[1]) and their consequences for relations between Poland and the USSR as well as the considerable improvement of the international climate.

The first attempts at conversion of certain military production facilities to civilian production were made at this time. However, in the latter half of the 1960s tendencies for further expansion and restructuring of the military sector

[1] In Oct. 1956 major social protests and anti-Stalinist demonstrations were held in Poland which led to a change of the ruling party leadership; Wladyslaw Gomulka was released from jail to become the Party leader.

reappeared. The 1970s in particular were marked by considerable modernization of the arms industry, which resulted in the introduction of such new kinds of weaponry such as T-54 and T-55 battle tanks, tactical ballistic missiles, and several new types of combat aircraft and helicopter primarily built on the basis of Soviet licences. In the first half of the 1980s this tendency accelerated, and as a result, production of the modern T-72 battle tank was launched. Aircraft and naval industries were also considerably modernized.

A perceptible down-swing in the dynamics of military production appeared again in the second half of the 1980s, owing to considerably improved international relations and the diminishing intensity of the East–West confrontation, which resulted in the launching of a qualitatively new stage in disarmament negotiations. Movement towards reducing military spending and limiting the scope of arms production was characteristic of both East and West.

One can distinguish at least two phases in these processes in Poland.

1. The first phase (1986–90) was linked with developments in the Soviet Union under the political impact of General Secretary Mikhail Gorbachev's policy of *perestroika*, which resulted in a profound evolution of the internal and external relations of the USSR and other countries of the so-called socialist camp and led to a considerable acceleration of the disarmament dialogue. For Poland, the most important were the negotiations on the reduction of conventional forces, which resulted in the 1990 Treaty on Conventional Armed Forces in Europe (CFE Treaty). Under its provisions Poland is obliged to reduce the number of its battle tanks by 1120 units, artillery systems by 690 and armoured infantry combat vehicles by 127. Only in the case of helicopters were the limits set in the Treaty put at a higher level than the number possessed by the Polish Army (by 110 units).

2. The second phase (1990–91) was connected with the collapse of the socialist system in Poland and the other countries of Central and Eastern Europe. As a result of the profound systemic transformations, the political, military and economic ties between these countries changed in substance and character. The entire military sector found itself in an completely new setup. The dismantling of the Council for Mutual Economic Assistance (CMEA) and the dissolution of the WTO nullified all the premises on which previous military doctrine and policy were based and changed the place and role of the military sector in the country's transforming economy. The scale and dynamics of these processes are reflected above all in the evolution of the Polish national military budget.

II. The military budget

A downward trend in Polish military spending appeared in the latter half of the 1980s. The last year in which Poland's military expenditure rose in absolute terms was 1986. In the following years it tailed off and then plunged very sharply in 1990 and 1991 (see table 11.1).

Table 11.1. Polish military expenditure, 1986–92
Figures are in billion zlotys.

Year	Military expenditure		Dynamics, in percentages	
	At current prices	At constant prices	Change compared to the previous year	Change compared to 1986
1986	411	56 963	. .	. .
1987	506	55 311	*–2.9*	*–2.9*
1988	768	51 550	*–6.8*	*–9.5*
1989	2 147	46 395	*–10.0*	*–18.6*
1990	14 637	41 756	*–10.0*	*–26.7*
1991	18 211	27 142	*–35.0*	*–52.4*
1992	24 021	24 021	*–11.5*	*–57.8*

Source: Firlej, E., 'Okreslanie poziomu wydatkow wojskowych Polski' [Estimates of Polish military expenditures], paper presented to the Conference on the Military Budget of Poland, organized by the Centrum Badan Konwersji i Rozbrojenia SGH [Centre for Conversion and Disarmament of the Warsaw School of Commerce], Warsaw, 9 Apr. 1992.

As can be seen from table 11.1, military expenditure fell in 1990 by 10 per cent. In 1991 the reduction was even steeper: 35 per cent. Projections for 1992 indicate that the real level of military expenditure (in constant prices) will be almost 60 per cent lower than in 1986. Such changes have undoubtly influenced the position and altered the role of the military sector in the economy as a whole. This is clearly illustrated by data on the share of military expenditure in the gross domestic product (GDP) and in the state budget (see table 11.2).

Table 11.2 calls for additional explanation because the statistics which it contains do not—at first glance—indicate the qualitative change in the importance of the military sector in the structure of the national economy. These statistics show that the share of military spending in GDP declined in the period from 1986 to 1989 (from 3.2 to 1.8 per cent). In 1990 the share again increased and in 1992 fell back to 1.8 per cent again. The increase in 1990 may be clarified by noting the following: (*a*) in 1990 some substantial modifications were introduced in the methodology of the national accounting system (starting in 1990, the gross national product (GNP) and the GDP are now computed by a methodology almost fully competible with the Western system); and (*b*) military expenditure is now expressed in more realistic terms. For instance, some outlays, taxes and payments directly connected with the military sector which were previously covered by civilian departments are now financed from the budget of the Ministry of Defence. This is why, in spite of a real drop in military expenditure, its share in the national budget increased from 6.4 per cent in 1989 to 7.5 per cent in 1991 (see tables 11.1 and 11.2).

It must be added that the evolution in the relative position of the military sector in the economy as a whole was accompanied by substantial transforma-

Table 11.2. Share of Polish military expenditure in the gross domestic product and the state budget, 1986–92

Figures are percentages.

Year	GDP	State budget
1986	3.2	8.3
1987	3.0	8.5
1988	2.6	7.7
1989	1.8	6.4
1990	2.4	7.6
1991	2.0	7.5
1992	1.8	6.0

Source: Firlej, E., 'Okreslanie poziomu wydatkow wojskowych Polski' [Estimates of Polish military expenditures], paper presented to the Conference on the Military Budget of Poland, organized by the Centrum Badan Konwersji i Rozbrojenia SGH [Centre for Conversion and Disarmament of the Warsaw School of Commerce], Warsaw, 9 Apr. 1992.

Table 11.3. Structure of Polish military expenditure, 1987–92

Figures are in b. zlotys, at current prices; shares are expressed as percentages.

	1987		1990		1992	
	Value in b. zlotys	Share in total military expenditure	Value in b. zlotys	Share in total military expenditure	Value in b. zlotys	Share in total military expenditure
Personnel	212.5	*42.5*	7 904.0	*54.0*	15 613.7	*65.0*
Operation and maintenance	106.3	*21.0*	2 341.9	*16.0*	4 323.8	*18.0*
Procurement	156.8	*31.0*	3 366.5	*33.0*	3 122.7	*13.0*
Construction	30.4	*6.0*	1 024.6	*7.0*	960.8	*4.0*
Total	**506.0**	***100.0***	**14 637**	***100.0***	**24 021**	***100.0***

Source: Firlej, E., 'Okreslanie poziomu wydatkow wojskowych Polski' [Estimates of Polish military expenditures], paper presented to the Conference on the Military Budget of Poland, organized by the Centrum Badan Konwersji i Rozbrojenia SGH [Centre for Conversion and Disarmament of the Warsaw School of Commerce], Warsaw, 9 Apr. 1992.

tions in the structure of defence expenditure. These changes, for 1987–92, are reflected in table 11.3.

Attention must be drawn to the fact that the biggest shift in the structure of defence spending in this period was in arms procurement (down from 31 to 13 per cent from 1987 to 1992). The share of outlays on operation, maintenance and construction decreased to a lesser extent. The only item which increased is the share of outlays on personnel. The reasons for these changes are obvious. Under the general pressure for rationalizing military outlays, curtailments in material supplies were much easier to achieve than reductions in salaries and other payments to personnel. The necessity to keep remuneration in line with growing inflation outweighed certain other factors influencing the structure of the overall decline in military spending, including a substantial reduction in the size of the army itself. Unfortunately, the lack of reliable data on research and development (R&D) spending (because of the profound restructuring of the military research infrastructure) makes it impossible to comment on changes in this field, but indirect data indicate that its share in overall military outlays does not exceed 1 per cent.

III. Arms production, the economic condition and the status of military industry

The sharp decline in budget appropriations for military purposes decisively influenced the functioning and economic position of the arms industry. From 1988 to 1991 the volume of military production fell (in constant 1991 prices) from 12.5 billion to 4.13 billion zlotys, that is, by 67 per cent.[2] In 1991 this downward trend continued. It is necessary to add that the drop in military production during the period 1988–91 substantially exceeded the decline in overall industrial production. As a result the share of military production in aggregate industrial output declined from about 3 per cent in 1989 to 1–1.5 per cent in 1990/91.[3] This was accompanied by a considerable reduction in the level of employment: it is estimated that, in 1991, 180 000 persons were employed in the arms industry,[4] of which about 40 000 in the production of final products.

Economic conditions and functioning of military industry

These changes in military expenditure, arms production and arms industry employment undoubtedly reflect positive trends resulting from the historical transformation of Poland's internal and external relations. However, in the short run they pose extremely difficult socio-economic problems for both the economy as a whole and the arms industry in particular. The economic situation of the latter has dramatically deteriorated in recent years. The conversion of military production to other types of productive activity cannot be accomplished in a short period of time. The restructuring process is as a rule costly and highly capital-intensive. This is why the economic performance of the majority of military establishments in 1989–91 was extremely poor and why a large number of establishments became credit risks and found themselves on the verge of bankruptcy.

Several factors have contributed to this situation.

[2] Information obtained from the Military Department of the Ministry of Industry and Trade, in a note of 9 Apr. 1992.

[3] For 1989, see *Rzeczpospolita* 18–19 Feb. 1989; for 1990–91, authors' estimates based on information in note 1.

[4] *Rzeczpospolita*, 14 Feb. 1992 and note 1.

1. The marketization of the entire economy required stripping the military sector of all the economically unjustified privileges which accounted for its artificially high level of profitability. As a result military enterprises ceased to enjoy absolute priorities in access to cheap credits, special technology, tax preferences and, last but not least, substantial subsidies from the state budget. Currently, military enterprises have obtained independent financial status and operate according to the same rules as other establishments in the civil sector (except for their foreign operations).

2. The marketization process coincided with a drastic decline in military procurement. This started in 1989–90 and led to the near total marginalization of military purchases by the Ministry of Defence in 1991. At the same time the Ministry of Defence was obliged to spend considerable funds to eliminate surpluses in military equipment and weapons. It is estimated that dismantling a single T-55 battle tank costs about $10 000–12 000. Vivid proof of the almost total marginalization of military procurement is the fact that in 1991 the Ministry of Defence purchased only 10 T-72 battle tanks and 40 Polish Star trucks and six radar installations.[5]

3. There is a dramatic decline in Polish arms exports. This is an important subject in itself (discussed in section V), but in this context it is necessary to point out that arms exports to some countries were in the past a source of hard currency earnings and a highly profitable economic activity. Suffice it to say that, in 1988, to obtain $1 in exports to the hard currency area of armoured vehicles or battle tanks, only 200–300 zlotys were required to cover the costs of production. In the automobile industry the same dollar was obtainable for twice the cost and in electronics for triple the cost.[6] This can be explained by the fact that productivity in the arms industry was much higher than in other branches of industry. Priorities in high-technology appropriations, combined with a more disciplined organization of labour as well as substantial R&D content in the productive inputs, decisively contributed to the quality of military production. The competitive position of arms industry on the external market was therefore much higher than that of the automobile or electronics industries.

The collapse of Polish exports to the USSR and other WTO countries and to the less-developed world destroyed one of the basic pillars on which the economic performance of the Polish arms industry rested. All the above-mentioned factors resulted in the first nine months of 1991 in the military industrial sector sustaining overall losses estimated at 1.4 billion zlotys.[7] Against this background it is understandable that utilization of the productive capacity of the

[5] *Gazeta Wyborcza*, 25 Sep. 1991; and Kiljanek, K., 'Wojsko na bezdrozu' [Army without a way], *Prawo i Zycie*, 18 July 1992.

[6] Dziadul, J., 'Produkcja czolgow—zly czy dobry interes' [Battle tanks production: good or bad business?], *Polityka*, 14 Oct. 1989, p. 9; Markiewicz, W., 'Cena zycia' [The price of life], *Polityka*, 3 Feb. 1990, p. 9.

[7] *Polska Zbrojna*, 6 Feb. 1992. Converted according to the official 1991 exchange-rate, this amounts to about $135 million.

arms industry is currently appraised at only 20–30 per cent.[8] The financial indebtedness of the 30 major arms-producing factories in March 1992 exceeded 9 billion zlotys, of which more than 1 billion zlotys originated during the first two months of 1992.[9]

The bankruptcy of any leading establishment of the military industry obviously affects the defence interests of the country. Thus, serious efforts were made to draw up a consistent restructuring programme for the Polish arms industry, taking into account at least three important aspects of the problem: economic efficiency, defence priorities and possible conversion of military productive capacities to civilian use—an important and urgent but extremely difficult task.

IV. Organizational and systemic foundations of the arms industry

Over 350 industrial enterprises are currently involved in the production of armaments and military equipment in Poland. Of these, 120 manufacture final products. The degree of concentration in the military sector is generally higher than in the other branches of industry. It is estimated that the bulk of military final products originates in the 30 largest enterprises. A list of 32 major arms producers in 1991 is presented in table 11.4). Until 1989, the status of arms industry establishment was granted to 128 enterprises in manufacturing, services and trade. Of these, 84 came under the Ministry of Industry and Trade, 36 under the Ministry of Defence and 3 under the Ministry of the Interior. The rest were controlled by state departments.[10] In 1990–91 a further restructuring of the administrative and ownership set-up went ahead. Currently, there are three types of enterprise in the military sector: state arms enterprises, joint-stock companies and private enterprises.

State arms enterprises

The status of state arms enterprise is given to establishments producing basic weapons. Current plans postulate that only 19 establishments will belong to this category (8 subordinated to the Ministry of Industry and Trade (items 1–8 in table 11.4, and 11 to the Ministry of Defence,[11] items 9–19). Ministry of Industry and Trade establishments produce such basic items as battle tanks, small and heavy arms (caliber 5.45- to 23-mm), ammunition (rifle and artillery), 5.45- to 120-mm anti-tank missiles, explosives and electronic equipment. The 11

[8] *Zycie Warszawy*, 2 Oct. 1991.

[9] *Rzeczpospolita*, 20 Mar. 1992. Converted according to the official 1991 exchange-rate, these figures amount to about $850 million and $95 million, respectively.

[10] Wieczorek, P., *The Polish Arms Industry in the New Political and Economic Reality*, PISM Occasional Papers (PISM: Warsaw, 1991), p. 4.

[11] *Gazeta Wyborcza*, 9 Oct. 1991.

enterprises subordinated to the Ministry of Defence are establishments performing major repairs and modernization of arms and military equipment.

Joint-stock companies with majority government interest

In joint-stock companies the state has a majority holding and therefore controls decisions on production and structure as well as trade policy. It is assumed that this status will gradually be granted to 24 enterprises (producing training, transport and combat aircraft, helicopters, aircraft, ship and automobile engines, motorcars, armoured vehicles, trucks, battleships as well as explosives, ammunition and electronic equipment).[12]

Private enterprises and limited liabilities companies

These enterprises are beyond the sphere of direct organizational and functional subordination to government departments and undertake production activity at their own risk. They are mainly suppliers of components to the major arms producers.

The share of strictly military production in the overall output of the above-mentioned types of arms industry plant is highly differentiated. In factories with the status of Ministry of Industry and Trade enterprises, this share varied in 1991 from a very high level (as, for instance, at Radwar, 84 per cent, or Bumar, 71 per cent) to a very low level (for instance, 3.9 per cent at Gamrat and 1.9 per cent at Niewiadow). Much more homogeneous and military-oriented are establishments under the Ministry of Defence. Of these 11 establishments, the share of military production in overall output is 80–90 per cent. However, this share can be expected to diminish because the majority of the above-mentioned enterprises are increasingly involved in repair services in the civilian economy.

The share of military production at establishments having stock company status is also highly differentiated, but some of them concentrate on arms production and in several cases the share of military output in their total production is 60–90 per cent (see table 11.4).

The degree of concentration of production in the group of major producers is usually higher than in the rest of the economy. It is worth noting that 11 of the 32 enterprises classified as major arms producers employ over 1000 workers each. The level of employment in six of them is 1700–7400. The remaining major arms producers also belong to the category of large or very large factories, but the share of strictly military production in their total output is relatively small.

[12] Wieczorek, P., 'Konwersja przemyslu obronnego Polski' [Conversion of the Polish military industry], *Wojsko i Wychowanie*, no. 12 (1991), pp. 70–71.

Table 11.4. Major arms producers in Poland, 1991

Company	Location	Type of production	Share in overall production	Employment in arms production	Alternative production (after conversion)
Mesko	Skarzysko-Kamienna	Ammunition, anti-tank missiles	*46.8*	1 820	Farm tractors
Lucznik	Radom	Small arms	*15.6*	740	. .
Niewiadow	Niewiadow	Ammunition mines	*1.9*	252	Truck trailers
Tarnow	Tarnow	Anti-aircraft guns, small arms	*34.7*	880	Refrigerants
Pronit	Pionki	Explosives	*16.0*	665	. .
Gamrat	Jaslo	Explosives	*3.9*	67	Explosives for mining
BUMAR-Labedy	Gliwice	Tanks, crawlers	*71.4*	1 520	Hoisting cranes, loading machines, heavy dump-trucks
Radwar	Warsaw	Electronics	*84.0*	1 011	. .
PZL-Mielec	Mielec	Aircraft (transport, training)	*53.8*	7 495	Golf carts
PZL-Swidnik SA	Swidnik	Helicopters	*81.0*	5 868	. .
PZL-Rzeszow	Rzeszow	Aircraft engines	*66.6*	5 707	. .
PZL-Kalisz	Kalisz	Aircraft engines	*65.4*	2 204	. .
Okecie	Warsaw	Light-weight aircraft	*31.8*	704	. .
PZL-Warszawa II	Warsaw	Gas-mask spare parts	*68.8*	1 070	. .
Hydral	Wroclaw	Semi-units for mechanized combat equipment	*58.9*	1 365	Refrigerating installations, power meters, lighning meters
Stalowa Wola	Stalowa Wola	Self-propelled howitzers, armoured personnel carriers	*8.6*	1 876	. .
Star SA	Starachowice	Trucks, spare parts	*15.1*	854	. .
Warszawa-Wola	Warsaw	Engines for tanks and armoured vehicles	*25.0*	1 010	. .
Stocznia Polnocna im. Bohaterow Westerplatte (Westerplatte Heroes Shipyard)	Gdansk	Ships (landing, reconnaissance, hydrographic)	*94.7*	930	Repair services

Wisla	Gdansk	Patrol craft, tugboats	*19.3*	26	Repair services
Dezamet	Nowa Deba	Aerial bombs	*52.5*	356	Agricultural machinery
Krasnik	Krasnik	Semi-units for mechanized combat equipment	*6.0*	300	. .
Pressta	Bolechowo near Poznan	Shell, components of missiles	*60.2*	555	Equipment for food processing industry, concrete mixers, gas cylinders
Belma	Bydgoszcz	Mines, detonators	*22.2*	129	. .
Nitro-Chem	Bydgoszcz	Explosives	*4.6*	379	. .
ERG-Tychy	Bierun Stary		*0.6*	45	. .
Nitron-ERG	Krupski Mlyn	Explosives	*0.5*	40	. .
Przemyslowe Centrum Optyki (PCO, Industrial Centre for Optics)	Warsaw	Telemeter, fire control systems	*90.0*	1 100	. .
Warel	Warsaw	Electronics	*81.1*	570	. .
Radmor	Gdynia	Electronics	*43.3*	420	. .
UNIMOR	Gdansk	Electronics	2.2	285	. .
Zaklady Radiowe im. M. Kasprzaka (M. Kasprzaka Radio Factory)	Warsaw	Electronics	*23.6*	293	. .

Sources: Information from the Military Department of the Ministry of Industry and Trade, note of 9 Apr. 1992; and articles in *Polska Zbrojna*, Oct. 1991–Feb. 1992.

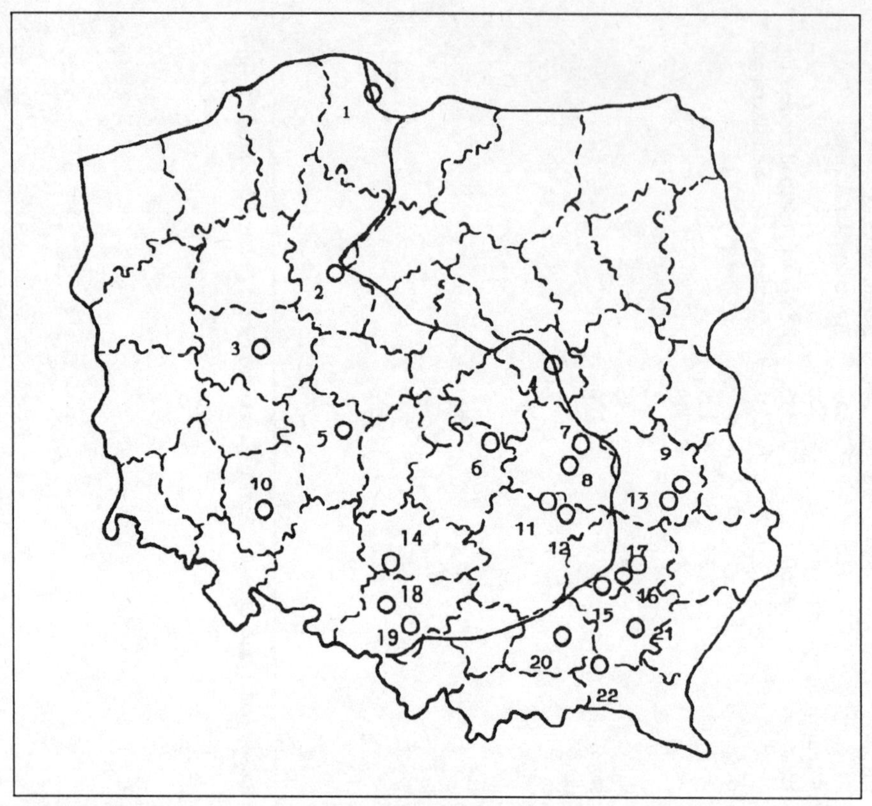

Figure 11.1. Location of the most important Polish military establishments, 1991

Key to the locations of the military establishments:

1. Gdansk, Gdynia:	Stocznia Polnocna im. B. Westerplatte; Wisla, Radmor, Unimor
2. Bydgoszcz:	Belma, Nitro-Chem
3. Bolechowo near Poznan:	Pressta
4. Warsaw:	Radwar, Warszawa-Okecie, PZL-Warszawa II, PCO, PZL-Wola, Zaklady Radiowe im. M. Kasprzaka, Warel
5. Kalisz:	PZL-Kalisz
6. Niewiadow:	Niewiadow
7. Pionki:	Pronit
8. Radom:	Lucznik
9. Swidnik:	PZL-Swidnik
10. Wroclaw:	Hydral
11. Skarzysko-Kamienna:	Mesko
12. Starachowice:	Star SA
13. Krasnik:	Krasnik
14. Krupski Mlyn:	Nitron-ERG
15. Mielec:	PZL-Mielec
16. Nowa Deba:	Dezamet
17. Stalowa Wola:	Stalowa Wola
18. Gliwice:	BUMAR-Labedy
19. Bierun Stary near Tych:	ERG-Tychy
20. Tarnow:	Tarnow
21. Rzeszow:	PZL-Rzeszow
22. Jaslo:	Gamrat

The motives for the specific distribution of military industry are determined by numerous factors of an economic, social and military nature. In Poland, as in other countries, the net result is a marked concentration of military industry in selected areas. Two-thirds of the establishments producing arms are deployed on the territory of the pre-war Central Industrial Region. This is clearly seen in the map in figure 11.1. The remainder of the military factories are located primarily around Gdansk, Warsaw, Bydgoszcz, Poznan and Wroclaw.

V. Arms exports

Export policy in the Polish military sector—as in other countries—is the result of numerous political, economic and military factors. Analysis of the determinants of Polish policy in recent years should take into consideration the following elements.

The field of manœuvre in this area is determined by the fact that the scale, structure and profile of the arms industry were formed in accordance with the participation of Poland in the Warsaw Pact, in which it played an important role in allied co-operation in arms production. Up to the end of the 1980s, 60–70 per cent of Polish arms exports were for other WTO member countries, particularly the USSR. The second major market was the Third World. In 1990–91 both markets collapsed. Nevertheless, former WTO and less-developed countries still absorb the bulk of Polish arms exports. It is estimated that in 1990–91 about 75 per cent of Poland's arms exports were directed to the Soviet Union. Czechoslovakia, Hungary, Bulgaria and Romania, in rank order, absorbed 10 per cent, and the rest was purchased by Third World countries.

As mentioned above, arms exports were regarded as the most profitable branch of export activity. The income from arms exports covered the cost of imports of both armaments and raw materials and licences for the arms industry. The largest arms export surplus (approximately $400–600 million yearly) was attained in 1978–80 and in 1984. In 1990 the positive balance in the arms trade was estimated at 275 million roubles (trade with the socialist countries) and $50 million (with the rest of the world) and in 1991 at $249 million.[13] Arms exports also made it possible to keep a considerable part of the industrial productive capacity in a state of so-called 'mobilizational readiness'. The national demand for arms accounted in 1988–90 for about 43–46 per cent of Polish arms production. In 1991 this share fell to 10–15 per cent.[14]

The dependence of the Polish arms industry on arms exports is relatively high. Before 1990, 48–52 per cent of arms production (so-called 'special production') was exported. The value of Polish arms exports in the period 1984–91 is presented in table 11.5.

[13] Information obtained from CENZIN; and Karkoszka, A., 'Socjalistyczny handel bronia' [Socialist arms trade], *Polityka*, 23–30 Dec. 1989.

[14] *Rzeczpospolita*, 21 Mar. 1991.

Table 11.5. Value of Polish arms exports, 1984–91

Figures are in m. roubles and US$, at current prices.

Values	1984	1985	1986	1987	1988	1989	1990	1991
In roubles	924.4	1 005.3	1 080.1	1 131.3	1 138.6	992.5	768.2	–
In US$	164.8	118.4	324.1	274.4	258.2	188.3	64.9	386.2

Sources: Zukrowska, K., *Organisation of Arms Exports in Eastern Europe and Prospects for Limitation of Arms Transfers*, PISM Occasional Papers, 1990, p. 15. For 1990–91, information obtained from the Ministry of External Economic Relations (CENZIN); and *Rzeczypospolita*, 28–19 Mar. 1992.

As can be seen from table 11.5, the highest value of exports came in the years 1986–88, as expressed in both transferable roubles and US dollars. Unfortunately, it is not possible to make credible estimations of total exports in convertible currencies. The data expressed in roubles indicate the trend in exports to the countries of the so-called 'socialist camp', whereas in dollars the values relate to the rest of the world (predominantly the less-developed countries). In both cases, a sharp decline is clearly noticeable from 1987. It must be added that in 1991 new rules were introduced under which trade among the former WTO member countries is carried out in convertible currencies. This is why in table 11.5 the values expressed in roubles disappear in the last column and the value of total exports is estimated only in US dollars.

According to the US Arms Control and Disarmament Agency (ACDA), the share of arms exports in total Polish exports was in 1989, 1.4 per cent; 1988, 3.9 per cent; 1987, 5.3 per cent; 1986, 6.4 per cent; and 1985, 7.3 per cent.[15] According to Polish sources, in 1990 the share was 1.6 per cent and in 1991 it increased to 3 per cent.[16] The increase in 1991 is explained by the fact that arms exports were diminishing less rapidly than overall exports.

The largest amount of Polish armaments were sold abroad in 1988. Since then the volume of arms exports has been gradually declining. The reasons are as follows.

1. The collapse of the WTO caused a radical break in co-operative ties between the industries of the member countries. All Central and East European states recorded a decline in military spending and dramatic curtailments in the demand for arms production, from both internal and external sources. For example, in 1988 Poland sold 300 An-2 aircraft and 200 MI-2 helicopters to the Soviet Union. In 1991 the USSR purchased only 19 aircraft and 17 helicopters.[17] A drastic deterioration in the economic terms of exports has also been caused by considerable delays in Soviet payments. The indebtedness of Russia,

[15] US Arms Control and Disarmament Agency, *World Military Expenditures and Arms Transfers 1990* (US Government Printing Office: Washington, DC, 1991), p. 104.

[16] Authors' estimates and information obtained from CENZIN.

[17] *Polska Zbrojna*, 23 Jan. 1992.

which assumed liability for Soviet debts in respect of Polish aircraft exports, reached the level of 1 billion zlotys.[18]

The difficult financial situation of arms-exporting enterprises limits their possibilities to sell on credit terms. Preferences for cash transactions always narrow the scope of external trade and have the same effect on Polish arms exports to the former CMEA countries.

2. The considerable number of less-developed countries which previously purchased appreciable amounts of Polish armaments now face similar economic difficulties and have stopped or limited their arms imports.

3. In the new international situation the Polish Government is granting concessions for arms exports in cases where there is certainty that the exported military equipment will be utilized for defence purposes only. Special responsibilities in this respect have been placed on the Ministry of Foreign Affairs. An integral part of any trade agreement in arms transactions is the so-called 'end-user certificate'. This is a document (warrant) issued by the importing state declaring that the purchased arms will not be transfered to third countries or organizations. The Polish Ministry of Foreign Affairs has drawn up a 'negative list' of countries to which arms deliveries are restricted or completely prohibited. In 1991 this list contained the following countries: Afghanistan, Burma, El Salvador, Iraq, Mozambique, Somalia, Sudan, Taiwan and Yugoslavia (total ban on exports); and Iran, Israel, Libya, South Africa and Syria (partial ban).[19] The list is subject to periodic review.

The attempt in March 1992 to export Polish military equipment to Iraq, which met sharp public criticism in Poland and abroad, induced the Polish Government to more closely scrutinize its international arms transactions. Preparations for a new law regulating the conditions of trade in arms and other strategic commodities are now in the final stage.

The 1991 Persian Gulf War also influenced Polish arms exports. Its disastrous outcome for Iraq strengthened the belief that the weapons and military equipment produced in Eastern Europe are much inferior to what is obtainable in the West.

4. A considerable proportion of the armaments exported by Poland are produced under licences obtained in the former Soviet Union. Export requires the permission of the licence seller which now is not so easy to obtain.

The scope, dynamics and structure of arms exports also depend a great deal on the organizational system of trade. Until 1990, arms deals were fully monopolized by the state. The only authorized institution was the Central Department of Engineering of the Ministry of Foreign Trade (CENZIN).[20] From March 1990 arms exports began to be granted also to some other institutions, including several arms-producing enterprises. Trade in arms by private persons and companies continues to be prohibited. However, private firms are

[18] See note 16.

[19] *Rzeczpospolita*, 14 Feb. 1992.

[20] As of 1990, CENZIN is a part of the Ministry of External Economic Relations.

allowed to engage in some training and advisory services. CENZIN itself has also been considerably transformed. It is now a holding company with a majority state capital stake (68 per cent of the stock is owned by the Ministry of Foreign Economic Co-operation and the remaining 32 per cent by 25 enterprises, mainly in the arms industry). CENZIN continues to be the main arms exporter and importer. In 1990–91 it accounted for almost 90 per cent of the Polish arms trade. The remaining 10 per cent went to a few other trading companies such as CENREX and PHZ-BUMAR. The domination of the arms market by CENZIN is due mainly to the specific problems of privatization in the arms industry.

VI. Conversion

The first modest attempts at military industry conversion to civilian production were undertaken in Poland in the mid-1950s. They were part of the changes carried out in the spirit of the events of the 'Polish October' of 1956. However, the failure of the socio-economic reforms and transformations of that period and the lack of deeper changes in international relations as a whole prevented much progress being made in this field. Conversion started again in 1989 in a different political atmosphere, conducive to more radical changes. No one doubts that there is no alternative and that it is inevitable for economic and political reasons. However, in spite of more favourable conditions for promoting the conversion process, it is much slower than desired and expected. There are serious political, economic and technical reasons for this. Political obstacles are created by the fact that debate on a new concept of national security and defence doctrine has not yet ended. As is usually pointed out, Poland has found itself in a certain 'security vacuum' in which old alliances have disappeared but new political and military options have not yet been institutionalized. At the same time new political challenges and threats have arisen. In these circumstances Poland's future requirements for military supplies both from internal and external sources are undefined. Needless to say, under such conditions it is not easy to formulate a comprehensive programme for development and conversion of the national arms industry.

The economic obstacles to conversion stem from the fact that it is very costly. The 'peace dividend' is a long-term prospect; in the short run any profound restructuring of industrial capacity is highly capital-intensive. In the existing military establishments only a part of the specialized technical equipment can be successfully utilized for civilian production. The rest has to be totally restructured or scrapped. Additional and even more important obstacles are created by the fact that conversion attempts coincide with a deep recession in the overall economy. Even if technical obstacles are surmounted (as is often the case), barriers in the market demand for new civilian products appear. A typical example is one of the largest arms producers—the BUMAR-Labedy factory—in which a considerable part of the production of battle tanks was converted into production of hoisting and loading equipment for the construction industry and mining. However, these industries are in a recession and thus do not create a sufficient demand for the expansion of new civilian products, especially since in the initial period of conversion these products are not technologically or economically competitive.

Better short-term prospects are attributed to the conversion of productive capacities in the field of repair services. This type of activity has been widely expanded in numerous military enterprises, including shipyards.

The specific features of military technology hamper the process of privatization in the arms industry. Neither its highly specialized stock of machine tools, which is therefore very costly and difficult to restructure, nor the specific location of the military establishments encourage private capital investment—which is oriented chiefly to quick profits. It should also be noted that for a very long period in the past the arms industry enjoyed specific conditions which were counter-productive to economic rationality. Military enterprises were not compelled to rationally calculate inputs (including investments), and over-costs tended to be easily forgiven if the production targets were met. As mentioned above, sales of their products were ensured, as was the supply of factors of production. The process of adaptation to new market conditions in this sector of the economy is naturally more difficult than in others and poses many specific social problems. A serious dilemma arises: on the one hand, slowing down conversion and scrapping military production is politically and economically unacceptable. On the other hand, speedy elimination of military productive capacities augments the existing problems of massive unemployment and creates dramatic social tensions, particularly in regions where arms industry is concentrated. Is there a way out of this dilemma?

Aside from solutions which may come from the general revitalization of the economy, great hopes are attached to large-scale development of industrial co-operation between Polish military establishments and Western companies. There are already some encouraging examples. The BUMAR-Labedy factory has entered into negotiations with Krupp and Volvo; if finalized, production of heavy dump trucks will be started in 1992. Some experts predict that production for the internal and external market may reach a level of 120 000–130 000 trucks per year.[21]

One may also hope that recently expanded contacts between NATO and the post-socialist countries of Central Europe will effectively contribute to the expansion of co-operation in this field. It has already been declared that one of the important platforms of NATO involvement in co-operation with these countries will embrace effective help in converting the productive capacities of military industry to civilian production. Conversion is one of the most convincing factors of the reinforcement of military stability on the continent. Real progress in this field will decisively narrow the possibilities that military concepts will reappear which pose new threats to international security on regional and global levels.

[21] *Polska Zbrojna*, 7 Jan. 1992.

12. Czechoslovakia: reductions in arms production in a time of economic and political transformation

Oldrich Cechak, Jan Selesovsky and Milan Stembera

I. The historical background

The history of the Czechoslovak arms industry dates from the mid-19th century, during the rule of the Austro-Hungarian monarchy. The arms industry was built up in parallel to general industrial development in the country during the so-called 'founder years'. It was expanded in the western part of Czechoslovakia only, particularly in Bohemia and Moravia, while the eastern part—Slovakia—was an underdeveloped region without any arms industry. At the end of the 19th century Bohemia and Moravia were the most developed industrial regions of the former Austro-Hungarian monarchy.

The Czechoslovak arms industry of that time was located predominantly in Plzen (Pilsen), with the production of artillery pieces, and Brno (Brunn), where infantry weaponry was produced. A number of smaller factories produced ammunition in other parts of the region, but in World War I the factories mentioned above were the most important suppliers of the Austro-Hungarian armed forces. Only a small portion of their production was exported.

At the foundation of Czechoslovakia in 1918 with the dissolution of Austria–Hungary, it inherited an effective arms industry, capable of both supplying its own armed forces and successfully competing in the export of armaments. A high level of technology and an experienced work-force were the backbone of the industry. The main commodities produced were artillery pieces (100-mm cannon and howitzers) and infantry weapons, in particular the Bren machine-gun, produced under licence from Enfield in England.

New perspectives for expansion of the Czechoslovak arms industry were opened by the founding of the so-called Little Alliance (*La petite entente*) between Czechoslovakia, Romania and Yugoslavia in the 1930s. Security considerations were not the main objective of this alliance, but good political and economic relations enabled Czechoslovakia to gain a dominant position in the arms trade with these countries. The export of armaments had a broader dimension and also extended to Turkey and on a smaller scale to Bulgaria. At that time foreign companies—such as the Schneider Creusot Group of France—invested in the Czechoslovak arms industry.

After 1918, the arms industry expanded mainly through the SKODA firm, which established a number of new enterprises and in 1935 founded OMNIPOL, an organization for the promotion of arms exports.

Czechoslovakia was in 1934–35 the world's largest exporter of weapons and ammunition (vehicles and aircraft are not included). Its share in the arms trade was 21 percent in 1934 and 24 per cent in 1935.[1]

After the German occupation in 1939 nearly the entire arms industry was integrated in the Hermann Göring Werke group. During World War II the country was part of the rear area, where until the spring of 1945 no military operations were conducted. More arms factories were therefore established here. During the war factories were established in the western part of Slovakia as well, particularly along the valley of the Waag (Váh) river, where not only infantry weapons and artillery pieces but also torpedoes, electronic equipment and more sophisticated weapons or parts were produced. Some of the factories were destroyed in the last weeks of the war, but the majority remained intact. The Czechoslovak arms industry was thus larger and more efficient at the end of the war than ever before.

The next period of development began in the 1950s, with the cold war. The industry, from then on continuously modernized, was made to serve the needs of a coalition strategy (organized according to the division of labour within the Warsaw Treaty Organization) and to follow strictly the principles of Soviet military doctrine. Thus it mainly produced large quantities of heavy weapons such as tanks, armoured vehicles, artillery weapons and large-calibre ammunition, most of which went to member states of the Warsaw Pact; the second largest customer was the Czechoslovak military forces, and the remainder was exported to Third World countries with a so-called 'progressive orientation' (see table 12.1).

From an economic and ecological standpoint, the performance of the arms industry was unsatisfactory. Moreover, it held back development in other important branches, in particular electronics and electro-optics. Today, products of these branches are in demand by the Czechoslovak armed forces and may be exported. Although tanks were exported in the 1980s and early 1990s, the only prospective export articles currently produced by the arms industry are trainer aircraft and radar systems.

In the early 1950s Czechoslovak arms production increased rapidly. Production increased between 1950 and 1953 by 374 per cent. The share of arms production in total machinery production in 1953 was 27 per cent.[2] New large plants producing heavy military equipment had been built in Slovakia after World War II.

[1] *Zbrojnivvroba v Ceskoslovensku, Hospodarske noviny* [Arms production in Czechoslovakia: economic papers], 5 Nov. 1991, p. 12.

[2] Federal Ministry of Economy, 'Defence conversion and armament production in the Czech and Slovak Federal Republic', paper presented at the NATO–Central and East European Countries Defence Industry Conversion Seminar, Brussels, 20–22 May 1992, p. 9.

In 1984–89 arms production was based on licences made available usually some time after the weapon systems had been introduced into the Soviet armed forces. Apart from the money that had to be paid for obtaining the licences, the weapons had to be modernized, for which more money was needed.

Another unsatisfactory feature affecting the arms industry, above all the heavy arms industry, was its concentration mainly in western and central Slovakia after World War II. The formal explanation for this heavy concentration of arms-producing enterprises was simply 'strategic reasons'; the true reason was the policy of the then party and state leaders of the Slovak Republic, who regarded this concentration of arms production under Soviet leadership as an opportunity to industrialize Slovakia rapidly.

II. The present Czechoslovak arms industry

The boom in the Czechoslovak arms industry culminated in 1987, when the value of the production of arms and other military equipment was over 29 billion CS crowns (see table 12.1), or nearly 3 per cent of total industrial production.[3]

Table 12.1. Selected official data on Czechoslovak arms production, 1987–91

Year	1987	1988	1989	1990	1991
Production					
Total armaments production (million CS crowns)	29 298	26 737	18 996	15 107	7 673
Index 1987 = 100	100.0	91.3	64.8	51.6	26.2
Regional distribution					
Czech arms production (million CS crowns)	11 557	12 331	10 587	7 515	2 417
Share (per cent)	*39.4*	*46.1*	*55.7*	*49.7*	*31.5*
Slovak arms production (million CS crowns)	17 741	14 406	8 410	7 592	5 256
Share (per cent)	*60.6*	*53.9*	*44.3*	*50.3*	*68.5*
Sales (shares, per cent)					
To Czech Army	*22.4*	*28.7*	*35.8*	*47.7*	*32.5*
For export:					
To former socialist countries	*58.2*	*56.6*	*58.8*	*41.7*	*20.6*
To other countries	*19.4*	*14.7*	*5.4*	*10.6*	*46.8*

Source: Federal Ministry of Economy, 'Defence conversion and armament production in the Czech and Slovak Federal Republic', paper presented at the NATO–Central and East European Countries Defence Industry Conversion Seminar, Brussels, 20–22 May 1992, p. 21.

[3] Note 1.

Table 12.2. Shares of Czechoslovakia in world arms exports, 1987–91

Figures are percentages.

Year	Czechoslovak share in world arms exports
1987	2.08
1988	2.36
1989	1.87
1990	3.29
1991	. .[a]

[a] According to information from the Czechoslovak Foreign Trade Ministry, Czechoslovakia exported arms valued at about half the value of its 1990 arms exports.

Source: *SIPRI Yearbook 1992: World Armaments and Disarmament* (Oxford University Press: Oxford, 1992), chapter 8, p. 272.

Until 1990, more than half of all arms industry production was exported to the socialist countries; the remainder was exported to the less developed countries. During the period 1985–89 Czechoslovakia was the seventh largest arms exporter in the world,[4] but in 1991 exports dropped dramatically (see table 12.2).

Sixty per cent of the arms industry was located in Slovakia and 40 per cent in Bohemia and Moravia in the 1980s. Weapons and other military equipment were produced in more than 100 enterprises, but not all these companies were dependent on arms production. Military production represented over 20 per cent of total production in only one-third of these companies.

The Czechoslovak arms industry was developed as an integral part of the division of labour of the WTO arms industrial base and was thus dependent on its allies, especially on the former Soviet Union's technology, innovation, development and co-operation in production. Nevertheless, a substantial number of weapons and other military equipment were designed and developed indigenously.

Because of reduced international tension in the late 1980s, the demand for arms decreased and a process of conversion of the Czechoslovak arms industry to civilian production was approved and initiated. After the November 1989 Revolution, the reduction of arms production accelerated. The volume of arms production is planned to be only 4 billion CS crowns (in constant prices) in 1992.

The extent and speed of the decrease in arms production (by 87 per cent in five years[5]) are unprecedented, and this has caused many complications in searching for alternative production programmes. Conversion to production of

[4] *Obranyschopnost, zbrojni vyroba a konverze, Studie ISS* [Defence, arms production and conversion, Studies of the Institute for Strategic Studies], (ISS: Prague, 1992), p. 167; *SIPRI Yearbook 1992: World Armaments and Disarmament* (Oxford University Press: Oxford, 1992), chapter 8, p. 272.

[5] Vrablik, J. and Kocevova, M., 'Konverze zbrojni vyroby' [Conversion of military production], *Ekonom*, no. 16 (1992), p. 19.

Table 12.3. Czechoslovak firms with the greatest share of arms production, 1987–88

Figures are percentages.

Firm	Share of production (average, 1987–88)
Slovak Republic	
ZTS Martin	60
ZTS Dubnica	70
Povazske Strojarne Povazska Bystrica	52
Tesla Liptovsky Hradok	57
Podpolianske Strojarne	50
Czech Republic	
Tesla Pardupice	73
Zeveta Bojkovice	60
Policske Strojirny Policka	80
Meopta Prerov	70
Adast Adamov	37
Pal Magneton Kromeriz	33
Ceska Zbrojovka Uhersky Brod	32
Blanicke Strojirny Vlasim	23

Source: *Obranyschopnost, zbrojni vyroba a konverze, Studie ISS* [Defence, arms production and conversion, Studies of the Institute for Strategic Studies], (ISS: Prague, 1992), p. 177.

civilian goods is being carried out in a situation when the Czechoslovak economy is experiencing a process of change from a centrally planned, bureaucratic and administrative economy to market relations. Systemic changes in the Czechoslovak economy are being realized parallel to the change of economic structure, which is to overcome the backwardness of the country's economy and technology.

The following products are the current core of the Czechoslovak arms industry: armoured personnel vehicles, tanks, military jet training aircraft, various types of radars and radio equipment, self-propelled howitzers, rocket launcher systems, various types of anti-aircraft systems, ammunition of all calibres, anti-tank guided missiles and infantry weapons.

Reductions in arms production are taking place in a time of economic and political transformation, not least the possible breakup of the country into two republics in 1993, in particular parallel to decreasing industrial production and a decrease in the gross national product (GNP).

Military expenditure

Czechoslovak military expenditure peaked in 1989, when it reached a level of 35 billion CS crowns (see table 12.4). Spending has substantially decreased since then as a result of both the favourable development of the international situation and especially the social and political changes occurring in the coun-

Table 12.4. Military expenditure of Czechoslovakia, 1989–92
Figures are in million CS crowns.

Year	GDP	Military expenditure	Share of GDP (%)
1989	947.6	35 062	*3.7*
1990	819.0	32 288	*3.9*
1991	977.8	26 500	*2.7*
1992[a]	1 056.8	27 914	*2.6*

[a] Figures for 1992 are estimates.

Source: *Obranyschopnost, zbrojni vyroba a konverze, Studie ISS* [Defence, arms production and conversion, Studies of the Institute for Strategic Studies], (ISS: Prague, 1992), p. 87.

try. In 1990 military spending amounted to 32 billion CS crowns, in 1991 to over 26 billion crowns and, for 1992, 28 billion crowns has been approved. Due to a 54 per cent rate of inflation in 1991 and an expected 10–15 per cent rate of inflation in 1992, the real value of military spending will not reach even half of the 1989 level of spending. The share of military expenditure in the GNP will have decreased from 3.7 per cent in 1989 to an expected 2.6 per cent in 1992. This decline is in accord with the new concepts to ensure the defence of the country.[6]

The transformation process of the Czechoslovak Army includes steps to transform the large armed forces that were formerly part of the Warsaw Pact coalition to a smaller army of a democratic state that will not be too heavy a burden on the state budget. Leaving the WTO and developing its own doctrine have entailed for Czechoslovakia a reduction, reorganization, structural change and relocation of the armed forces. The reduction of military expenditure is a precondition for the reallocation of financial resources for economic reform.

The decline in the Czechoslovak defence expenditure budget for 1990–92 has been so rapid that a lack of financial resources has stopped weapon development projects and led to substantial restraints in the procurement of military equipment and the reduction of Czechoslovak Army troop training.

Employment

In 1987, in the period of its greatest prosperity, the arms industry employed 73 000 workers. Another 70 000–75 000 people in support industry—which supplied equipment, material, spare parts, semi-manufactured products, and so on—depended indirectly on arms procurement. This figure of almost 150 000 arms industry employees represented about 2 per cent of the working population and about 5 per cent of those employed in industry.[7]

[6] Fucik, J., 'The Czechoslovak armament industry', *Military Technology*, no. 7 (1991); *Obranyschopnost, zbrojni vyroba a konverze, Studie ISS* (note 4).

[7] Note 5.

Table 12.5. The largest arms producers in Czechoslovakia, 1991

Figures are in million CS crowns, at 1991 prices.

Enterprise	Type of production	Total production
1. Turcianske strojárne Martin	Heavy equipment	1 500
2. ZTS Dubnica	Heavy equipment	450
3. ZVS Dubnica	Heavy-calibre ammunition	400
4. Podpolianske strojárne Detva	Heavy equipment	400
5. Provázské strojárne Povázska Bystrica	Aircraft engines	200
6. Tesla Pardubice	Communications technology	150
7. Blanicke Strojirny Vlasim	Ammunition	70
8. Zeveta Bojkovice	Ammunition	60
9. TATRA Koprivnice	Heavy trucks	50
10. Vitkovice s.p. Ostrava	Engineering technology	40

Source: *Obranyschopnost, zbrojni vyroba a konverze, Studie ISS* [Defence, arms production and conversion, Studies of the Institute for Strategic Studies], (ISS: Prague, 1992), p. 210.

In 1987, 44 000–46 000 employees worked in the Slovak arms industry and 26 000–28 000 in Czech industry. By the end of 1991 the work-force of the arms industry had decreased to 22 000–24 000 in Slovakia and to 14 000–16 000 in the Czech Republic as a result of restrictions on military production. On the basis of initial conversion programmes, it is calculated that a further reduction of arms industry employment by 16 000 (7 000 in the Czech Republic and 9 000 in the Slovak Republic) will take place in the next few years.[8]

The highest density of arms industry employment was in 1987 in central Slovakia: 32.6 per cent of all employees of the Czechoslovak armament industry worked there. Next comes southern Moravia (23 per cent) and the Prague area (17.4 per cent).[9]

Conversion has resulted in the loss of about half of the jobs in the arms industry. The rapid arms production decline secondarily threatens another about 60 000 jobs in support organizations. Conversion programmes are expected to create about 30 000 new jobs—about 9000 in Bohemia and Moravia and 21 000 in Slovakia—or less than 50 per cent of the redundant jobs.[10]

The concentration of arms production varies among the Czechoslovak enterprises. In the boom period of the 1980s, the share of military orders in total production in the 100 most important plants was about 25 per cent. The highest concentration of arms production was in ZTS Dubnica (70 per cent) and

[8] Note 4, p. 202.

[9] *Nektere poznatky, zavery a doporuceni z hodnoceni prubehu konverze sbroji vyroby v CSFR, Vyzkumny ustav provedeckotechnicky rozvoy* [Information, conclusions and recommendations from an assessment of the course of coversion of military production in Czechoslovakia], (Institute for Science Technical Development: Prague, 1991), p. 1.

[10] *Obranyschopnost, zbrojni vyroba a konverze, Studie ISS* (note 4), p. 221.

Turcianske strojarne Martin (about 60 per cent) of total production (see table 12.5 for the largest arms producers in 1991).[11]

III. Conversion

It is difficult to sell civilian products in Czechoslovakia today because of the state of the economy, and the economic recession has caused the bankruptcy of many enterprises. Funds for new investments are scarce. It is therefore very difficult to compensate for the decline in the procurement of weapons with a demand for non-military products.[12]

Table 12.6. Czechoslovak Government subsidy to conversion, 1989–92

Figures are in million CS crowns, current prices.

	1989	1990	1991	1992
Total	**685**	**1 200**	**1 500**	**1 000**
Czech Republic	68	388	300	..
Slovak Republic	617	812	1 200	..
The purchase of surplus supplies	..	2 600	..	..

Source: *Usneseni vlády CSFR z let 1989-92, Vláda CSFR* [According to the Government Resolutions in 1989–92, Government of the CSFR], (Federal Ministry of Economy: Prague, 1992).

The Federal Government has undertaken a number of activities to support arms industry conversion. In 1989 arms-producing enterprises were subsidized with 685 million crowns, in 1990 with 1.2 billion crowns, in 1991 with 1.5 billion crowns, and it has been calculated that 1 billion (see table 12.6) crowns are planned to be spent to stabilize the industry and develop alternative production programmes in 1992.[13] For example, the tank producer ZTS Martin was reoriented to the production of construction machinery and diesel engines; ZTS Dubnica, the producer of armoured vehicles, turns out mobile machinery (earth moving equipment); and Provázské strojárné B. Bystrica produces agricultural machinery, small motorcycles and equipment for environmental protection.

The conversion process of the arms industry is an important part of the structural changes of the Czechoslovak economy and is closely linked to the development of Czechoslovak arms production in the 1990s. On the basis of the new Czechoslovak Defence Doctrine adopted in 1990, the Federal Government accepted a new concept for the arms industry. It was a compromise, taking into account not only the defence demands but also the limitations of the economic,

[11] *Obranyschopnost, zbrojni vyroba a konverze, Studie ISS* (note 4), p. 222.

[12] *Conversion of Time Line of Basic Indices in Industry in the Years 1948–1990* (Federal Statistics Office: Prague, 1991); Federal Ministry of Economy, 'Defence conversion and armament production in the Czech and Slovak Federal Republic' (note 2).

[13] *Usneseni vlady CSFR z let 1989–92, Vlada CSFR* [According to the Government Resolutions in 1989–92, Government of the CSFR], (Federal Ministry of Economy: Prague, 1992).

scientific and technical resources of the country, the tradition of Czechoslovak arms production and possible future co-operation with foreign partners.

Among the advanced indigenous military industrial branches that will probably be developed are those which produce jet trainer aircraft, hand-held anti-tank weapons, ammunition, small arms, airport radars, radar reconnaissance stations, trailers, pontoon bridging equipment, logistic road bridges, gas masks, gas mask filters, detection tubes, NBC (nuclear, biological and chemical weapon) protection, and others. However, the process of conversion to non-military production will continue in other sectors, especially in tank and armoured vehicle production. Production of tanks and infantry fighting vehicles will probably cease entirely by the second half of the 1990s.

A controversy on questions of conversion of Czechoslovak military production began in 1989 and intensified in 1990, as a reaction to implementing Presidium resolutions 84/1989 on slowing down and then stopping tank production and 42/1990 on the production, export and conversion of military production. In this controversy there were two basic opposing views: those of the initiators of conversion (as a rule high officials of the administration) and those of its adversaries (military industry, representatives of municipalities in concerned regions, some theoretical and macro-economists, and persons in weapon and technical support of the army, etc.).[14]

The protagonists of rapid conversion were largely recruited from the post-November 1989 wave of so-called intellectual politicians who had accepted conversion even before the Revolution and started to implement government conversion measures. They justified the acceleration and extension of conversion by arguing the need for a credible Czechoslovak peace policy, citing the reduced threats to the country, European trends towards co-operation, the decline in the security importance of military power, and over-capacities and the unprofitability of military production.[15]

The arguments of the adherents of gradual and selective conversion cited the unpreparedness of the macro- and micro-economic spheres for this large undertaking, insufficient resources for its support, the social impacts of conversion, especially the negative employment effects, the instability of the East and Central European region, the unique standard of development of some sectors of the Czechoslovak arms industry, the possibility for increasing foreign co-operation on the basis of the arms industry's high technological standard and the need for arms exports to earn hard currency. This debate, and primarily the difficulties involved in conversion and the intricacy of economic transformation, have required corrections in the conversion plan.

The declarations of the political parties which won the 1992 election indicate that there may be a more circumspect course in the strategy of reducing military

[14] See note 13; Vrablik and Kocevova (note 5).

[15] Selesovsky, J. and Krc, M., 'Ekonomika a nova vojenska doktrina' [Economy and new military doctrine], *A-revue*, no. 10 (1991); Stefkeje, J., 'Ekonomicke aspekty vystavby armady' [Economic aspects of military production], *Vojenske rozhledy*, no. 4 (1992).

production and its future development in new, independent Czech and Slovak Republics.

IV. The Czechoslovak arms industry in the Central European region

According to those who advocate retaining the current level of the arms industry, the inherited military–economic structure of the country remains an important part of its defence capabilities. The indigenous arms production and some research and development capacities are a component of military–economic power.

The whole of Czechoslovak society, including the economy and armed forces, is at the present time engaged in a chaotic transformation. Reduction and reorganization of the armed forces are taking place, the defence budget is being reduced, and the armed forces are gradually losing their power. However, the country continues to produce most of its military equipment requirements indigenously. This is considered necessary in order to face possible, not precisely defined, threats. Under present conditions, there remains a need for a sufficient, modern and effective arms industry. However, the main prerequisite is not its size, but its technological potential.

It is argued that a weakened arms industry could lead to the weakening of the geostrategic position of the state, mainly with regard to the successor states of the former Soviet Union, whose policies are not transparent internally or to the rest of the world. There is also a possibility that these states will try to retain their domestic arms industries.

In this connection, and in the situation of the existing security risks in the region, it is argued that the necessary arms industry capacities should remain in Czechoslovakia. This in turn means that only some of these industries could be converted to civilian production.

Exports

President Vaclav Havel and Foreign Minister Jiri Dienstbier, who belonged to the intellectual opposition before the 1989 Revolution, were elected to the highest state posts, where they tried to implement a restrictive policy on arms exports and a conversion programme.

It is necessary to understand their policy as primarily a gesture of the political will of both these politicians and a signal of the new orientation of Czechoslovak foreign policy. However, it later transpired that they did not base their programmes on an analysis of the technological, economic, social and military factors involved in the radical conversion process. This is why the withdrawal of this internationally acclaimed goal cannot be understood only as a result of the counter-attack by the arms industry lobby and political opposition. It was rather primarily a result of the enormous difficulties posed for the arms industry conversion programme by the adverse economic conditions in the country. Furthermore, the other major limiting factor for the conversion programme was its extent and the speed of progress envisaged, which could not be realized.

Regarding the economic difficulties, the most important was the distintegration of central power and the ensuing weakening of the state economic planning bodies. In addition, the lack of ownership rights and of a functioning market made conversion virtually impossible, and it became necessary to subsidize the industry. The state was not, however, able to compensate for the millions of crowns lost by cutting arms procurement and limiting arms exports through non-military orders: in the period 1987–91 arms exports declined by a factor of five.

Regarding the envisaged scope and speed of arms industry reductions, the early effects were severe and the country was ill prepared: the value of arms production fell from about 29 billion crowns in 1987 to about 4 billion crowns in 1990 (in constant prices). Alternative programmes were not devised on an adequate macro- or micro-economic level; thus the management of arms-producing factories was unprepared in the new situation. The effects included high enterprise debts, an increase in unemployment, a decrease in wages, and social, economic and political effects in particular regions of the country. As a consequence, by 1992 the government had to abandon its programme for restrictions on arms production and exports.

Today both republics admit the failure to reach the declared objectives. However, the Czech Republic, more than the Slovak Republic, is characterized by a pragmatic position that arms exports must be permitted 'to a reasonable extent', a position held by both representatives of industry and many theoretical economists. The Slovak Republic is more universally of the opinion that conversion efforts have badly affected the republic and must be reversed.

It is generally believed that if the Federation breaks up into two republics both would adhere to a pragmatic position, although it is also possible that arms production may even be stepped up. It is also reasonable to expect future co-operation between the two republics as well as an increasing effort to co-operate in other European arms programmes.

Czechoslovakia will attempt to maintain and develop production of military jet trainer aircraft and radars for export. It remains one of the largest producers of jet trainer aircraft in the world: air force pilots of many states have been trained on these aircraft. The modernized type L-39 MS, for example, is regarded as a technically advanced aircraft.

Radar technology is in a similarly favourable position. Mobile airfield radars have been successfully used by many countries, and their quality has been confirmed under the most exacting climate and working conditions. Similarly, good results have been obtained with a passive radio communication reconnaissance system with a potential for use in the implementation of arms control measures and thus appropriate for the arms control verification market.

There were also favourable results in the field of electro-optics for various battlefield purposes, but their export importance was only modest in comparison with the commodities mentioned above.

The export of tanks and armoured vehicles is declining, and in the foreseeable future it might be stopped totally. The same will probably be done in the field of artillery systems. However, infantry weapons and ammunition, as well as arms for police and other civilian purposes, may find a market abroad.

The legal framework for Czechoslovak arms exports was codified in Government Decision No. 147/91, under which application for export licences is made to the Ministry of Foreign Trade. On 1 January 1992, however, foreign trade was liberalized in the Commercial Code no. 513/91Sb, which replaced a number of arms export regulations. A temporary decision was taken to entrust the Council of Defence with the authority to grant arms export licences. A final decision is still pending at the time of writing (November 1992). If Czechoslovakia breaks up into two republics, this authority will revert to the new republics.

V. The influence of a breakup of the federation on the arms industry

A breakup of Czechoslovakia into two independent countries and two national armies would affect the arms industries. For more than 70 years of co-existence between the Czech and Slovak Republics, such close economic relations have been created that their economies are tightly integrated. About 60 per cent of all Czech enterprises are dependent on deliveries from Slovak enterprises and, conversely, about 75 per cent of Slovak enterprises are dependent on deliveries from the Czech Republic.[16]

The Czech Republic contributes 71 per cent of the industrial production volume, and 74 per cent of the Czechoslovak GNP is produced in the Czech Republic. In some basic commodities, a mutual strategic dependence exists between the two republics. This is the the case for oil and gas transport across Slovak territory to the Czech Republic and for deliveries of coal and electricity from the Czech to the Slovak Republic.

Concerning the arms industrial base, about two-thirds of the arms production capacity is concentrated in Slovakia. While in the Czech Republic arms industry the machines, equipment and assembly lines of the arms industry are valued at 26–28 billion CS crowns, the equivalent figure for the Slovak Republic is 40 billion crowns. The production of heavy equipment (tanks, infantry combat vehicles and artillery weapon systems) is concentrated in Slovakia. The Czech Republic produces mainly small arms, ammunition, auto-

[16] Czech and Slovak confederation of trade-union associations, Prague, 1992; Barak, O., 'Moznosti materialne technickeho zabezpeceni Cs. armady', *Vojenske rozhledy*, no. 4 (1992); Capek, A., 'So-called specification of Slovak economy', *Political Economies*, 1992.

mobiles, means of radio communication and reconnaissance, and aircraft assembly.

The Slovak arms industry is equipped with investment machinery that has been produced largely in the Czech Republic. The Czech arms industry depends on delivery of semi-finished products from, and therefore on co-operation with, the Slovak Republic.

A breakup of the Federation will entail the division of a relatively unified arms industry potential. It will be necessary either to continue co-operation or to look for foreign partners. The latter would entail each republic having to specialize in certain types of military production and research, and co-operation would have to take place on the basis of convertible currencies.

VI. Conclusion

Arms production has developed in Czechoslovakia under conditions of total state ownership, in an environment without competition and in a technologically closed system (isolated from the rest of the economy). In the current initial period of privatization, it still retains these characteristics.

A new state concept for the development of arms production and conversion to production of civil goods is being prepared in which state participation will continue. If the state supports military industrial facilities, these facilities will have better prospects for survival and may also begin to export arms again.

A substantial improvement in the technological level of Czechoslovak arms production and its competitiveness on the world market will depend on the research, development and production co-operation of Czechoslovak enterprises with well-developed foreign partners. With the present Czechoslovak foreign policy, there is a strong orientation towards co-operation with countries of the European Community and a hope that the arms industry could benefit from such co-operation.

13. Turkey: the arms industry modernization programme

*Gülay Günlük-Senesen**

I. Introduction

In an era of significant disarmament undertakings, Turkey is an exceptional case because it is arming while large parts of the world are disarming. The implications of recent statistical indicators, in combination with Turkey's current defence policy, call for a closer look at the major factors underlying the trends of the past decade in the militarization of Turkey. The main dimensions of Turkey's militarization are its arms imports and domestic arms production.

The average annual growth rate of Turkish military expenditure in the period 1982–91 is 4 per cent. Noting that the European NATO total increased by an average annual rate of only about 0.6 per cent in this period, Turkey stands out as an exception (see table 13.1). The comparison is even more striking in procurement expenditure for major weapons: the total for European NATO diminished annually by 0.1 per cent, while Turkey's expenditures on arms procurement increased by over 10 per cent per year. Although these financial indicators are striking, they none the less seem to underestimate the flow of weapons into Turkey in real, quantitative terms: Turkish imports of major weapons, measured by the SIPRI methodology,[1] increased by about 14 per cent annually, while the European NATO annual average growth rate for the same period was 0.4 per cent. This implies that the amount of arms acquired by Turkey in this period was much greater, a fact which does not show up in the offical Turkish import statistics. In terms of arms import value, SIPRI data on imports of major weapons also show that for the period 1987–92 Turkey ranks as the highest among the NATO countries and third in the Middle East region,[2] after Saudi Arabia and Iraq.[3]

[1] For a note on the SIPRI methodology, see note 3, chapter 1 in this volume.

[2] It should be noted that SIPRI does not include Turkey in the Middle East region in its arms trade statistics, but in Europe. In other sources Turkey is variously included in the regions of Europe, the Middle East, the Third World, etc.

[3] *SIPRI Yearbook 1992: World Armaments and Disarmament* (Oxford University Press: Oxford, 1992), table 8.2, p. 273.

* The author wishes to acknowledge the grant given by the Swedish Institute for this research at SIPRI during September–October 1992 and the stimulating environment which the SIPRI staff provided. Special thanks are due to Ian Anthony, Paul Claesson, Elisabeth Sköns, Siemon Wezeman and Herbert Wulf, whose comments on an early draft very much helped to improve the text.

Table 13.1. Military expenditure, arms procurement expenditure and imports of major weapons by Turkey and the European NATO countries, 1982–91

Figures for military expenditure and arms procurement expenditure are in US $m., at constant (1988) prices; figures for arms imports are SIPRI values, in constant (1990) prices.

	Military expenditure		Arms procurement expenditure		Imports of major weapons	
Year	Turkey	European NATO[a]	Turkey	European NATO[a]	Turkey	European NATO[a]
1982	2 528	144 589	271	30 197	434	6 337
1983	2 393	147 235	241	32 473	404	4 543
1984	2 325	148 977	304	33 317	683	4 805
1985	2 467	150 360	336	33 417	737	3 257
1986	2 772	152 131	496	33 769	593	4 009
1987	2 647	156 735	559	35 403	1 203	4 731
1988	2 664	156 092	600	35 026	1 419	5 968
1989	3 082	155 811	530	33 630	1 138	6 937
1990	3 725	156 228	745	30 477	1 067	5 242
1991	3 870	154 250	711	29 817	1 558	6 611
Change (%), 1982–91	*53.08*	*6.68*	*162.36*	*– 1.26*	*258.99*	*4.32*
Average annual rate of change, *r* (%)	*4.35*	*0.65*	*10.13*	*– 0.13*	*13.63*	*0.42*

Note: r is calculated by (1991 figure/1982 figure) = $(1 + r)^{10}$.

[a] Turkey is included in the figures for European NATO.

Sources: For military expenditure, *SIPRI Yearbook 1992: World Armaments and Disarament* (Oxford University Press: Oxford, 1992), table 7A.2 (European NATO total calculated), p. 259; for arms procurement expenditure, *SIPRI Yearbook 1992*, table 7.16, p. 229; and for arms imports, the SIPRI data base, 1992.

However speculative it might be, there might be several explanations for this discrepancy: foreign military assistance, combined with 'free' arms transfers under the 1990 Treaty on Conventional Armed Forces in Europe (CFE),[4] account for some of the difference. Long-term credit or loan arrangements which do not require immediate, high payments might be a contributing factor as well as subsidized sales by foreign suppliers to benefit their own industries; it may also be due to the fact that national income accounts data do not include military transactions. However, there is no documented evidence for these assumptions. The question of which party, Turkey or the suppliers, incurs the financial costs or what these costs are remains unanswered.

[4] The CFE Treaty ceiling on certain weapon categories has led to a 'cascade' within NATO: weapons that must be withdrawn from the Treaty area are—in accordance with the Treaty—'cascaded' or transferred to other NATO countries, among them Turkey, which in turn have agreed to destroy older equipment. The main suppliers are the US forces in Europe and Germany. See *SIPRI Yearbook 1991: World Armaments and Disarmament* (Oxford University Press: Oxford, 1991), pp. 426–27.

Under the 10-year arms industry modernization programme which was initiated in 1985, Turkey is taking significant strides towards developing a role as an arms producer in a variety of areas. According to Brauer,[5] for the period 1975–84 Turkey was classified as a country with consistent arms production with limited sophistication and quantity. With regard to its industrial diversification base, Turkey ranked fifth among 32 arms-producing developing countries in this period, suggesting that prior to the modernization programme the local industry provided a good infrastructure base for starting industrial modernization. The main incentive for the modernization programme was to achieve arms import substitution so that Turkey would eventually not have to rely on foreign military equipment. The fulfilment of this goal, however, faces serious challenges which are discussed below. Although a modern Turkish arms industry is still in the initial stages of installation, this ambition has brought Turkey to the attention of international arms producers, which are facing a seriously diminished demand for weapons in the Western world and are looking for new opportunities.

This chapter is concerned mainly with Turkey's arms industry modernization programme: the political and financial set-up as well as developments in the aerospace industry, which is so far the most developed sector. Section II presents the main motives behind Turkey's militarization attempts. The context of the programme and its implementation are evaluated in sections III and IV, respectively. Section V reviews the aerospace production experience. General conclusions, including the challenges to the self-sufficiency goal, are discussed in the final section.

II. Incentives for militarization

Notwithstanding the significance of other factors reviewed later in this section, the dominant incentive behind Turkey's active militarization endeavours is its geopolitical position.[6] Turkey is the only NATO country located in the Middle East region. It offered the landing and maintenance facilities on its southern NATO base to the allied forces during the 1991 Persian Gulf War against Iraq. Combined with current tensions in the region, this fact alone explains why the Turkish Government has been importing large quantities of weapons over the past few years.

Turkey does not envisage a diminishing role in Europe for NATO in the future, regardless of the end of the cold war era. NATO is perceived by Turkey as the 'core of dialogue among the "Sixteen Nations"'.[7] In addition, Turkey expects to continue to play a crucial role in European NATO. Thus, according

[5] Brauer, J., 'Arms production in developing nations: the relation to industrial structure, industrial diversification, and human capital formation', *Defence Economics*, no. 2 (1991), pp. 165–75.

[6] *Nato's Sixteen Nations*, Special Edition, 1989/1990, p. 10; Wyllie, J., 'Turkey—adapting to new strategic realities', *Jane's Intelligence Review*, Oct. 1992, p. 450.

[7] Interview with H. Dogan (then Minister of National Defence), *Nato's Sixteen Nations*, Special Edition, vol. 36, no. 2 (1991), pp. 18–20.

to official policy, 'the developments towards disarmament throughout the world does not require a modification of the Turkish defence policy'.[8] Apart from the obligations attached to its membership in NATO, such as the pressure to assume more of its own defence burden, Turkey's militarization attempt is also linked to its alertness to prolonged unrest in the region and hence to arms buildups in the countries bordering on Turkey.[9]

The yet unresolved problem of the PKK (People's Army for the Liberation of Kurdistan) insurgency, mainly in the south-eastern and eastern parts of Turkey, is one of these problems. The Turkish Land and Air Forces are currently conducting operations against the Kurdish guerrillas in the south-eastern part of Turkey and in northern Iraq.[10] The allegedly tolerated mobility of PKK camps in Iran, Syria, Cyprus and, to a more proven extent, in Iraq creates tense diplomatic relations from time to time. The tension is even more exacerbated by the conflict over the water supply from the Tigris and Euphrates rivers, the sources of which are controlled by Turkey. The Turkish South East Anatolia Development Project, initiated in 1983, comprises the massive Atatürk Dam and a series of dams on the Tigris and Euphrates. This comprehensive project is not due to be completed for at least a couple of decades but has raised Syrian and Iraqi anxieties over the availability of water. Moreover, in spite of Turkey's assuring policy, Syria and to a lesser extent Iraq fear that they might encounter a threat of a cut-off of the water supply; this seems to be on the future agenda of the countries concerned.[11]

Turkey has so far been reluctant to interfere in the clashes between Azerbaijan and Armenia in Transcaucasia. Its efforts to establish ties with the five Islamic republics of pre-revolutionary Turkistan in the former Soviet Union is mainly for economic reasons; a substitute for the frustration of delays in the approval of its full membership in the European Community (EC).[12] The Black Sea Economic Co-operation Pact, initiated by Turkey in 1990 and signed by Albania, Armenia, Azerbaijan, Bulgaria, Georgia, Greece, Moldova, Romania, Russia, Turkey and Ukraine in June 1992, is an indication that Turkey is seeking a new role in the region, which, with this diversity of countries, cannot be attributed to primarily military goals. However, these should not be overlooked.

Turkey has occasional diplomatic disputes over territorial water zones in the Aegean Sea or over Turkish minority rights with Greece, but the most impor-

[8] Interview with Turkish National Defence Minister Nevzat Ayaz, *Nato's Sixteen Nations*, no. 4 (1992), pp. 57–62.

[9] Interview with Turkish National Defence Minister Nevzat Ayaz, *Defense News*, 31 Aug.–6 Sep. 1992, p. 46; *Nato's Sixteen Nations*, no. 4 (1992), pp. 57–62.

[10] *Financial Times*, 11 Sep. 1992; *The Economist*, 10 Oct. 1992, p. 35.

[11] Starr, J. R., 'Water wars', *Foreign Policy*, no. 82 (1991), pp. 17–36; Anderson, E., 'Water coflict in the Middle East—a new initiative', *Jane's Intelligence Review*, May 1992, p. 227; Wyllie, J. 'Turkey—adapting to new strategic realities', *Jane's Intelligence Review*, Oct. 1992, p. 451.

[12] Krause, A., 'Turkey looks beyond the EC', *International Herald Tribune*, 14 Nov. 1990; Rugman, J., 'Turkey hopes its ship is coming in', *The Guardian*, 3 Feb. 1992; 'The Turkish Question', *The Independent*, 1 Apr. 1992, p. 26; Erginsoy, U., 'Turkey renews friendships with Turkic neighbors', *Defense News*, 31 Aug.–6 Sep. 1992, p. 10; *The Commonwealth of Independent States and the Middle East*, vol. 17, no. 6 (1992), pp. 10–12.

tant dispute between Greece and Turkey is over the partitioning of Cyprus, the northern part of which has been controlled by Turkish troops since 1974.[13]

III. The context of the modernization programme

The Cyprus Operation of 1974 was a turning-point for the Turkish armed forces in raising the issue of the modernization of its arms inventory, since Turkey faced a US embargo on arms deliveries until 1978 because it had used NATO arms in Cyprus against the Greek Cypriots.[14]

Aeronautics, artillery and ammunition production in Turkey along with maintenance and overhaul facilities, although with a weak heavy industry infrastructure, were established during the early years of the Republic (i.e., over half a century ago). However, production stagnated during World War II and Turkey's post-war NATO membership, which resulted in the provision of military equipment mainly by the USA, slowed down arms industrialization efforts and Turkey continued to depend on arms imports from abroad.[15]

The Cyprus Operation carried out in retaliation for a pro-Greek *coup d'état* in Cyprus was the first active involvement of Turkish armed forces on the battlefield since the Korean War. It was also a test of Turkey's military capabilities and shortcomings. This created an awareness of the need to become self-sufficient in arms production, to avoid the restrictions attached to military aid and quickly to replace out-moded equipment through local production since reliance on imports entails the risk of an embargo or a severe scarcity of foreign exchange. Turkey's pride in having the second largest army in NATO after that of the USA was fading away; having a modern domestic arms industry and revenues from arms exports were seen as a way to help ease the burden on the defence budget.[16] Subsequently, the initial steps were taken towards capital accumulation in order to create an independent national arms industry. Donations and grant aid, indicators of public support, flowed into various foundations established to develop the Army in general, and the Air Forces, Land Forces and the Navy in particular, and hence provided the core of funds to be invested. These funds were in 1987 combined under the name of the Turkish Armed Forces Foundation.

The political and economic instability of the late 1970s forced Turkey to postpone its aspirations for a national arms industry until the military take-over on 12 September 1980.[17] During the military rule, which lasted until 1983, the

[13] *Jane's Defence Weekly*, 14 Dec. 1991, p. 1171.

[14] Recent provision of US credits to finance co-production of 200 general-purpose helicopters in Turkey is also based on the condition that the aircraft concerned would not be used in Cyprus. *Defense News*, 7–13 Sep. 1992, p. 32.

[15] Erdem, V. (head of DIDA/UDI since its inception), 'Defence industry policy of Turkey', *Nato's Sixteen Nations*, Special Edition, 1989/1990, p. 10; Erdem, V., 'History of the Turkish defence industry', *Nato's Sixteen Nations*, Special Edition, vol. 36, no. 2 (1991), p. 28.

[16] de Briganti, G., *Defense News*, 14 May 1990, p. 17; *Jane's Defence Weekly*, 28 July 1990, p. 126; Erdem (note 15).

[17] The Turkish Army, by definition, has undertaken responsibility for safeguarding the Turkish Republic, inside and outside, since the establishment of the state. The military coups of 1960, 1971 and

armed forces initiated preparations for a comprehensive modernization plan with full support from Turkish bureaucrats.[18] The publication of the negotiations on the production of the F-16 fighter aircraft in Turkey in late 1983 marks the first step in the implementation of the modernization plan and the commitment behind it. The F-16 indirect offset financing programme included investment in the communications and tourism sectors, in mining of copper and cobalt as well as in marketing of Turkish traditional export goods by the subsidiary companies (in Turkey) of General Dynamics of the USA.[19] This was the first time that the economic and thus welfare benefits which the installation of military complexes would bring to the country was presented to the public.[20]

The administrative basis of the modernization programme

The modernization goal, with its vast scope beyond arms-producing factories owned and oeprated by the military, called for a new organization to centrally plan, finance, implement and supervise the steps towards fulfilment. Both the implementing body and the Defence Industries Support Fund (see below) related to it resemble parallel organizations in Saudi Arabia and Egypt. However, since the former have weak ties with local industry and the latter lack central funds dedicated to the defence industry, the Turkish model seems to be unique.[21]

The Defence Industries Development and Support Administration (DIDA, or UDI)[22] was formed at the end of 1985 to administer the 10-year, $10 billion modernization programme. This administrative structure is summarized in table 13.2.

1980 were carried out on the grounds of restoring internal peace. Hence, the army has traditionally always been involved in internal politics. This involvement was institutionalized with the National Security Council (NSC), formed under the 1961 Constitution. Its position was strengthened with the 1982 Constitution. The NSC is chaired by the president; the prime minister, ministers of national defence, interior and foreign affairs, along with the Chief of General Staff and the Commanders of the Land, Air, Naval and Gendarmery Forces, are members. The NSC overviews almost all aspects of the current situation of Turkey but is beyond supervision by the government; its decisions have priority on the agenda of the government.

[18] There are no power conflicts between the branches of the Turkish armed forces, as is the case in for instance Argentina, among the less-developed countries.

[19] *Cumhuriyet*, 21 Sep. 1983.

[20] When the conflict over the implementation of the indirect offset agreement only partially arose in 1989, the confidence in military priorities prevented it from becoming a critical issue. *Armed Forces Journal International*, June 1989, p. 58. This led to the adoption of a more flexible offset policy subject to revisions. By the end of 1991, however, the UDI announced the Industry Offset Guidelines, which were taken by the candidate contractors as rather restrictive and disadvantageous. *Jane's Defense Weekly*, 9 Nov. 1991, pp. 883–90; Silverberg, D., 'U.S. contractors say Turkey misfires with offset obligations', *Defense News*, 24 Feb. 1992, p. 16.

[21] Sayigh, Y., *Arab Military Industry, Capability, Performance and Impact* (Brassey's: London, 1992), pp. 167, 210.

[22] The status of the DIDA was changed to the Under-secretariat for National Defence Industries (UDI, or SSM in Turkish) under the auspices of the Ministry of National Defence in 1989.

Table 13.2. Administration of the Turkish military modernization programme, in order of decision-making and implementation, since 1989

Body	Task	Members
DISBC: Defence Industry Supreme Board of Co-ordination	General defence policy; overall co-ordination, planning and funding	Prime Minister, Chief of General Staff, Commanders of Land, Air, Naval and Gendarmery forces, Minister of National Defence, 4 ministers, 3 under-secretaries
DIEC: Defence Industry Executive Committee	Decision-making body on the basis of the general policy of the DISBC; procurement and funding decisions	Prime Minister, Chief of General Staff, Minister of National Defence, Under-secretary of the UDI
UDI: Under-secretariat for Defence Industry	Monitoring and implementation of decisions of DIEC; management of the DISF	A partially independent body, with its own budget, under the Ministry of National Defence
DISF: Defence Industries Support Fund	Provides funding for investments in the arms industry	Under the auspices of the Central Bank; managed by the UDI; audited by a special committee

Source: Office of the Turkish Prime Minister (T. C. Basbakanlik), *Savunma Sanayii* [Defence Industry] (T. C. Basbakanlik: Ankara, 1990).

The UDI is the policy-implementing body for the modernization programme. Its primary functions, to be carried out in line with the decisions of the Defence Industry Executive Committee (DIEC), include the following:

– To ensure the execution of the large scale projects of the Turkish Armed Forces involving supply of defence equipment, within the framework of cooperation between national industry and foreign technology and capital in Turkey;
– To organize and integrate the existing national industry in line with the needs and requirements of the defence industry;
– To encourage and guide any new investments in keeping with the current requirements;
– To finance any research and development activities, prototype manufacturing and similar works as well as any related investments and operational activities.[23]

In short, the UDI is the office which calls for tenders, evaluates projects, monitors adherence to implementation schedules, provides updated and detailed information, and controls the Defence Industries Support Fund.

[23] Erdem (note 15), p. 17; Office of the Turkish Prime Minister (T. C. Basbakanlik), *Savunma Sanayii* [Defence Industry] (T. C. Basbakanlik: Ankara, 1990).

Financing the modernization programme

The Defence Industries Support Fund (DISF) was established in connection with the UDI under Law 3238 to provide continuous and stable support for the investments in arms production. Along with the initial assets, the main sources of income of the Fund are transfers from the foundations mentioned above, and taxes levied on income (5 per cent goes to the Fund), on fuel consumption, (10 per cent goes to the Fund) and on alcoholic beverages and tobacco consumption. Certain shares of the national lottery, gambling, payment for exemption from military service, bank interest revenues, the National Defence Ministry budget and the General Budget are also passed on to the Fund. The Fund is unique in Turkey since it is nearly entirely exempt from Turkish accounting and bidding laws in order to ensure secrecy and rapid procedures. It is audited and supervised by a three-member board, with the Prime Minister, Ministry of National Defence, and Ministry of Finance and Customs equally represented.

The UDI also initiated the formation of the Defence Industry Producers Association (SASAD) by major local firms in 1990. Membership in the SASAD is restricted to those firms which fit the definition of arms producer. The number of SASAD members increased from 21 in 1991 to 33 in 1992, marking a growing interest in arms production by local firms. These are listed in table 13.3, with their main areas of activity. In consequence, the co-ordination of main defence industry-related activities (bidding procedure, project evaluation, financial matters, finalizing decisions, etc.) was achieved.

These developments might appear to be organizational matters, but they in fact reflect a radical deviation from the former public sector monopoly which characterized Turkish defence policy for some 50 years. This drastic policy change has two main implications: (*a*) the public sector is taking over the planning, co-ordination and financing of defence production and is becoming indirectly involved in production through partnerships; and (*b*) although dominant in the aerospace industry, the USA is not automatically assumed to be the major arms supplier and/or co-producer for Turkey, with tenders announced internationally. The contractors listed in table 13.3 reflect this diversification.

As explicitly stated in Law 3238, top priority is assigned to the installation of a domestic, high-technology arms industry through co-operation with the public and private sectors and, more importantly, with foreign producers. In the words of Vahit Erdem, 'It is believed that the cooperation of the Turkish private sector with foreign partners will contribute to the establishment and implementation of the basic industrial infrastructure that will allow Turkey to develop its own defence industry technology over the long term and a considerable emphasis is accordingly being placed on the Turkish private sector's participation in this field'.[24]

[24] Erdem (note 15), p. 23.

Table 13.3. Leading arms producers in Turkey, on the basis of SASAD membership

Company	Year founded	Major activity
Aremsan	1945	Diesel generator sets for military purposes
Aselsan	1976	Military communication, electronics for F-16s
Asil Celik	1974	Steel
Baris	1986	Tubes for rocket launchers, helicopter blades
CAM IS MAKINA	1970	Glass molds, mechanical systems
Coskunoz	1973	Hydraulic and mechanical presses, automotive spare parts
ERMEKS-ER	1988	Parts for CASA aircraft
FMC-Nurol (FNSS)	1988	Armoured (combat) infantry fighting vehicles
Hema Elektronik	1986	Electronic equipment
Hema Hidrolik	1973	Hydrolic systems
Kale Kalip	1969	Various moulds and spare parts
KOSGEB	1990	Small- and medium-scale industry support administration
Marconi (MKAS)	1988	HF-SSB Radio Communications
Mercedes-Benz	1967	Tactical vehicles
MES	1965	Precision parts, rocket motors, tubes, cartridge cases
Mikes	1987	Electronic warfare sets for F-16s
MKEK	1950	Artillery, small arms ammunition, anti-tank rocket launchers, machine-guns
MKEK-AV Fisek	1930	Ammunition
MKEK-Barutsan	1989	Explosives, propellants
NETAS	1967	Communication electronics
NUROL	1982	Electronics
Otokar	1963	Diesel engines, Land Rover chassis
Roketsan	1989	Propellants and rocket motors
SGS-PROFILO	1988	Mobile telephones
SIMKO	1955	Electronics
STFA-Savronik	1986	Fire control and secure communications
TUSAS-TAI	1984	F-16 aircraft
TUSAS-TEI	1985	F-110 engine components for F-16s
TDI	1991	Association of a group of exporting firms in Defence Industry
Teletas	1984	Communications electronics
TESTAS	1976	Electronic components
TRANSVARO	1988	Night vision systems
TSKGV	1987	Funds defence projects

Sources: Milli Savunma Bakanligi (MSB, Ministry of National Defence), *Savunma Sanayii Envanteri* [Defence Indusry Inventory], SAGEB/1 (MSB: Ankara, 1988); information supplied by SASAD and the companies listed in the table.

In order to establish a defence industry which is 'dynamically structured, has a certain export potential, which can easily adapt to new technologies and is capable of updating itself in line with technological developments and improvements and thus ensures balanced co-operation with other, particularly,

NATO countries',[25] a call for foreign companies was considered inevitable. Moreover,

> Difficulties created by technology transfers, under license arrangements oriented to new developments and improvements, have led to the adoption of a "joint venture" model instead of license production. With the new approach, joint venture partners on both the foreign and national sides, are held jointly responsible at each level of production, and advantages accruing from foreign investment, in terms of quality control, cost and offset commitments, flow into sales to third-party countries.[26]

The final comment points out the possible international spill-over effects, which can be interpreted as another indication of Turkey's ambition to become a part of the international market.

Domestic spill-over expectations

It has been officially expected that domestic spill-over effects from the modern arms industry would include more diverse industrial production, more efficient production, product quality improvement, foreign exchange savings, acceleration of economic growth, increased value added, less unemployment, increases in the overall technology level, and improvement of the quality of the labour force and university education, especially engineering.[27]

These points have been on the agenda of economists for a long time. Although econometric findings from time series data sometimes contradict those from cross-sectional research, it is undebated that the cost of military activities—public goods for security needs—is in the long run paid at the expense of social welfare.

A macro-econometric simultaneous equation model for Turkey estimated for the period 1964–85 indicates that defence expenditure has no significant effect on the growth of investments and that the overall impact on the economic growth rate is negative.[28] Similar models estimating the impact of military expenditure on the growth rates of potential arms industries for the period 1972–86 show that the impact will be adverse, except for the basic metals (iron

[25] Erdem (note 15).

[26] Erdem (note 15).

[27] Interview with V. Erdem, *Ankara Sanayi Odasi Dergisi*, no. 110 (1991), p. 19; Gencler, R., *Savunma Sanayi Sektör Raporu* [Sectoral Report for the Defence Industry], 1991 Sanayi Kongresi [1991 Congress on Industry], no. 149-3 (TMMOB–MMO), p. 209; *Armed Forces Journal International*, June 1989, p. 58. The causality regarding spill-over effects has in fact gone in both directions, as has long been discussed in the related literature. The Turkish Chief of General Staff made an interesting evidential statement of the reverse link in 1989: in order to achieve full modernization of the arms inventory in the near future, the army demanded that the inflation rate should be curbed, the population growth rate and unemployment decreased, and the economy grow at an optimal rate with a balanced budget, supported by increased exports and decreased foreign debt. *Cumhuriyet*, 17 Sep. 1989.

[28] While military spending is not for arms production activities alone, it can be an indicator of the magnitude of a new potential demand to be injected into the economy, as in the case of Turkey. Günlük-Senesen, G., 'Yerli Silah Sanayiinin Kurulmasinin Ekonomiye Olasi Etkileri' [Probable Economic Impacts of the Installation of the (Turkish) Domestic Arms Industry], *1989 Sanayi Kongresi Bildirileri*, I, MMO/134, 1989, pp. 267–74.

and steel) industry.[29] Noting that all of these industries are key industries in the Turkish manufacturing sector and also the most import-dependent ones, the statistical analysis provides evidence that contradicts the optimistic expectations. Specific inputs of the defence industry, such as basic ferro and aluminium alloys as well as rubber, are totally imported. Turkey lags far behind in terms of space aviation and electronics technologies.[30] In addition, the present state of research and development (R&D) activities (with a 0.24 per cent share of R&D expenditures [public plus private] in the gross national product [GNP]) far from provides a promising environment for the spill-over effects of innovations. A recent experience speaks for itself: the oldest state enterprise in ordnance and ammunition production, MKEK, could not compete in the arms market in the Middle East during the Persian Gulf crisis and had to withdraw because other tenders had included the MKEK-type of products as bonuses in their offers.

Nevertheless, the determination of Turkish authorities to continue the modernization programme disregards its possible shortcomings. It should also be noted that there is consensus on this subject in the Turkish Parliament as well as between the armed forces and the government.

IV. The modernization programme in practice

It might be considered too early to make an overall evaluation of the implementation of the modernization programme since related investments started in late 1989, following a period of calling for bidders and evaluation of projects. However,an overview of the initial projects undertaken and the stage of their progress might shed light on the recent emergence of Turkey as a significant arms importer and as an aircraft producer.

Tables 13.4 and 13.5 summarize the stages of the implementation of the modernization programme by the UDI. Turkey's industrial modernization takes a two-pronged approach: production of equipment locally wherever possible; and imports of both modern equipment and technology where necessary. The initiation in 1987 of Turkey's first international arms fair, the International Defence Equipment and Avionics Exhibition (IDEA), is an expression of this policy—of trying to involve Turkish arms-producing companies and simultaneously trying to attract foreign arms producers. The IDEA '89 fair hosted 306 participating companies in 1987; peak attendance was reached in 1989, with the participation of 700 exhibitors. The reversal of this trend in 1991, with 300 participating companies, seems to indicate that Turkey has become less attractive as an arms market. Hence, by the end of 1991 doubts had arisen concerning the original

[29] Günlük-Senesen, G., 'An econometric model for the arms industry of Turkey', paper presented at the 11th European Congress on Operational Research, 16–19 July 1991, Aachen, Germany; Günlük-Senesen, G., 'An evaluation of the arms industry in Turkey', poster paper presented at the Economics of International Security Conference, 21–23 May 1992, The Hague, Netherlands.

[30] Çakmakçi, A., *Savunma Sanayii* [Defence industry], (Seminar/Lecture Notes), (I.T.Ü. Isletme Fakültesi [Istanbul Technical University, Faculty of Management]: Istanbul, 1989).

Table 13.4. List of military projects initiated in Turkey as of mid-1992

Project	Local producer (licenser)	Date of contract
Armoured Infantry Fighting Vehicle	FMC (USA) Nurol	15 Aug. 1989
Propellants and rocket motors (for Stinger (under licence), MLRS, Maverick)	Roketsan ARC	14 June 1989
F-16 electronic warfare	Loral (USA) Mikes	20 Sep. 1989
HF/SSB radio communications system	Marconi (UK) HAS CIHAN ELIT	9 Jan. 1990
Basic trainer aircraft (SF-260D)	Agusta (Italy) TAI KIBM	21 Mar. 1990
Mobile Radar Complex Project		
for C^3	AYDIN (USA) Hema Elektronik	8 Oct. 1990
for radar (TRS-22XX)	Thomson-CSF (France) Tekfen	
Light Transport Aircraft (CN-235M)	CASA (Spain) TAI KIBM	Feb. 1990
Multiple Launch Rocket System	LTV (USA) Roketsan	Feb. 1990
General-purpose helicopter (UH-60)	Sikorsky (USA)	21 Sep. 1992
Unmanned air vehicles (UAV)	AAI (USA) General Atomics (USA) TAI	Oct. 1992

Sources: Office of the Turkish Prime Minister (T. C. Basbakanlik), *Savunma Sanayii* [Defence Industry] (T. C. Basbakanlik: Ankara, 1990); information supplied by UDI, 1992.

expectation that Turkey would become 'one of the world's most lucrative defense markets'.[31]

There have been complaints about the lack of determination of decision makers, not in terms of the final goal but in terms of forwarding defence equipment requirements. Foreign contractors expressed their resentment of the fact that competitive tenders were delayed, cancelled or begun anew after a winner had been officially selected and announced. In addition, the scaling

[31] de Briganti, G. 'Turkey begins to lose shine as arms market', *Defense News*, 11 Nov. 1991, p. 14.

Table 13.5. Military projects under negotiation as of mid-1992

Low Level Air Defense System
35-mm Anti-Aircraft Fire Control System
MCM Vessels
Coast Guard Vessels
Advanced Technology Industrial Park
Aviation Center and Airport Construction

Sources: Office of the Turkish Prime Minister (T. C. Basbakanlik), *Savunma Sanayii* [Defence Industry] (T. C. Basbakanlik: Ankara, 1990); *Nato's Sixteen Nations*, vol. 36, no. 2 (1992), (special issue); *Jane's Defence Weekly*, 26 Sep. 1992, p. 5.

down or repeated delay of procurement decisions has caused contracting firms to hesitate to make major investments without future procurement guarantees.[32]

Furthermore, major decisions, including decisions on arms procurement, were delayed until after the October 1991 elections. After the Gulf War experience, Turkey became more determined to restructure the armed forces into a mobile, technologically sophisticated, professional force with superior firepower and increased capabilities in air defence, communications, electronics and electronic warfare. This restructuring involves reductions in the Army as well as in the compulsory conscription periods, put into effect in 1992.[33] Hence, an overall review of the procurement plan was found necessary, which obviously contributes to the delays in finalizing decisions.

Conformity to NATO Allied Quality Assurance Publications (AQAP) (officially started in 1988 in Turkey for the first time), industrial security measures, and working to tighter tolerances and standardization present difficulties to Turkish industry. In addition, the inadequacy of R&D activities and the shortage of state funds allocated to R&D are among the complaints of the local producers, who also request protective measures for local producers, state-guaranteed long-term financial funding and accelerated procedures. There seem to be profitability concerns as well; since the present market is limited to a single buyer, the Turkish armed forces, undergoing large-scale projects might not be as profitable as was expected, if the dim prospects for exports are also taken into account. Hence, it is proposed that establishing companies on the basis of specialization in components might be an alternative to the present practice of forming companies on a project basis.[34]

Present doubts concerning the continuity of the modernization aspirations and the effectiveness of the UDI notwithstanding, armoured infantry fighting vehicles, Scimitar HF/SSB radio sets and radar production programmes are the

[32] de Briganti (note 31); Aris, H., 'Creating a defence industry', *Nato's Sixteen Nations*, May/June 1991, p. 81; *Jane's Defence Weekly*, 1 Feb. 1992.

[33] Interview with Turkish Defense Minister Nevzat Ayaz (notes 8 and 9).

[34] Aris (note 32), p. 78; Minutes of the discussions held between industrialists and UDI representatives, *Ankara Sanayi Odasi Dergisi*, no. 110 (1991), pp. 21–35; Tosun, A.,'Savunma Sanayii ve Düsündürdükleri' [Thoughts on the defence industry], *Ankara Sanayi Odasi Dergisi*, no. 110 (1991), pp. 50–51.

major ongoing programmes. The Turkish Armed Forces Integrated Communications Systems (TAFICS) project is an emerging ambitious one which aims at modernizing the outmoded military communications network in compliance with NATO standards.

Contrary to this bustling of tenders and negotiations concerning the Air and Land Forces, the upgrading of naval equipment seems to be following a more settled and better defined route. This might be because of the collaboration with German shipyards since the late 1970s. FRG shipyards (mainly Blohm & Voss, HDW and Lürssen Werft) have so far sold naval equipment for DM 6300 million (approximately $4000 million) to Turkey. These orders, consisting of frigates, submarines and fast attack craft (FAC), form a significant part of related German companies' overall business which are supported by the German Government and banks. Although second-hand warships were previously received in large numbers from the US Navy at no or a nominal cost, Turkey's warships are largely German-designed and built locally with technical assistance from the parent yards in Germany.[35] In addition, the Turkish Navy has firm plans for ordering six Tripartite minehunters from France with an option for two more, some of which to be built in Turkey. Hollandse Signaal of the Netherlands will be providing weapon control and combat information systems to Turkish FAC.[36]

Two navy yards, Gölcük and Taskizak, are owned by the Turkish Navy and are involved in defence production and maintenance. Gölcük produces frigates and submarines, while Taskizak produces FAC and landing craft.[37]

Turkey will spend $1.59 billion from the defence budget and $760 million from the DISF on arms procurement in 1992 and a combined total of $2.7 billion in 1993.[38] Turkey's determination in advancing its aerospace industry deserves special attention, however, as evidenced by the ongoing F-16 fighter production by TAI and recently started production of light transport aircraft and trainer aircraft in collobaration with CASA (Spain) and Agusta (Italy), respectively. TAI is also involved in prototype design for the Unmanned Air Vehicle (UAV).

V. The aerospace industry

The formation of the Turkish Aerospace Industries (TAI) in May 1984 is concrete evidence of Turkey's aspirations for a domestic aircraft industry, an area in which technological requirements are universally the most demanding. TAI's

[35] *Defense and Foreign Affairs Weekly*, 5–11 Feb. 1990, p. 5; *Jane's Defence Weekly*, 28 July 1990, p. 131; *Military Technology*, no. 4 (1991), pp. 15–16; *Naval Intelligence*, 3 July 1992, p. 4; *Wehrtechnik*, no. 9 (1992), p. 54.

[36] *Asian Defence Journal*, no. 10 (1991), p. 117; *U.S. Naval Institute Proceedings*, Mar. 1992, p. 119; *Jane's Fighting Ships, 1992–93*, 95th edn (Jane's Information Group: Coulsdon, UK, 1992), p. 873.

[37] Todd, D., 'Mediterranean naval shipbuilding, challenges and prospects', *International Defense Review*, no. 10 (1992).

[38] *Defense News*, 31 Aug.–6 Sep. 1992, Pp. 6, 12.

significance lies in the fact that it is this company model which encouraged and led other companies to enter defence production in Turkey.

As mentioned above, TAI was formed before the introduction of the modernization programme, as a joint venture with a 51 per cent Turkish share (49 per cent of TUSAS-TAI, 1.9 per cent of Turkish Armed Forces Foundation and 0.1 per cent of the Turkish Air League) and a US 49 per cent share (42 per cent of General Dynamics and 7 per cent of General Electric). As of the end of 1991, TAI employed 2266 people, with a US General Manager.[39]

The first F-16 Fighting Falcon fighter assembled under licence by TAI was delivered in late 1987. By early April 1992, TAI had passed the half-way mark in the 152-plane F-16 Peace Onyx fighter programme, which it expects to complete on schedule in 1994. A total of 96 F-16s were delivered to the Turkish Air Force, in a combination of Block 30 and 40 configuration. Parts of the F110-GE-100 jet engine which powers the F-16 are produced, assembled and tested by TUSAS Engine Industries (TEI), which was formed in 1985 by TUSAS (51 per cent) and General Electric (49 per cent), also as a joint venture. TAI started to build the aft fuselage section in 1988 and the centre fuselage section in 1989, previously supplied by Sonaca and Fokker, respectively. With some parts, the forward fuselage, still coming from these companies, the cockpit and the fins are to be supplied by General Dynamics. In mid-1992, of each F-16 aircraft, 70 per cent of the airframe (aft and centre fuselage sections, as well as the wings) is domestically produced. The local content is planned to increase to 90, 92 or 95 per cent (the share varies according to the source, including TAI) in the second phase of the Peace Onyx programme, with 160 aircraft by 1994. The realization of this stage depends on the provision of a total of $3.5 billion over five years, by Saudi Arabia, Kuwait, the United Arab Emirates and the USA as compensation for Turkey's economic costs during the Gulf War.[40] When the export of 46 Turkish-made F-16s to Egypt by 1995 was announced in 1991, Turkey's hopes for entering the international market were revived. In addition, this order might be considered as confirmation of the sought after significant role in the market, with the expectation that the Turkish F-16 plant might become a major international supplier of F-16 fighter aircraft plus parts and components after the USA stops production in the mid-1990s.[41]

The recently resolved debate between Turkey and the USA over an electronic countermeasures system (ECM) to be mounted on the F-16s should be taken as a signal of the fragility of the aspiration to achieve the self-sufficiency goal in defence production. The USA had rejected Turkey's demand to be able to

[39] Company records; 'Türk Havacilik ve Uzay Sanayi (TAI)' [The Turkish aerospace industry], *Ankara Sanayi Odasi Dergisi*, no. 110 (1991), pp. 37–39.

[40] Aris, H., 'A new player in the making, Turkish defence industry', *Military Technology*, no. 4 (1991), pp. 11–15; *Jane's Defence Weekly*, 12 Oct. 1991; *Defense News*, 11 Nov. 1991; *Military Technology*, no. 4 (1992), p. 95.

[41] 'Where East meets West', *Armada International*, no. 2 (1989), editorial, p. 2; Boyle, D. and Salvy, R., 'Turkish defense modernisation', *International Defense Review*, no. 6 (1989), p. 847; *Defence Industry Digest*, Oct. 1991, p. 16; Enginsoy, Ü., 'Turkey seeks more F-16s with Arab, U.S. funds', *Defense News*, 11 Nov. 1991, p. 12; Sariibrahimoglu, L., 'Building an industry', *Jane's Defence Weekly*, 9 Nov. 1991, p. 881; *Arms Sales Monitor*, no. 13–14 (Mar.–Apr. 1992); *Military Technology*, no. 4 (1992), p. 95.

modify the device's US software and develop a national ECM system. Although the USA called this an unlawful transfer of technology, it was announced in August 1992, in a letter from the US State Department, that Turkey will be permitted to modify the device's software, the extent of which is not revealed.[42]

VI. Conclusions

The Turkish experience of indigenous arms production, although still in an early phase, seems bound to suffer from the timing, since it started in an era of re-evaluations of world defence and security requirements. Nevertheless, the determination of Turkish authorities, both the armed forces and politicians, combined with the search by international producers for new markets, backed by their governments, should be expected to carry the aspirations of modernization further, although in a development embodying a number of contradictory factors.

The flow of armaments into Turkey as a consequence of NATO's reductions under the CFE Treaty and the 'cascade' programme[43] along with increased arms supplies to Turkey after the Gulf War have already served to upgrade Turkey's outmoded equipment. The USA and Germany are the two major suppliers of military equipment to Turkey. These large imports might act as a counter-incentive for local production.

On the other hand, the CFE Treaty excludes the south-eastern part of Turkey, which is a battlefield at the present time, with the Turkish Land, Air and Gendarmery Forces carrying out uninterrupted operations against the PKK. This has hastened the possible acquisition of modern high-technology arms, especially helicopters and armoured vehicles, which are to be bought from Russia without the imposition of territorial restrictions.[44] This should be taken as an indication of the possibility that a preference for off-the-shelf purchases and/or purchases on the market might weaken the improvement of local production, the gestation period of which is not sufficiently short to meet the urgent requirements of 'hot' conflicts.

According to the Defence Industry Inventory published in 1988 by the Ministry of National Defence to provide information to investors, 983 establishments are located in 45 cities in Turkey.[45] More than one-third of these are situated in the Marmara Region, the most industrialized part of Turkey. The

[42] Consequently, the installation of Loral AN/ALQ-178 Rapport III integrated electronic countermeasures systems on Turkish Air Force F-16s started in July 1992 at the TAI-Mürted factory in Ankara. *Jane's Defence Weekly*, 9 May 1992, p. 796; Erginsoy, Ü., 'Pratt, GE vie for Turkish F-16 engine contract', *Defense News*, 31 Aug.– 6 Sep. 1992, p. 8.

[43] 'Arms windfall dilemma for Turkey', *Financial Times*, 27 June 1991; Hitchens, T., 'Nato arms transfers benefits Turkish military', *Defense News*, 31 Aug.–6 Sep. 1992, p. 7.

[44] Some suppliers put restrictions on the import of arms regarding where they may be used. In the negotiations with Russia which were taking place in the autumn of 1992, this did not appear to be a condition that Russia would set. *Financial Times*, 11 Sep. 1992, p. 5.

[45] Milli Savunma Bakanligi (MSB, Ministry of National Defence), *Savunma Sanayii Envanteri* [Defence Industry Inventory], SAGEB/1 (MSB: Ankara, 1988).

trade union in this sector, HARB-IS, has 41 500 members (not all of whom work in arms production), but, as noted above, with the exception of the few leading, mostly new, establishments, the overall technological capability is limited. High-technology companies, on the other hand, are provided with technology by foreign partners (e.g., the TAI). The Turkish case, at this stage, is another example of the contradiction between an independence incentive for the installation of a local defence industry and the inevitably vital dependence of this production on foreign technology and resources. It is unknown how long this dependence will last.

The armament aspirations of Turkey will no doubt continue to put pressure on the economic capabilities of the country. The share of defence expenditures in the gross domestic product (GDP) was 4.9 per cent in 1991: Turkey ranks third in NATO, following Greece and the USA. On the other hand, Turkey's per capita defence expenditures are the lowest in NATO, but this is also true for the GDP per capita (approximately one-half that of of Portugal and one-third that of Greece). These data apply to the 1980–91 period.[46] The dilemma of setting priorities between defence and major social welfare items such as health and education seems to have been resolved for the near future in favour of the former, the long-term effect of which might be detrimental even to defence spending.

The last but not the least contradiction which the Turkish case illustrates is that the so-called ongoing 'arms reduction' and 'conversion' of arms industries in the industrialized world appear to be compensated for by the militarization of the less-developed countries, such as Turkey. Agreements on stocking of conventional weapons, qualitative and quantitative improvement of local inventories, and encouragement of local production might be serving to upgrade the global inventories. Complementary to this, it should be noted that, steered by new public incentives, the Turkish industry is reorganizing towards 'conversion in the opposite direction': civilian industries are being militarized. Needless to say, 'arms reduction' is not on the Turkish agenda, as shown in this chapter.

The probable negative economic and social consequences of this militarization trend are not matters of explicit public discussion in Turkey, but this does not appear to be discussed in the international arena either. The Turkish case appears to be challenging the disarmament prospects of the Western world. However, it is an irony that, while preaching and practising arms control and gradual disarmament at home, the Western countries (those which have developed arms industries) are assisting Turkey in its modernization effort. The security-concerned circles, then, must include this neglected dimension on their own agendas before it is too late.

[46] *Nato Review*, no. 1 (1992), pp. 33–35.

Part V
Other countries

14. The People's Republic of China: arms production, industrial strategy and problems of history

*John Frankenstein**

I. Problems of history

Chinese discussion of military modernization and the future of Chinese defence industries is today carried out in terms of professionalization, technology transfer, conversion of industries to civilian production and the economics of public administration. Before examining contemporary Chinese defence industries, this chapter reviews some of the salient historical issues in Chinese military development. These issues relate to China's traditional military system and what happened in the mid-19th century when the industrializing West, fired by the twin ideologies of free trade and evangelism, and imperial China, holding to the historic (and persisting) imperatives of the supremacy of state interests, collided head-on in the Opium War of 1839–42. These issues are of more than historic interest since the present-day Chinese leadership often invokes them in policy discussions and debates. By raising these historically resonant issues, the Chinese may also be expressing an awareness—or even an anxiety—that there may be tides or patterns in both the Chinese past and present from which escape is difficult, if not impossible.

The traditional system

The Chinese state has always maintained a large military establishment. The Board of War was one of the six key central government ministries of the traditional Chinese state. Confucian protestations to the contrary, the emperors of China well understood the need for a state imperiled from outside—by nomadic marauders who sometimes seized power, most spectacularly in the case of the Mongol Yuan (1261–1368 AD) and Manchurian Qing (1644–1911) dynasties—and from within—by rebellion—to maintain a strong army. Indeed, the political culture of China puts great stress on dealing harshly with first internal and then external enemies. Song Yingxing (Sung Ying-hsing), a scholar in the late Ming dynasty (1368–1644), reflecting on these matters, asked: 'How would it be possible for any intelligent king and sacred emperor to exist without

* The author would like to acknowledge the invaluable research assistance of Paul Frankenstein and Susan Turney in the writing of this chapter.

military power? Bows and arrows were traditionally used for controlling the country.'[1]

China has historically directed substantial resources to the military. For instance, the Song (Sung) emperors, pressured by the northern barbarians, devoted up to 80 per cent of state expenditure to military purposes. As economic historian Mark Elvin has noted, the Song dynasty maintained 'the most formidable military machine the world had yet seen', with a standing army of 1.25 million troops and a military industry capable of enormous 'standardized mass production' of weapons.[2]

Military technology

China also had an extensive history of military innovation. Gunpowder and rocket arrows (often fired *en masse* from devices that look remarkably like modern rocket launchers) are the best known of these. Others included huge arcuballista firing three quarrels per shot, self-loading crossbows, repeating muskets, poison gas grenades and various kinds of cannon and mortars, some firing fragmentation charges. The Chinese also adapted technologies from outside. Stirrups and the short compound bow were borrowed early from the nomadic Xiong-nu tribes of the steppes. Much later, during the Ming and Qing dynasties, Jesuit missionaries were welcomed by the Inner Court not so much for their spiritual message as for their skills in astronomy and cannon casting.

Questions of leadership

The Chinese military was not always successful in defending the state. After all, the standard mode of dynastic change for over 2000 years was either invasion or rebellion. Despite technical innovations and large numbers, the traditional Chinese military establishment appears to have been just as prone to bureaucratic decay as other parts of the imperial system. Military leadership also entered into the equation. Imperial officials were chosen on the basis of an examination that tested their literary skills and knowledge of the Confucian classics. The best and the brightest did not opt for military careers; indeed, there were no institutional rewards for those who would consider a career

[1] Song Yingxing (Sung Ying-hsing), *T'ien Kung K'ai Wu: Chinese Technology in the 17th Century*, (E. Z. Sun and S. C. Sun, trans.), (Pennsylvania State University Press: University Park, 1966), p. 261. Compare 'Persist in performing domestic functions of the People's Army', *Liberation Army Daily*, 27 Mar. 1990: the mission of the PLA is to 'defend state power, maintain social stability and safeguard peaceful labor of the people'; 'while class struggle still exists within a certain scope, the existence of this people's army can scare the hostile forces . . .'; *Foreign Broadcast Information Service China Daily Report* (hereafter FBIS), 25 Apr. 1990, pp. 35–38.

[2] Elvin notes: 'Even at the beginning of the dynasty, the Bow and Crossbow Department at the capital was turning out 16.5 million arrowheads a year. By 1160, the yearly output of the Imperial Armaments Office . . . came to 3.24 million weapons. Body armor was manufactured in three regulation styles to the extent of several tens of thousands of sets annually.' Elvin, M., *The Pattern of the Chinese Past* (Stanford University Press: Stanford, 1973), p. 84. See also Fairbank, J. K. and Reischauer, E. M., *East Asia: The Great Tradition* (Houghton Mifflin: Boston, 1965), p. 205.

outside the traditional bureaucracy. This was a serious systemic problem—statesmen tasked with defending the empire would routinely find themselves without the kind of officer corps that could cope with 'warfare under [then] modern conditions'. In China today, under the reforms of Deng Xiaoping, the centralized press speaks of the need for civilian and military officials who have 'emancipated their minds' from 'ossified', ideological thinking.[3]

The crisis of the Opium War

All of these matters came to a crisis point in the Opium War of 1839–42 and in the revolutionary 'century of humiliation' that followed. Westerners, with superior weapons and tactics, crushed the Chinese forces sent against them. The Chinese response was varied. Some Chinese felt it was necessary to reform the system completely (a political *tendance* which ultimately led to Marxism). Others, sensing the dangers to the imperial tradition, responded in a reactionary, xenophobic manner. Some—the 'Self-Strengtheners', as they came to be known—felt that the traditional system could be retained and that, at the same time, certain useful things from the outside world, particularly military technology, could be adapted to Chinese requirements. In a brilliant passage that echoes even today, Feng Gueifen, one of the intellectuals behind this movement, wrote that to repel the barbarians China would have to disdain 'empty bravado' and use 'the instruments of the barbarians'.

> Some have asked why we should not just purchase the ships and man them with [foreign] hirelings, but the answer is that this will not do. If we can manufacture, repair and use them, then they are our weapons. If we cannot manufacture, repair and use them, then they are still the weapons of others . . . In the end, the way to avoid trouble is to manufacture, repair and use weapons by ourselves. Only thus can we pacify the empire; only thus can we become the leading power in the world; only thus can we restore our original strength, redeem ourselves from former humiliations, and maintain the integrity of our vast territory so as to remain the greatest country on earth.[4]

[3] This problem has deep historical roots. The Song (Sung) dynasty reformer Wang Anshi denounced the examination system as producing scholars who have no military expertise whatsoever and who, 'when finally appointed to office . . . [do] not have even the faintest idea of what to do'; cited in DeBary, W. T., *Sources of Chinese Tradition*, vol. 1 (Columbia University Press: New York, 1964), pp. 416 ff. The Ming scholar Song wrote that an ambitious person would 'undoubtedly toss this book [about technology] onto his desk and give it no further thought; it . . . is in no way concerned with the art of advancement in officialdom'; Song (note 1), p. xiv. In the autumn of 1985, the *Guangming Daily* published a short article, aimed at enemies of the reform programme, entitled '"Stressing military force" and not "belittling military force"—on a basic national policy of the Northern Song'. The article points out that 'because the idea of strictly following "the rules of the ancestor" fettered the minds of many people', China's military problems were not solved, and thus 'when the . . . kingdom of Jin invaded the Northern Song, the Northern Song could not help but resign itself to extinction'. FBIS, 19 Sep. 1985, pp. K20–21.

[4] Cited in DeBary (note 3), vol. 2 (1964), p. 46. Compare the following, which deals with Iraq's defeat in the 1991 Persian Gulf War: 'In the recent regional war, the Iraqi Air Force had more than 800 fighter planes, which formed a powerful air force . . . However, as the planes were all purchased from abroad . . . the Air Force was not yet sufficiently prepared for air combat. Some highly difficult ground maintenance work had to be done by foreign experts [who withdrew during the war] . . . [Iraq] was entirely unable to

The Self-Strengthening movement was summed up in the famous slogan of Zhang Zhitong, another Chinese official: *'Zhongxue wei ti, Xixue wei yong*; Chinese learning as the base, Western learning for use'. Thus the first modern Chinese forces and arsenals came to be established by the Self-Strengtheners, but even these were unable to stave off the ultimate collapse of the old regime.

To begin with, the Self-Strengtheners were constantly starved for resources. Li Hongzhang, one of the most important officials of the late Qing dynasty, argued that military spending was crucial for the survival of the Dragon Throne.[5] At the same time, Li and the other Self-Strengtheners recognized that military strength had to have a basis in industry. 'From the early 1880s,' writes historian Wellington Chan, 'Li . . . argued that guns and gunboats alone did not make a nation strong; their operation required the support of industry in manufacturing, mining and modern communications; industry would create new wealth—a further source of national wealth.'[6] The Court did not heed Li's words. Li's Northern Fleet, which was to meet defeat by the modernized Japanese Navy in the 1905 Sino-Japanese War, had to resort to commercial shipping to support itself.[7]

The recurring issue of leadership was also a problem. China needed engineers and a professional officer corps, not *literati*, to cope with the new challenges. Feng Guifen complained that men 'of intelligence have spent their entire lives and exhausted all their energies on such useless matters as the eight-legged essay [a style of writing required for the Imperial Examinations] and on calligraphy of a prescribed style'. Rather, he proposed, honours and position should be given to 'craftsmen who distinguished themselves in arsenals and shipyards'. Li Hongzhang suggested that a special technical section be added to the examination system so that 'talents' will see the new category as their road to 'wealth, rank and honor'.[8]

Even though arsenals were established and 'talents' set to work in them, institutional change came with difficulty. Provincial rivalries plagued modernization: in northern China, Li Hongzhang bought warships from Europe because he could not depend on support from shipyards in the south. Chinese shipyards and arsenals depended on expensive foreign technology and expertise and, at great cost, acquired equipment that was often beyond the limited technical capacity of the workers to use. The cost of producing a ship in China was twice that in Britain. Development was uneven: in the 1870s, while the Tianjin shipyard was experimenting with electric torpedoes, some arsenals were still mak-

bring the Air Force's fighting power into play.' *Liberation Army Daily*, 22 July 1991, in FBIS, 9 Aug. 1991, p. 33.

[5] Li wrote: 'All other expenditures of our nation can be economized, but the expenses for supporting the army . . . should be all means never be economized. If we try to save funds . . . we shall never be strong . . . The amount which has already be spent will . . . become sheer waste.' See Teng, S. Y. and Fairbank, J. K., *China's Response to the West* (Antheneum: New York, 1954), p. 109.

[6] Chan, W., 'Government, merchants and industry to 1911', in eds D. Twitchett and J. K. Fairbank, *The Cambridge History of China: Late Ching*, vol. 11 (Cambridge University Press: Cambridge, 1980), p. 418.

[7] Rawlinson, J. L., *China's Struggle for Naval Development, 1839–1895* (Harvard University Press: Cambridge, 1967), p. 144.

[8] Twitchett and Fairbank (note 6), vol. 10 (1978), pp. 499, 503.

ing muzzle-loading muskets.[9] In sum, Qing China was faced with the management problems typical of incomplete technology transfer; trapped in technological dependence on foreign sources, and prisoners of the Confucian past, it was unable to innovate and respond to the forces that ultimately brought the imperial system crashing down.

In the final analysis, it was China's ambivalence towards modernization and the outside world that determined the outcome. The Self-Strengtheners' *ti-yong* slogan cited above assumed that China could graft selected aspects of industrial technology to the agrarian Confucian root without damage. Chinese powerholders at the end of the 19th century, unlike their Japanese contemporaries, could not admit, much less imagine, the nature of changes required for survival in the new industrial world.

One hundred years later, the same forces continued to be expressed in Chinese policy debates and power struggles:

1. The call to open up to the outside world, while upholding the 'Four Cardinal Principles' of Party dictatorship and struggling against 'bourgeois liberalism', reflects a mind-set that has its roots in the *ti-yong* strategy.
2. The question of ideological foundation—Confucian or communist—continues to be opposed to the matter of expertise in the distinction between 'Reds' or 'experts' or 'Leftists' and 'Rightists'.
3. The question of self-reliance or international interdependence (with its overtones of technological dependence) has yet to be fully answered.

In studying China's modern-day defence production, these issues should be kept in mind: Chinese decision-makers, who tend to keep a nervous eye on the past, are certainly aware of them.

Structure

When Mao Zedong and his colleagues declared at Tiananmen on 1 October 1949 that China had finally stood up, they were announcing the end of a 'century of humiliation'. However, the country was devastated after decades of foreign intervention, revolution, civil war, the disgrace of the warlords, Japanese invasion and civil war. The warlords and the defeated Guomindang (Kuomintang, or KMT) regime of Jiang Jieshi (Chiang Kaishek) had their arsenals and even aircraft manufacture, but the new regime had to start over. Not surprisingly, the heavily militarized industrial structure that they built was modelled after that of the Soviet Union, but with Chinese characteristics.

Since then the formal structure of military production has changed a number of times. Today, the complex system includes both military enterprises run by the PLA and industries that fall under the industrial ministries of the State Council.

[9] Rawlinson (note 7), pp. 44, 81; Twitchett and Fairbank (note 6), vol. 10 (1978), p. 522.

Table 14.1. China's military–industrial commercial complex in the 1990s

CENTRAL MILITARY COMMISSIONS[a]					STATE COUNCIL[b]			*CHINA INTERNATIONAL TRUST AND INVESTMENT CORP. (CITIC)*
GENERAL STAFF DEPT (GSD)[c]	GENERAL POLITICAL DEPT (GPD)	GENERAL LOGISTICS DEPT (GLD)[c]	PEOPLE'S ARMED POLICE	COMMISSION ON SCIENCE, TECHNOLOGY & INDUSTRY FOR NATIONAL DEFENSE (COSTIND)	MINISTRY-LEVEL CORPORA-TIONS:	MINISTRY OF ENERGY RESOURCES (MER)	MINISTRY OF MACHINE-BUILDING & ELECTRONICS (MMBEI)	MINISTRY OF AEROSPACE INDUSTRY (MAS)
Air Force: Lantian Corp	Kaili (Carrie) Corp. (Communica-tions equip-ment, publica-tions)	Xinxing Corp. (Clothing, food, construction materials, fuels, vehicles)	(Joint *Public Security Bureau* command)	**Xinshidal Corp.*** (Exchanges, exhibitions, publications, planning; marketing plans)	China Electronics Industry Corp. (Chinatron: R&D, systems, consumer goods)	China Nuclear Instrumentation & Equipment Corp. (CNIEC)	**China National Electronics ImpEx Corp. (CNEIC)*** (Electro-optics, radars, crypto, EW) (ex-4th MBI)	**China National Aero-Technology ImpEx Corp. (CATIC)*** (ex-3rd MBI) (aircraft, missiles, engines)
Navy: Xinghai Corp.			Jingan Equipment ImpEx Corp. (Police/ Security equipment)			Rainbow Development Corp. (Nuclear power)		
2nd Artillery (Strategic Forces)		**Zonghe Technology Corp.*** (Missiles, military satellites)		Xiaofeng Technology & Equipment Corp. (computers, robotics, advanced technology)	**China Ship Construction Corp.*** (ex-6th MBI) (naval weapons, technology)	**China Nuclear Energy Industry Corp.*** (NBC protec-tion equipment)	China National Machinery & Equipment ImpEx Corp. (CMEC) (Armour, vehicles) rockets, medical equipment	**China Precision Machinery ImpEx Corp. (CPMIC)*** (ex-8th MBI) (Optics, missiles & tools, appliances)

Equipment Dept. Polytechnologies (Arms trading; connections with *CITIC)* & Ping He Electronics (Military technology; has connections with *Everbright)*	**China North Industries (NORINCO)*** (small arms, armour, artillery, radars) (ex-Ordnance Ministry) (ex-5th MBI)	Great Wall Corp. (space launch, space equipment)
Communications Dept. China Electronic Systems Zhihua Corp. (communications, connections with *Ministry of Foreign Economic Relations & Trade*)		Beijing Wan Yuan Industry Corp. (launch, space services)
PLUS PLA enterprises under the 7 Military Regions & Local Authorities (including administrations of Special Economic Zones) (The Military Regions are: Beijing, Chengdu, Guangzhou, Jinan, Lanzhou, Nanjing, Shenyang)		Beijing CHang Feng Ind. (spacecraft)
		Chinese Academy of Space Technology (CAST) (satellites)

ImpEx = Import–Export.

[a] Constitutionally, there are two Military Commissions charged with directing the PLA: the Party CMC and the State Council CMC. However, the senior memberships are virtually identical, and the Party CMC appears to have more clout, as appropriate in a Leninist system.

[b] The Ministry of Defence does not have operational control of the armed forces, although the Minister is on the CMCs; rather, the Minister appears to act as a co-ordinator and spokesman for military matters in the State Council.

[c] The GSD and the GLD oversee and co-ordinate, with and through COSTIND, external sales of the industrial ministries.

Note: The table indicates reporting relationships within the Chinese military–industrial complex. The Military Commissions have primary authority over the military enterprises in columns 1–5; the State Council has primary jurisdiction over the Ministries in columns 6–9. However, COSTIND and Xinshidai Corp. provide guidance and co-ordination to the ministries and their enterprises and act as a bridge between the military authorities and the ministries. CITIC also provides input at the State Council level. However, as with any organization chart of this sort, not all connections can be shown: informal relationships based on family or personal or factional ties may often transcend the formal chain of command. Furthermore, the table does not attempt to resolve to the enterprise level (for instance, it does not name the many enterprises that fall under NORINCO). Accordingly, it does not show either enterprise-level linkages between the defence and civilian sectors or linkages with organizations and enterprises outside China. In a few important cases, cross-reporting relationships are indicated in italics. Organizations whose names are in bold were identified by a PLA General Logistics Department document as being the 'Big 8' in the arms trade.

Sources: China's Defense Industrial Trading Companies, Defense Intelligence Reference Series (VP-1920-271-90) (US Defense Intelligence Agency, DIA: Washington, DC); 'China's military procurement organizations', *China Business Review*, Sep.–Oct. 1989, p. 31; 'Chinatron: ghost of MEI?', *China Business Review*, Jan.–Feb. 1992, p. 30; and for those marked with an asterisk (*), 'Zhonggong de Qiangpao Waijiao' [Chinese Communists' foreign arms traffic], *Jiushi Niandai* [The Nineties], May 1991.

A co-ordinating body, now called the Commission on Science, Technology and Industry for National Defence (COSTIND),[10] operates at the ministerial level but reports to the Chinese Central Military Commissions (CMCs).

It is in fact at the top of the military system (see table 14.1) that the complexity begins, where there are two CMCs: one nominally part of the state apparatus, and the other an organ of the Party Central Committee, COSTIND, and the Ministry of National Defence (MND). The Communist Party CMC is a 'central organization' of the party and is at the top of the chain of command. In 1982 the State CMC was created, as part of the general effort of the Deng reform programme to make a clear distinction between the administrative and operational responsibilities of state organizations and the less well-defined 'guiding' functions of party organs. The State CMC has the responsibility to 'direct the armed forces of the country' in all its aspects and oversees the various general departments of the armed forces. COSTIND co-ordinates and manages weapon research and development (R&D) in the industrial ministries. and, among other functions, also serves a marketing role.

The role of the MND, a State Council body, is to 'direct and administer the building of national defense'[11]—that is, to oversee the industrial base for the PLA. Thus the MND is not in the direct chain of command; rather, it acts as a spokesman for and representative of the PLA in the State Council and among the industrial ministries that report to the State Council. The standing of the MND relative to the CMCs is indicated in the Chinese State Constitution, which reserves a separate section (Section IV) for the CMC and simply includes a paragraph mention of national defence responsibilities in Section III, devoted to the State Council. However, this may be a false distinction. To begin with, the senior memberships of the two CMCs are almost identical: the Minister of Defence is on the State CMC. Furthermore, all of the senior positions at the top of the military system are included in the party *nomenklatura* system. This results in overlapping assignments (see table 14.2) and the frequent reassignment of senior personnel. Finally, the fluidity in personnel assignments is reflected in the fluidity of the structure itself. The military–industrial ministries have merged, separated and changed designations several times.

The lack of organizational clarity is typical of the Chinese regime and underlines three important features of the nature of power and the military in China:

1. Power in the system is weakly institutionalized and highly personal; formal position is hardly a guarantee or sign of authority.

2. The PLA is not subordinate to the state structure; rather, it is the third pillar of the Chinese Party–State–Army system of rule and, although under party direction, is a distinct and separate element.

[10] The Soviet analogue would have been the Military–Industrial Commission (VPK).

[11] Constitution of the People's Republic of China (1982), Article 89(10), in Wang, J. C. F., *Contemporary Chinese Politics* (Prentice Hall: Englewood Cliffs, N.J., 1992), p. 391.

Table 14.2. Overlapping memberships at the top of the Chinese military system, 1991

Name	Politburo	Central Committee	Party CMC	State CMC	Other/title
Jiang Zemin	x	x	Chairman	Chairman	Chairman, CPC
Yang Shangkun	x	x	x	x	President, PRC
Gen. Yang Baibing		x	x	x	Director, PLA Political Dept.
Gen. Liu Huaqing			x	x	CMC Vice-Chairman
Gen. Qin Jiwei	x	x		x	Minister of Defence
Gen. Chi Hotian		x		x	PLA Chief of Staff and Party Secretary
Gen. Cho Nam Qi		x		x	Director & Party Secretary, General Logistics Dept
Gen. Ding Henggao			x		Minister and Party Secretary, COSTIND

Note: The commanders and party secretaries of the PLA Air Force and Second Artillery are all members of the Central Committee; the commander of the Navy is a Central Committee alternate member.

Source: Drawn from *China Directory 1992* (Radio Press: Tokyo, 1992).

3. The national security system prizes security and secrecy; thus the names of organizations are often misleading and relationships are often deliberately obscured, even if they are also built on family ties.[12]

II. The evolution of the military–industrial system

It would be correct, but incomplete, to say that the Chinese defence industries produce (or have produced) a small number of strategic nuclear weapons plus a full range of conventional munitions and weapon systems—small arms, tanks and other armoured vehicles, artillery, tactical missiles, bombers, fighters and naval vessels from nuclear submarines to in-shore patrol craft, plus an array of defence electronics; that these systems, obsolescent when compared to the state-of-the art, were manufactured for the Chinese People's Liberation Army (PLA); and that many of these conventional weapon systems and perhaps some nuclear technology as well have been exported.

The present organization of industrial ministries evolved from the several numbered 'machine-building' industries (MBIs) that China established in the

[12] Polytechnologies, one of China's major arms traders, serves as an example. It reports to the Equipment Department of the PLA General Staff. and is headed by He Ping, a PLA senior colonel who is Deng Xiaoping's son-in-law; He Ping is reported to be next in line to take over the Equipment Department. The sons-in-law of other top PRC officials—President (and General) Yang Shangkun and PRC Vice-President Wang Zhen among them—are also officials with the company; many of the employees come from high-ranking military families. However, the bright, modern offices of the enterprise are hardly the scruffy, standard-issue Chinese Government bureaus. All the young staffers this author saw on a brief visit in 1989 to the enterprise's offices in the CITIC Building in Beijing, one of Beijing's most prestigious business addresses, were dressed in sharp-looking civilian clothes: no 'PLA green' was in sight.

1950s and 1960s, following the Soviet model. In 1949 the PRC had inherited a scattered and run-down assortment of arsenals and factories—hardly a promising base from which to build.

The Korean War provided considerable impetus to start reconstruction. A Military Industry Commission under Zhou Enlai and an Aviation Industry Commission were set up in 1951, and in 1952 the Second MBI, with responsibility for military production, was set up. During the period of the First Five-Year Plan (1953–57), during which considerable amounts of Soviet industrial assistance flowed into the country, the military industries were further developed. In the general reconsolidation and reorganization that followed the catastrophe of the Great Leap Forward (1958–61), a joint system of R&D, supervision and co-ordination through the PLA National Defence Science and Technology Commission (NDSTC) and the State Council National Defence Industries Office was set up, and work on nuclear weapons was accelerated.[13]

By the mid-1960s, on the eve of the Cultural Revolution, there were six numbered military MBIs that dealt with: nuclear weapons (originally Number 3 MBI when established in 1956, but renamed No. 2 in 1958); aircraft (No. 3 MBI); electronics (No. 4); ordnance (No. 5); naval vessels (No. 6); and ballistic missiles (No. 7). The factories of these ministries not only were located in the major industrial centres (Beijing, Shanghai, Tianjin, Shenyang, and so forth) but also, in a massive, accelerated construction effort starting in 1964, were built in the remote and primitive interior (Sichuan, Yunnan, Guizhou, Gansu, Qinghai, Ningxia and the western portions of Hunan, Hubei, Henan and Shaanxi), in a programme of supposed strategic dispersion.

These 'Third Front'[14] factories had several functions. They were built so as to be out of range of US attack but were also intended to bring self-sufficiency in military production to the country's military regions and, in addition, were to provide some technological diffusion to the backward localities in which they were built. The effort and investment were huge since both the factories, designed with typical Stalinist giganticism, and the infrastructure (roads, energy and railways) had to be built.

[13] This account is drawn from *China Report 1949–1989* (Influxfunds: Hong Kong, 1989); Jencks, H., *From Muskets to Missiles* (Westview Press: Boulder, Colo., 1982); Ostrov, B. C., *Conquering Resources: The Growth and Decline of the PLA's Science & Technology Commission for National Defense* (M. E. Sharpe: Armonk, N.Y., 1991); Shambaugh, D. L., 'China's defense industries: indigenous and foreign procurementp, in ed. P. Godwin, *The Chinese Defense Establishment: Continuity & Change in the 1980s* (Westview Press: Boulder, Colo., 1983); Latham, R., 'People's Republic of China: the restructuring of defense–industrial policies', in ed. J. Katz, *Arms Production in Developing Countries* (Lexington Press: Lexington, Mass., 1984); and Latham, R., 'China's defense industrial policy: looking toward the year 2000', in ed. R. H. Yang, *SCPS PLA Yearbook: 1988/89* (Sun Yatsen University: Kaohsiung, Taiwan, 1989).

[14] The term has a Maoist military flavour and refers to the rear area towards which Chinese forces would retreat should the first and second defence fronts fail. Lin Biao is reported to have used the term in 1962, when KMT attacks on Shanghai (The First Front) and Suzhou (the Second Front), backed up by US forces, were feared. Naughton, B., 'The Third Front: defence industrialization in the Chinese interior', *China Quarterly*, no. 115 (Sep. 1988), p. 352. Such a plan, with its tactics of strategic retreat and 'luring deep', is classic People's War strategy.

Investment

The precise magnitude of the PRC military industrial investment is difficult to judge, but it was huge. According to Chinese statistics, 11 per cent of total Chinese state investment in the period 1969–71 was in military industries; a report from Sichuan indicated that almost one-quarter of the investment in Sichuan in the period 1950–81 went to the armaments industry. According to figures collected by Barry Naughton, over half (52.7 per cent) of Chinese national investment was plowed into the Third Front alone in 1966–70; just over 40 per cent went to the Third Front in 1970–75. However, if these percentages are recalculated relative to industrial investment, Naughton estimates that in 1966–75 the Third Front consumed about two-thirds of national investment in industry. Only about 20 per cent went 'directly to military industries'. The other 80 per cent had to be sunk into infrastructure, which, among other things, included the construction of a railway line between between Chengdu and Kunming that went, literally, through the mountains—about a third of its 1083-km length are tunnels.[15]

According to one Chinese source, the Third Front built, among other things, 'more than 1000 large and medium sized enterprises, research institutes, specialized schools and communications and post and telegraph projects' and 30 industrial cities;[16] *China Daily*, reporting the move of some of these industries to industrial centres in eastern China in 1991,[17] put the number of firms set up during the drive at 2000. A Chinese source says that the Third Front cost 'more than 100 billion yuan'; Naughton estimates the investment to be about 140 billion yuan. Salisbury cites Chinese sources to the effect that, with 75 per cent of China's nuclear weapon plants and 60 per cent of its aerospace facilities included, total Third Front investment came to around 200 billion yuan.[18]

The Third Front not only required huge immediate outlays but also had significant opportunity costs. Since the factories were in remote areas, away from good communications and other industrial centres, the output relative to investment was low. Naughton cites Chinese estimates that productivity in the coastal areas is almost four times greater than in the inland areas and calculates that, if the same investment had been made in eastern rather than western China, Chinese industrial output could have been 14 per cent higher per year. In 1991 *China Daily* simply termed the distribution of the Third Front plants as 'irrational', adding, significantly, that 'conditions are hard, and unattractive to college graduates'.[19]

[15] These figures are from Naughton (note 14).

[16] Li Yue, *Zhongguo Gongye Bumen Jiego* [Chinese Industrial Structure], 2nd edn, (Zhongguo Remin Daxue Chubanshe [Chinese People's University Press]: Beijing, 1987), p. 218.

[17] *China Daily*, 5 Dec. 1991, p. 1.

[18] Li (note 16), p. 218; Naughton (note 14), p. 379. See also the account in Salisbury, H., *The New Emperors: China in the Era of Mao and Deng* (Little, Brown: Boston, 1992), pp. 128 *passim*, who draws on interviews with officials in China.

[19] Naughton (note 14), p. 379; *China Daily*, 5 Dec. 1991, p. 1.

It is important to understand that, as in the Soviet case, the military industries came to form a virtually separate economy, with first call on talent and resources. The Chinese themselves acknowledge this, referring repeatedly in press commentary to the closed nature of the military–industrial system and to its superior technology and personnel. The importance of the military effort could be seen in the person of the chairman of the NDSTC, Marshal Nie Rongzhen. When the NDSTC was set up in 1958, Nie held key party and state positions: he was both a vice-chairman of the Party Central Military Commission and, as head of the Science and Technology Commission, a vice-premier as well.

Although the military industries were put under the dubious protection of the Cultural Revolution Group during the Revolution and work on nuclear weapons and other advanced systems continued relatively unimpeded, it would be misleading to suggest that the military industries were islands of tranquillity during that chaotic time. The ministries all emerged from the Cultural Revolution with military men at the top. There were conflicts between the PLA, the NDIO and the NDSTC; ministries had been factionalized; and there were old scores to settle.

Reorganization

Despite their insulation, the military industries were also not immune from the general programme of reform advanced by Deng Xiaoping. In 1982, in a wholesale bureaucratic reform, the NDIO, the NDSTC and the Military Commission's Science and Technology Equipment Commission were merged to form COSTIND. Formally, COSTIND was put under the State Council, but it receives direction from the Party CMC. COSTIND absorbs the R&D functions of the NDSTC and the co-ordinating functions of the NDIO.

In the same move, the numbered MBIs were renamed and the 6th MBI became a ministry-level corporation, the China Ship Construction Corporation.

Table 14.3. Evolution of the Chinese military industry system to ministries, 1982–88

Number	Name of ministry in 1982	Ministry into which it merged in 1988
2	Nuclear Energy	Energy Resources (MER)
3	Aviation	Aerospace Industries (MAS)
4	Electronics	Machine Building & Electronics Industry, partially corporatized as the China Electronics Industry Corp. ('Chinatron') in 1991
5	Ordnance	MMBEI
6	Ship Construction Corp.	No change
7	Space Industry	MAS
8	(Missiles	Merged with No. 7 in 1981)

Source: Author's archive.

In 1988, in another shuffle, the nuclear MBI joined other energy-related ministries in the Ministry of Energy Resources (MER), the ordnance and electronics ministries merged to become the Ministry of Machine-Building and Electronics Industry (MMBEI), and aviation and space-related activities joined to form the Ministry of Aerospace Industry (MAS). The commercial trading arms of the old ministries—such as CATIC and NORINCO—simply followed their old organizations into the new ministries (see table 14.3).

III. Arms production

Several questions, some only partially answerable, need to be addressed in order to examine the military equipment this industrial system produces. First, there is the matter of listing the weapon systems and munition types. Then it is useful to explore the issue of design origin and modifications to the original design, which also involes the matter of foreign technology transfer. Lastly, there is the question of production quantities: the size of the inventory and trends in its development.

Weapon types

Given the close political and economic co-operation between the PRC and the USSR in the 1950s, it is not surprising to find that most Chinese weapons are based on Soviet designs of that period. However, over time, China has introduced its own modifications to these systems and has incorporated Western technology into many of its weapon platforms.

A fairly full picture of the weapon types produced and modifications made can be obtained from public sources.[20] It should be recognized that the exact

[20] For this broad view, the author draws upon the slick *Directory of Chinese Military Equipment* (CONMILIT Press: Hong Kong, 1987), which covers everything from ICBMs to hand grenades and which looks very much like an export brochure, albeit without prices; it also omits a number of obsolete weapon systems formerly produced in China (such as the Chinese version of the MiG-15 jet fighter and some naval vessels, including older submarine types), and lists the military trading companies as the manufacturers. The CONMILIT Press also publishes in Hong Kong a magazine on military matters entitled in English *CONMILIT*, and in Chinese *Xiandai Junshi* [Contemporary Military Affairs]. Despite their Hong Kong origin, these publications can be considered authoritative; the magazine recently began to identify its editorial office as being the China Defence Science and Technology Information Centre, a research organization subordinate to the Commission on Science, Technology and Industry for National Defense (COSTIND), which in turn is directly under both the Central Military Commission and the State Council. The CDSTIC also publishes an internally circulated bi-weekly newspaper, *Defense Industry Conversion News [Jun Zhuan Min Bao]* on defence industrial issues. In any case, it was evident earlier that *CONMILIT* was a useful source, as it was clear that it was associated with a Chinese business called Xinshidai (New Era), described by the *China Business Review* as 'the commercial arm' of COSTIND. See Gillespie, R., 'The military's new muscle', *China Business Review*, Sep.–Oct. 1989, p. 31. For a more detailed examination of selected equipment, information is taken from the appropriate directories published by the Jane's Information Group, London. Both of these sources are supplemented with information taken from the annual series *The Military Balance*, published by Brassey's for the International Institute for Strategic Studies (IISS); and *Shijie Junshi Gongye Gailan* [Survey of World Military Industries] published in Beijing. See IISS, *The Military Balance* (IISS: London, various years); and *Shijie Junshi Gongye Gailan* [Survey of World Military Industries], edited by the Technology Information Office of COSTIND

Table 14.4. Contemporary Chinese arms production

Category	No.	Descriptions/designations of types
Strategic weapons		
Launch vehicles (manufactured by CSTF)	3	CZ-1, CZ-2, CZ-3: Dongfeng (East Wind) series
SSBN	1	Xia Class, in trial, carries 12 SLBMs
Ground forces (manufactured by NORINCO unless noted otherwise)		
Long-range tactical missile (manufactured by CPMIEC)	1	Conventional warhead, truck-launched, with 600-km range; under development
Tanks	4	Main/medium battle tanks: T-80/85, -69, -59, -54
	2	Light/amphibious: T-62, T-63
Armoured personnel carriers, other armoured vehicles	18	Includes infantry fighting vehicles, command vehicles, ambulances, tank recovery vehicles, bridge layers
Howitzers/guns	3	Self-propelled (122- to 152-mm)
	7	Towed (85- to 155-mm)
Multiple rocket launchers	6	Self-propelled (includes 1 APC-mounted, 40 122-mm tube system, 1 4-tube FROG system on a truck, other systems 107- to 130-mm)
	1	Towed, 107-mm
Mortars	2	Self-propelled systems (82-, 120-mm)
	7	Conventional (60-, 82-, 100-, 120-mm)
Anti-aircraft missiles (manufactured by CPMIEC)	1	Shoulder-launched (HN-5A)
	1	Self-propelled on tracked chassis (HQ-61)
	1	Fixed site (HQ-2J)
Anti-aircraft guns	2	Self-propelled (Twin 57, 37-mm) guns
	8	Towed and mounted (includes machine-guns (twin and quad 14.5-mm), guns (twin 23-, 37-mm guns, 57-mm gun)
	2	Tripod-mounted 12.7-mm machine-guns
Anti-tank missiles	1	Self-propelled system (with 8 Red Arrows) on an APC chassis
	2	Tripod-mounted (Red Arrow HJ-8 TOW/Milan type, HJ-73 (Sagger type); both wire-guided)
Anti-tank rockets	2	40-, 62-mm, including versions based on Soviet RPG-7
Recoilless rifles	1	Self-propelled (105-mm)
	1	Tripod-mounted (82-mm)
Mines	3	Anti-tank
	1	Anti-personnel
	1	Mine-laying rocket system (Type 74), truck-mounted
Hand grenades	5	Includes 2 stick-type grenades
Machine-guns	4	Heavy, tripod-mounted, 7.62-mm
	4	Light, 7.62-mm
Assault rifles	8	Various 7.62-mm models, incl. 5 AK-47-based versions

(Guofang Kexue Jixu Gongye Weiyuanhui Keji Qingbaoju), Jin Zhude *et al.* (eds), (Guofang Gongye Chubanshe [National Defense Industry Publishing]: Beijing, 1990.]

Category	Number	Descriptions/designations of types
Pistols	6	7.62-mm semi-auto models, including pocket and silenced versions
	2	9-mm semi-auto models, including machine pistol
Naval vessels (manufactured by CSSC unless otherwise noted)		
Submarines	1	Nuclear (Han)
	1	Diesel
Missile Frigates	3	Main armament surface-to-surface missiles; one with AA missiles
Corvettes	2	Surface-to-surface missiles
Small craft	3	Missile boats
	2	Torpedo boats
	1	Coastal patrol craft
Anti-ship missiles (manufactured by CATIC, CPMIEC)	5	Includes Silkworm, Exocet type missiles
(manufactured by CPMIEC)	1	Anti-submarine missile (CY-1)
Torpedo	1	CST-2, 1.3-km range
Naval guns	2	Twin 57-mm
(manufactured by NORINCO)	1	Twin 30-mm
	1	Twin 25-mm
Aircraft (manufactured by CATIC)		
Fighters	3	J-8II: latest fighter, was Peace Pearl technology recipient; factory base: Shenyang
	..	F-7M Airguard: export model of J-7 (MiG-21 derivative); factory base: Chengdu
	..	J-6: MiG-19 derivative
Attack planes	1	Q-5: based on J-6, exported as A-5; upgrades based on French (Q-5K) and Italian (A-5M) avionics; factory base: Nanchang
Bombers	2	H-5: based on Soviet Il-28 Beagle H-6; based on Soviet Tu-16 Badger
Amphibians	1	SH-5: turbo-prop; Chinese design; built in Harbin
Helicopters	2	Z-8, based on Aérospatiale Super Frelon; factory base: Jingdezhen, Jiangxi
	..	Z-9, based on Aérospatiale Dolphin; anti-tank prototype flown
Trainers	2	JJ-5: Jet fighter trainer, based on MiG-17; factory base: Chengdu
	..	JJ-7: J-7 trainer. Factory base: Anshun, Guizhou
Transports	1	Y-8: 4 turbo-prop model, based on Antonov AN-12B;. factory base: Chengdu, Shaanxi
RPVs, drones	1	'Chang Kong' 1C
Air-to-air missiles	4	PL-2, -2A, -5B, -7; all appear to use IR homing

Table 14.4 *contd*

Category	Number	Descriptions/designations of types
Aircraft guns	4	23-mm models
(manufactured	1	30-mm model
by NORINCO)		
Bombs, aircraft rockets	2	90-mm unguided anti-tank rockets
(manufactured	11	Various bomb types (blast, anti-tank, fragmentation,
by NORINCO)		marking, cluster; largest is 3000-kg bomb)

Source: Directory of Chinese Military Equipment (CONMILIT Press: Hong Kong, 1987).

number of systems actually deployed is not known, and some of the more advanced systems may be only at the prototype or test stages (see table 14.4).

IV. The technology acquisition process

It is apparent from table 14.4 that Chinese arms production has a wide technical mix. Following a suggestion by Wendy Frieman, the modes of technology acquisition can be arranged in a continuum that ranges from indigenous R&D and design to licensed production, reverse engineering, extensive modification (including incorporation of foreign technology) of existing designs and co-production.[21] The boundaries between these modes of acquisition are not clear-cut: indigenous R&D certainly draws upon exposure to foreign technology and, in addition, weapon systems are purchased outright from abroad and factor in the experience gained from arms co-production with other countries outside China.

1. According to Frieman, weapon systems emanating from *indigenous R&D and design* include strategic missiles, warheads and satellites; nuclear submarines, some warships (Luda and Anshan Class destroyers and patrol craft); some field and machine-guns; and some armour (light and amphibious tanks, and some armoured personnel carriers).

2. *Licensed production* includes weapon systems obtained from the Soviet Union and more recently from the West. It is difficult to separate this mode from *reverse engineering,* since some spill-over is inevitable. Weapons produced in this latter mode include the J-6 and J-7 fighter series, missiles of various sorts (surface-to-air, air-to-air and anti-ship missiles) and the main battle tank series (T-59s). This mode is of some importance, given the massive Soviet aid given to China through the 1950s.

3. From reverse engineering to *extensive modification (including incorporation of foreign technology) of existing designs* is a short conceptual step, although the realization may take years. Examples from the aircraft industry—the Q-5 fighter bomber and the J/F-7 series—are perhaps the best known. European

[21] Frieman, W., 'Foreign technology and Chinese modernization', in eds C. D. Lovejoy and B. W. Watson, *China's Military Reforms* (Westview Press: Boulder, Colo., 1986).

equipment (French, German and Italian) has also been incorporated in Chinese naval vessels and other systems.

4. *Direct foreign assistance and purchase* lie at the end of the spectrum. Examples include the purchase and co-production of French helicopters. More recently, there have been reports that China is in the market for Soviet Su-27s, an advanced fighter plane several generations ahead of anything made in the PRC. In the late 1970s and mid-1980s the PLA did quite a bit of window shopping in the world arms bazaar but made few purchases: according to the *SIPRI Yearbook 1985*, China examined over 50 types of advanced systems from the West, including anti-aircraft radars and fire control systems; jet fighters, transport planes and helicopters; submarines; anti-tank, anti-aircraft and anti-ship missiles; tanks; and communications equipment. However, only a few purchases were made: helicopters from France, Germany and the USA; the Spey jet fighter engine from the UK (which never was put into Chinese production); lorries from the USA; and European avionics.[22] However, much foreign assistance was cancelled or otherwise reconsidered after the Tiananmen Square events of June 1989, including the 'Peace Pearl' J-8 upgrade programme.

5. Finally, at the extreme end of the technology process there are *clandestine transfers*. The sketchy information on these kinds of transfers—almost always denied or deniable—generally involve exotic technologies. It is believed that China has sought advanced computer software, electro-optics, advanced munitions, aero-space technologies, and special materials such as composites for aircraft and missiles or advanced manufacturing techniques and advanced materials for integrated circuits. The capability of Chinese technologists to work on these technologies in the laboratory should not be doubted—the real question is whether China can field them.

Tanks and aircraft

It is possible to get a better idea of Chinese defence production and technology absorption capabilities by looking at tanks and aircraft in more detail.[23]

The T-59 (the Chinese designation for the Soviet T-55, which the USSR supplied to China in the 1950s) is the basic battle tank of the PLA and is the basis for further Chinese designs. In the early 1980s, China showed the T-69 tank, an enlarged T-59 model, equipped with infra-red sights, laser sights and either a smooth bore or (in the T-69II) rifled 100-mm gun. The T-69 chassis has also been used as the basis for twin 37-mm and twin 57-mm self-propelled anti-aircraft guns as well.

[22] *SIPRI Yearbook 1985: World Armaments and Disarmament* (Taylor & Francis: London, 1985), p. 258.

[23] The discussion here is drawn entirely from Foss, C. F. (ed.), *Jane's Armour & Artillery, 1991–92*, 12th edn (Jane's Information Group: Coulsdon, 1991); Lambert, M. (ed.), *Jane's All the World's Aircraft, 1990–91* (Jane's Information Group: Coulsdon, 1990); and the CONMILIT *Directory of Chinese Military Equipment* (note 20).

Today the most advanced main battle tank produced in China is the T-80/T-85, a descendant of the T-59, which carries a 105-mm gun and includes a computer for calculating ballistic trajectories, a laser range-finder and a gun stabilization system. It weighs about 38 tons (compare the weight of the US M1 Abrams, about 58 tons, and the Soviet T-80, about 42 tons). The 105-mm gun has been reported to be a version of an Israeli tank gun; further modifications of the basic T-80, called the T-85, include a welded rather than cast turret, which allows the attachment of advanced composite and reactive armour.

The basic T-59 is so numerous that NORINCO has developed eight different retrofit packages to upgrade the chassis; these have probably entered the export market as well, as they would fit the popular (in the less developed countries) Soviet T-55s. These retrofit packages include 'mix and match' combinations of new guns, fire control and safety systems, and upgraded armour, including reactive armour. The T-59 was also the basis for upgrade packages from Western firms, including a never-finalized engine upgrade by a US engine manufacturer, and a 'new' low-cost ($1 million) export model tank to be called the Jaguar and to be developed by Cadillac Gage of the USA and the China National Machinery and Equipment Import Export Corporation. The upgrades—a new gun, a new armour overlay and a new fire control system—were all of either US or British origin. Although the project was shelved after the 1989 events in Tiananmen Square, the British Jane's Information Group reports that Cadillac may be pursing the retrofit market.[24]

China's light tanks are also based on Soviet designs. The T-62 is a scaled-down T-59, weighing 21 tons and sharing a 85-mm gun with the T-63 amphibious tank, which appears to be modelled on the Soviet PT-76. These light tanks appear to be ideally suited for the hilly and difficult terrain in southern China.

Other Chinese armoured vehicles show the same pattern: a mix of foreign (usually Soviet) and Chinese designs. Various armoured personnel carrier (APC) chassis are used as the platforms for anti-tank, anti-aircraft, self-propelled artillery and other systems. For instance, the WZ501 infantry fighting vehicle is closely modelled on the Soviet BMP-1 and is said to have been developed after Egypt provided China with one example out of their Soviet-supplied inventory; its chassis is used for a wide variety of vehicles, including a missile launcher, command vehicle and armoured ambulance. Other models include the Type 77, based on the Soviet BRT-50, and the type YW 534, used as the foundation for anti-aircraft and anti-tank systems. In 1984 a wheeled APC, apparently based on the Belgian SIBAS design, was shown. China has developed its own six-wheel WZ 551 APC, which can be used as a cannon and missile platform as well. Foreign companies—Vickers, GIAT and FMC—have been involved in jointly upgrading Chinese chassis. However, one item about Chinese design that may be special to the PRC is worth noting: photographs of Chinese armoured ambulances all show that they carry not only the internationally recognized white circle and red cross but also top-mounted machine-guns.

[24] *Jane's Armour & Artillery, 1991–92,* 12th edn (note 23).

The same combination of Soviet design, Chinese modification and foreign technology can be seen in the aircraft produced in the PRC. The developmental history of the Jian-7 aircraft, based on the Soviet MiG-21, is instructive. The plane was first licensed for assembly and manufacture in China out of knocked-down kits in 1961. According to Jane's and other sources, the plane was not assembled until 1964, and the first flight came two years later. A subsequent model, the J-7 II, which included a Chinese engine, compass system and new ejection seat, began development in 1975, with the first flight at the end of 1978. The F-7M programme to develop an export model began three years later; the aircraft features, among other things, British General Electric Corporation avionics, a better Chinese engine and air-to-air missile capability. At the same time, Chinese engineers were working to further upgrade the plane—the J-7 III and J-7 E models, which feature a new wing, additional fuel capacity, and improved avionics and weapon capabilities. Prior to the 1989 Tiananmen Square events, Grumman Aircraft was involved in an upgrade programme to produce an improved Super-7 export model, with a new wing, Sidewinder air-to-air missiles and a new cockpit based on the Northrop F-20.

The Q-5 fighter-bomber is an example of extensive Chinese modification of a Soviet design. Based on the MiG-19 supersonic day fighter, the Q-5 is virtually a new plane, with internal bomb bays in the original version, air intakes moved to alongside the fuselage, the cockpit moved forward and a distinctive extended nose cone. According to Jane's,[25] the original design proposal came from the Shenyang aircraft factory in 1958, but responsibility for the prototype was shifted to the Nanchang works. The programme was reportedly cancelled in 1961, but was revived in 1963; the first prototype flew in 1965. The Cultural Revolution must have delayed series production, which eventually was approved at the end of 1969, and deliveries began in 1970. Subsequently, as noted above, the aircraft has undergone further modification, including the installation of Western avionics and other sub-systems. Photographs of the aircraft show it configured with a variety of weapons, including air-to-air missiles, air-to-ground bombs and rockets, anti-ship missiles and electronic countermeasure (ECM) pods.

The Peace Pearl project to upgrade the J-8, although ultimately scrapped for political reasons, was a similar modification project undertaken on a Chinese airframe. The basic aircraft, a twin-engined delta-wing fighter, has gone through several versions which converted it from a clear-weather day interceptor to an all-weather fighter-bomber. In 1987 the US Grumman Aerospace Corporation, Westinghouse and Litton, under a US Air Force Foreign Military Sales (FMS) contract, undertook to provide a new avionics infrastructure for the aircraft, which would have included an inertial guidance system, heads-up display, and new fire control radars and computers. The project was ultimately

[25] *Jane's All the World's Aircraft 1988–89* (Jane's Information Group: Coulsdon, Surrey, UK, 1988), pp. 39–40.

cancelled, but at least two aircraft were delivered (to the USA) for the upgrade and flight-test.

It should be noted that foreign involvement with the Chinese aircraft industry has given rise to a number of unconfirmed reports over the years about new aircraft of uncertain design, capabilities and designation. In late 1991, the Hong Kong press reported that a new 'J-9' was flown before dignitaries in Sichuan; about the same time, there were press reports that China is buying Su-27s from Russia. Jane's has reported a swept-wing, canard-configured 'J-?' in development at the Shenyang works, and a new multi-role attack plane, the H-7, similar to the European Tornado, planned by the Xian works (a model of the aircraft was shown at the Farnsworth Airshow in 1988). In 1988 the *Sunday Times* reported Israeli–PRC co-operation in producing a fighter based on the scrapped Israeli Lavi project. The tentative nature of these stories is indicative of China's active interest in foreign military technologies, but what will count is whether these new systems actually enter into production.[26]

Numbers

To say that information about the size of production runs of these weapon systems is imprecise would be to understate the nature of open data. However, given the size of the Chinese armed forces—around 3 million men (down from over 4 million in the early 1980s)—the inventories are large. Table 14.5 gives 1990 data published by the London-based International Institute for Strategic Studies (IISS) on PLA manpower and major systems numbers.[27]

Any attempt to use public numbers to estimate production rates is not quantitatively satisfactory: the numbers vary in an inconsistent manner and are simply a reminder of the imperfect knowledge in this area. However, the trend towards increasing sophistication of weapon systems is reflected in what little information is available. In 1975, the IISS counted no intercontinental ballistic missiles (ICBMs) in the Chinese inventory; by 1980 there were four, and by 1990 there were eight (plus 60 intermediate-range ballistic missiles, IRBMs). Over the same 15 years the number of strategic bombers accounted for doubled, from 60 to 120; the number of destroyers went from 4 to 18; and the number of J-7 fighters went from 50 to 500.

While there has been growth in Chinese major weapon systems, it is equally important to note the increasing number and sophistication of perhaps less immediately spectacular smaller military systems, particularly in electronics and munitions. For instance, China has, not surprisingly, always shown intense interest in Western anti-tank missile technologies and in the mid-1980s was known to have acquired versions of the TOW, Milan and HOT missiles. Israel

[26] For the J-9, see FBIS, 3 Oct. 1991, p. 27. See also *Sunday Times*, 3 Apr. 1988; this article also reported on a secret Israeli mission to the PRC which agreed to provide China with advanced anti-armour and artillery munitions.

[27] International Institute for Strategic Studies, *The Military Balance 1990–91* (Brassey's: London, 1991), pp. 148 ff. Figures in this section are drawn from various years of this annual publication.

Table 14.5. Manpower and major weapon systems of the PLA, 1990

Type of force	Numbers
Ground forces	
Manpower	2.3 million
Main battle tanks	7 500–8 000 (of which 6000 were T-59s)
Light tanks	2 000 (of which 1200 were T-63 amphibious tanks)
Towed Artillery	14 500 pieces
Multiple rocket launchers	3 800
Air Force	
Manpower	470 000
Fighters	4 000 (of which about 3000 J-6, 500 J-7, 50-100 J-8 versions)
Fighter-bombers	500 Q-5s
Bombers	470 (of which 120 H-6s, 350 H-5s)
Air defence artillery	16 000 pieces (100 SAM units with HQ-2, -61 SAMs)
Navy	
Manpower	260 000 (includes 6000 Marines, 25 000 Naval Air Force troops)
SSN	4 Han Class
Conventional submarines	87 (Ming and Romeo Class)
Destroyers	18 (includes 16 Luda Class, with missiles; 2 Anshan Class with missiles and guns)
Frigates	37
Patrol/coastal craft	About 915 (includes 235 missile boats; 160 torpedo boats; 540 other patrol craft)
Other	Includes 52 mine warfare craft, 58 amphibious warfare ships, including 16 LSTs
Naval aircraft	600 various fighters (J-5, -6,-7,-8) 50 Q-5s, 30 H-6, 150 H-5)
Strategic Rocket Forces (Second Artillery)	
ICBMs	8
IRBMs	60
SSBN	1 Xia Class boat, with 3 believed to be under construction

Source: International Institute for Strategic Studies, *The Military Balance 1990–91* (Brassey's: London, 1991), pp. 148 ff.

is said to have extensive dealings in anti-tank munitions, rockets, guns and other systems; indeed, Israel may be China's leading supplier of foreign military technology.[28] Italy has supplied naval electronics and Germany machine-gun technology. Where the Chinese press discusses 'advances' in military technology, it almost always includes examples of improved command and control systems, radars, communications and the like.

What is important here is not so much the specifics of this kind of technology acquisition but rather the long-term implications for the Chinese military. It

[28] See Chugani, M., 'Israel in trouble over China connection claims', *South China Morning Post*, review section, 21 Mar. 1992, p. 5.

seems clear that China, within severe resource limitations, has embarked on a long-term programme of technology acquisition. The short-term consequences of such a programme can be criticized: the results are apt to be scattered, waste is inevitable, the hardware and technology acquired may be obsolescent or inappropriate, dependencies may develop, and the initial outcomes may be marginally incremental or crude. The long-term effects, however, may be profound: the exposure of China's military and engineering elite to a wide sample of advanced technology that can be drawn upon for China's own development. What is known about Chinese military budget allocations, which show increases in R&D (see below), reinforces this conclusion. It is a strategy not unlike that embarked upon by Japan at the close of the 19th century. In 1983—in words that echo those of the earlier Self-Strengthener Feng Gueifen—then Defence Minister Zhang Aiping wrote in *Hongqi*, the party theoretical journal:

> [I]t is not realistic or possible for us to buy national defense modernization from abroad. We must soberly see that what can be bought from foreign countries will at most be things which are advanced to the second grade. This cannot help us attain the goal of national defense modernization, nor will it help us shake off the passive state of being controlled by others. Depending or modeling one's weaponry on others is not a way of realizing national defense modernization either. At the onset it is necessary to obtain some technology that can be imported and model some weaponry on that of others. However, if we are content with copying, we will only be crawling behind others and still be unable to attain our anticipated goal. The fundamental way is to rely on ourselves. . . . Since our country has a vast territory, a long border and complicated geographic and weather conditions, only by developing—through self-reliance and in a realistic light—sophisticated military equipment that can be adapted to various conditions can we satisfy our Army's needs . . . [29]

V. Reforms

There is more to China's defence production story than just a list of weapons and foreign technology transfer. During the decade of the 1980s, the entire country underwent what some writers called a 'Second Revolution' of economic, political and social reform. Not surprisingly, the defence industries were caught up in these reforms.[30]

Perhaps the most crucial task facing the post-Mao leadership in the late 1970s was the restoration of government authority and party legitimacy, both severely compromised by the political chaos of the Cultural Revolution and its 'Gang of Four'-dominated twilight (1966–76) and by the stagnation of the national econ-

[29] From FBIS, 17 Mar. 1983, pp. K2–7.

[30] For a useful summary of the industrial reforms in the larger economy, see Solinger, D., 'The future of China's industrialization program: why should the United States care', ed. W. Tow, *Building Sino-American Relations* (Washington Institute/Paragon: New York, 1991), pp. 95–124. The reforms she discusses—formation of enterprise groups, exports, contracts and the coastal zone development strategy—are all reflected in the defence sector. For an experienced businessman's frank assessment of the reforms in practice, see Stepanek, J. B., 'China's enduring state factories: why ten years of reform has left China's big state factories unchanged', *China's Economic Dilemmas in the 1990s* (Joint Economic Committee, US Congress: Washington, DC, 1991), pp. 440–54.

omy. It took Deng Xiaoping, himself both the instigator and victim of purges, almost three years to re-assert himself after Mao's death, but by 1979 Deng and his supporters were clearly in dominant positions and were ready to try—indeed, compelled to try—new solutions.

Deng decreed that the 'main contradiction' facing China was no longer class struggle. Rather, the problem was economic—inadequate production in the face of rising economic demand. The solution, applied in fits and starts and still, in 1992, the object of contention, was economic reform of China's Stalinist system of central plans, rigid and bureaucratized vertical enterprise structures, allocated resources, production quotas, subsidies and administered prices. Economic reform thus required structural and organizational reform as well. Rebuilding and reforming the economy to make it more efficient became the central task.

The reforms occurred under party leadership and direction. Deng's 'Four Cardinal Principles' of keeping to Marxism–Leninism–Mao Zedong Thought, proletarian dictatorship, the socialist road and leadership of the party saw to that, but economic reform entailed a certain degree of social relaxation and depolitization of daily life. It also called for 'opening up' to the outside world and not only engaging in foreign trade but even allowing investment from the capitalist world.

Economic reform also required a re-assessment of China's international situation. While China previously saw itself surrounded by enemies, under threat of sudden attack and, in any case, awaiting the Marxist–Leninist Armageddon of the inevitable showdown between capitalism and socialism, later war was not considered an immediate threat. The international situation had eased. There were, of course, remaining dangers—US imperialism and Soviet hegemonism—but 'the eagle and bear had blunted claws'.[31]

Deng's reform programme—the 'Four Modernizations' of agriculture, industry, science and technology and national defence—required strategic redefinitions of both the domestic and international situations. However, if the country was no longer in peril, defence modernization would have to follow reconstruction of the economy—the proposition was that defence needs could only be built on the basis on a strong economy, the 'material foundation' of national defence.

The reformulation did stimulate debate: there were those in the military and elsewhere in the bureaucracy who continued to see China in danger and who put defence needs above economic reconstruction; some argued that national defence modernization could proceed at the same time as economic rebuilding. Ultimately, it was the the reformer's line that carried the day, but the tactical implementation of the strategy was not without its difficulties.[32]

[31] See Godwin, P., 'Chinese defense policy and military strategy in the 1990s', *China's Economic Dilemmas in the 1990s* (note 30), for a useful discussion of this issue.

[32] It should be added that this argument had been engaged in before. Many of the policies associated with the Deng reforms of the 1980s had been given airings in the 1950s, and the disagreement about economic priorities was not new. In the first years of the PRC, top leaders, including the veteran generals Zhu De, Ye Jianying and Peng Dehuai, debated the issue publicly. Mao Zedong eventually settled the issue—

How did the new general line affect the military and military industries? To begin with, if the country was not under immediate threat, resources could be re-allocated from the defence to the civilian sectors. Military budgets would be reduced. At the same time, it would be possible to rebuild and modernize the military, just as it was necessary to rebuild and modernize the civilian sector. So a double-edged process of reform and reduction began in the PLA and in the defence industries.

PLA reform took several directions. One of the most dramatic steps came in the mid-1980s, with a troop reduction of about 1 million, but there may be less to this move than meets the eye: many troops were simply transferred to the People's Armed Police and many veterans were retired. Indeed, the reduction in force, if anything, actually increased PLA strength by getting rid of less effective personnel and allowing some consolidation and rationalization of equipment distribution.

The PLA also took the more important move to upgrade the quality of the officer corps with the establishment of the National Defence University and the expansion of training. In other words, in the US phrase, the PLA moved to become 'leaner and meaner'. No doubt some of the lessons learned in China's none-too-successful border war with Viet Nam were important in this regard.

The defence industry was also deeply affected by these changes in both military and civilian policy:

1. With economism the watchword and with a smaller, more professional force to supply, the industry could consolidate its structure and rationalize production.

2. Along with the national emphasis on economic reconstruction came substantial pressure from the centre for the military industries to use its surplus capacity and superior technology to produce civilian and import-substitution goods for the domestic market, and to make products for export. There were incentives to both industry and the PLA as well: enterprises could retain profits and a portion (and sometimes all) of any hard currency earnings (rather than having to remit these funds back to the centre); and this retained income could be used to cover budget shortfalls.

3. Because of these financial incentives and the new policies encouraging trade—which included the authority to set up trading companies independent of the Ministry of Foreign Trade (later the Ministry of Foreign Economic Relations and Trade)—China's military industries could move into the commercial international arms market.

at least temporarily—in his 1956 speech 'On the Ten Major Relationships', in which he argued for the primacy of economic construction. The speech is reprinted in Howe, C., *China's Economy* (Basic Books: New York, 1978).

VI. Conversion of the arms industry: 'swords into plowshares'?

Almost immediately upon regaining power in late 1978, Deng Xiaoping and the reform group kicked off an expanded policy of using military factories for the production of civilian goods. No less than Deng promoted the policy: 'The military industry has two strong points. One is that it has advanced and adequate equipment. The other is that it has a strong technical force. The military industry should make use of these two strong points to serve the country and contribute to economic reform in a still better way'.[33]

At the same time, military production was not neglected but rather, with more efficient use of resources, enhanced. This policy was summed up in a '16 Character Slogan', which in English is a little longer: 'Military–civil integration, peacetime-wartime integration, priority to production of military goods, and maintaining the war industry with profits from civil-purpose production'.

There appear to be some subtle differences here, and a debate has surfaced about the appropriate weight that should be given to the military or civilian component. However, even mentioning the potential of the military industries for civilian production was a great change from previous policies, which had essentially prohibited any discussion of military industrial policy.[34] Still, compliance has been difficult, and it is not clear that everyone is committed. A review of press commentary since the policy was announced shows that the centre has had to urge, indeed, *order* the military industries repeatedly to get in line.

For instance, in January 1986, seven years after the conversion policy was announced, the Ministry of Ordnance decreed that 'all factories . . . switch part of their management capacity to civilian production without any exceptions no matter how heavy the military commitments might be'.[35] Nearly six years later, the party newspaper *Renmin Ribao* (*People's Daily*) admitted that 'at present, the arms industry is encountering some temporary difficulties during the process of applying military technology to civilian uses'.[36] In response, however, the ministries and industries have submitted reports that indicate enthusiastic compliance. This is typical of Chinese 'campaign' style: the centre will set the theme, and the periphery—the provinces and ministries—will chorus back the appropriate variations.

Since the almost canonical dialogue has been going on for almost 15 years, one can wonder about the success of the policy. Part of the problem appears to lie in the independence of the defence industries and in foot-dragging. There are also indications that the defence industries could not cope with the change in management outlook required to shift from mandatory to guidance planning to provide for the civilian market. A national-level industry meeting in Xian in

[33] Quoted in *Wenhui Bao* (Shanghai), 28 Nov. 1984, in FBIS, 4 Dec. 1984, p. K9.

[34] See Latham, 'China's defense industrial policy: looking toward the year 2000' (note 13), which contains the best analysis of these issues in print.

[35] *China Daily*, in FBIS, 7 Jan. 1986, pp. K14–15.

[36] FBIS, 15 Nov. 1991, p. 45.

1986 noted that it was 'essential to bring [the defence industries] into the orbit of the entire national economy, breakdown the closed-style management system that has formed over a long period ... [and] dismantle the barriers between defense and civilian industries'.[37]

China is not alone in having difficulties with the conversion of defence industries to civilian production. The issue is equally problematic in the former Soviet Union and the United States. Conversion is an exceedingly complex process, involving questions of manpower planning, technology adaptation and social impact, and is one that rarely meets with unqualified success. Indeed, even US defence contractors have troubles moving into the civilian arena. Their problems suggest that defence industries throughout the world have a great deal in common, regardless of their immediate political context. As *International Business Week* noted, 'Most [US defence] contractors have trouble learning to do the extensive market research that commercial ventures require, instead of following the very specific directions the Pentagon gives them'.[38]

Defence industry conversion involves changing the way business is conducted—that is, changing corporate culture, a difficult and sometimes wrenching process. In China's case, there are also the severe economic and cultural strains of moving from a planned economy to a mixed economic system that combines some aspects of central planning with features of a more liberal system, such as contracts, a certain degree of competition and a focus on economic results. The process is also at the heart of current political debate in China—which can literally be a matter of life or death—which provides a better understand of the ambiguous signals emanating from the PRC on this matter.

However, the structural and productivity problems of the defence industries, particularly the dispersed Third Front industries, appear to play a role as well, as noted above. There is additional evidence for this judgement: in 1986 the vice governor of Shaanxi Province complained that the defence industries there accounted for 25 per cent of the industrial assets of the province and employed 20 per cent of the work-force—but produced less than 10 per cent of the province's output value, had a return on assets 'two-thirds less' than local enterprises and attained profits at only half the provincial average.[39] Still, the defence industries claim some progress. Factories have moved into the manufacture of all kinds of civilian goods: sewing-machines, clothing, cameras and other photographic equipment, radios, white goods, automobiles, lorries,

[37] See a report on the problems of shifting from mandatory to 'guidance' planning in Shaanxi, FBIS, 8 June 1981, p. Q1; for the Xian meeting, see FBIS, 21 May 1986, p. K13.

[38] See 'From bullets to bullet trains: it won't be easy', *International Business Week*, 20 Apr. 1992, p. 66. The same issue contains commentary on two recent books on the subject: Weidenbaum, M., *Small Wars, Big Defense* (Oxford University Press: Oxford, 1992); and Markus, A. and Yudken, J., *Dismantling The Cold War Economy* (Basic Books: New York, 1992). See also Gansler, J., *The Defense Industry* (MIT Press: Cambridge, Mass., 1980); and Gansler, J., *Affording Defense* (MIT Press: Cambridge, Mass., 1989). For the Soviet conversion efforts, see SIPRI, *SIPRI Yearbook 1990: World Armaments and Disarmament* (Oxford University Press: Oxford, 1990), chapter 8; for press comment, see 'Converting Soviet arms factories', *The Economist*, 15 Dec. 1990, pp. 19–21; and Murray, A., 'Soviets tapping defense firms for commerce', *Asian Wall Street Journal*, 27–28 Sep. 1991. There is a growing bibliography on the subject.

[39] FBIS, 25 Apr. 1986. p. T2.

motorcycles (in 1984, the Ordnance Ministry claimed that the 500 000 motorcycles it made represented two-thirds of the national output).[40]

A 1991 catalogue issued by the China Aero-Space Civil Products Corporation (CASCPC), a branch of the Ministry of Aerospace Industry (MAS), is indicative of the highly diversified range of products and the large scope of production of Chinese industrial ministries. According to the catalogue, the corporation engages more than 800 000 employees in over 200 factories. It has assets worth more than 10 billion yuan (approximately $1.8 billion).[41]

The CASCPC claims to have set up 160 production lines for civilian products, and to make more than 7000 types of product. The catalogue gives a sense of the Ministry's diverse production—it would appear that the Corporation provides a full line of services.[42]

The catalogue also includes pictures of hovercraft, windmills, rocket launchings and the assembly of McDonnell Douglas MD-82 jet transports. MAS makes nose cones and other aircraft parts for civilian aircraft manufactured by Boeing and McDonnell Douglas as well. The corporation also provides various metal castings and manufactured a huge bronze Buddha image, said to be the largest in the world, for a monastery on Lantao Island in Hong Kong.

There are other claims of massive civilian contributions, but data released at the end of 1991 indicate that while 65 per cent of military industry output was for civilian use, that output came from a minority—40 per cent—of military factories.[43] Simple arithmetic shows that the productivity of these 'civilianized' factories is almost three times higher than that of those remaining in strictly military production.

Exactly how great a financial contribution these factories make is unknown. The sources for the figures from which the above percentages are drawn are unknown, but the figures would have to be small given the results, since, if the repeated claims that civilian production by the military industries growing at 20 per cent or so per year are correct, that would produce a six-fold increase over a

[40] *China Daily*, in FBIS, 7 Jan. 1985, p. K14.

[41] [Untitled catalogue], (CASCPC: Beijing, 1991); the numbers on GM and Boeing are from the 1991 'Fortune 500' listings, in *Fortune International*, 20 Apr. 1992, pp. 130–31. By way of comparison, the largest US manufacturer, General Motors, with assets worth $184.3 billion, had 756 300 employees; Boeing, the largest aircraft manufacturer, had assets of $15.8 billion and 159 1000 employees.

[42] The catalogue has the following goods on offer: industrial control computers; geosynchronous and meteorological satellites; satellite receiving stations; typewriters; 4 kinds of passenger buses; 11 models of light trucks; 8 models of heavy trucks, including fuel and garbage trucks; 12 kinds of motorcycles and motorbicycles, plus accessories and parts; various textile equipment, including 3 looms, thread spinners and cone winders; 9 kinds of food packaging machinery, including tea-bag packing and soft-drink filling machines; medical equipment including X-ray systems, titanium and cobalt alloy artificial joints and an ultrasonic 'kidney stone crusher'; turbine engines; industrial robots, lathes and five-axis milling machines; colour television sets; 3 models of refrigerators, 6 models of air conditioners, plus compressors; portable radios and tape players; fans; vacuum cleaners; teflon-coated cook ware; watches and clocks; drills and cutters; automobile body jigs; and various hand-held pneumatic drills and drivers. The catalogue also has pictures of hovercraft, windmills, rocket launchings and the assembly of McDonnell Douglas MD-82 jet transports. MAS makes nose cones and other aircraft parts for civilian aircraft manufactured by Boeing and McDonnell Douglas as well. The corporation also provides various metal castings and manufactured a huge bronze Buddha image, said to be the largest in the world, for a monastery on Lantao Island in Hong Kong.

[43] *Renmin Ribao* [People's Daily], in FBIS, 7 Nov. 1991, p. 32.

decade. In one of the few instances of results being cited, Beijing television reported in 1990 that civilian exports from the military industries over the past decade were worth 'more than US$10 billion'.[44] Even making allowances for this imprecise figure, it would represent only a small fraction (3–4 per cent) of China's exports over the period.[45]

The Chinese press has revealed other problems in the conversion effort. In the race for quick profits, the output of some military enterprises came into competition with that of purely civilian industry. In 1982 parts of the Jiangsu MBI were criticized for choosing only 'very profitable jobs', putting on 'arrogant airs' and not being 'enthusiastic' in servicing customers.[46] Most seriously, some of the output simply cannot be sold: *People's Daily* reported in 1989 that 'the development of military-civilian combinations is not balanced; some civilian products are unmarketable and seriously overstocked'.[47]

There are signs that the leadership is not pleased about the lack of progress and the problems in civilian conversion. In October 1989 the Xinhua news agency reported that Premier Li Peng had warned the military enterprises that they would be allocated more resources for military production only after the development of civilian goods: 'Only after the defense industry adapted its products according to state policies', the report went on, 'could it have a way out.'[48] Under the reforms, new enterprise forms and relationships were encouraged: enterprises formed joint ventures and groups that crossed ministerial and locational lines, contracts replaced allocation, and a number of Third Front enterprises took the opportunity to move to the booming south-east—Guangdong, Fujian and the Special Economic Zones there. In December 1991 *China Daily* reported that 'more arms factories are to be moved from China's mountainous hinterland to industrial centres' and that $566 million had been set aside for '115 major military firms' to 'readjust and renovate'.[49] Other moves affecting the military industry included: (*a*) the 1988 CITIC take-over of an old

[44] FBIS, 26 Jan. 1990, Supplement, p. 36.

[45] FBIS, 26 Jan. 1990, Supplement, p. 36. It should also be noted that some parts of the military industrial system have a greater portion of their production in civilian goods than others—for instance, about one-third of the nuclear industry's output was in the civilian sector in 1988 (Zhongguo Xinwenshe, in FBIS, 16 Feb. 1989, p. 26), while the reported proportions of others—aerospace, electronics, ordnance—was considerably higher, in the 60 per cent range. In particular, the aerospace industry, which offers commercial satellite launch services and is building high-valued added aircraft components for Western firms, is in an enviable position. For instance, the deputy manager of the Shenyang aircraft works told a reporter from *Aviation Week & Space Technology* in 1989 that 70% of the factory's revenues came from non-military production. This included overseas contracts worth over $40 million for parts and components and a $4.2 million contract for cargo doors for Boeing 757s. See Fink, D., 'China aviation: Shenyang focuses on commercial projects as military Aircraft Requirements Shrink', *Aviation Week & Space Technology*, 11 Dec. 1989, pp. 70–75. See *Jane's All the World's Aircraft 1990–91* (Jane's Information Group: Coulsdon, Surrey, UK, 1990); and *SIPRI Yearbook 1991: World Armaments and Disarmament* (Oxford University Press: Oxford, 1991) for additional details on civilian production in the aircraft industry.

[46] FBIS, 9 Aug. 1982.

[47] FBIS, 12 Oct. 1989, pp. 50–51.

[48] FBIS, 16 Oct. 1989, pp. 30–31.

[49] *China Daily*, 1 Dec. 1991, p. 1. Indeed, these firms have moved not only to the high growth Special Economic Zones and other 'windows' on the coast but have also set up offices in Hong Kong and overseas.

tank factory, with the aim of converting it, at the cost of 320 million yuan, into a factory making power stations and motor vehicles; (*b*) a move to put the civilian production of the military industries into the state plan; (*c*) formation of a joint venture between the PLA's leading vehicle plant, the Songliao factory in Liaoning, and a civilian enterprise, to make automobiles; and (*d*) in order to accelerate the transfer of military technologies to the civilian sector, COSTIND has set up provincial technology transfer centres. Press reports are weak on the details of the transfers. A 1989 Xinhua news agency story, reporting the pending declassification of 2300 'items' and the actual declassification of 210 'items' in the previous year, was typical. Another figure in the same report should also be noted: 10 per cent of the 210 released technologies found civilian application.[50]

The People's Liberation Army

The PLA also entered into civilian production with the formation, in 1984, of Xinxing Corporation under the General Logistics Department, and even local units began to engage in civilian business. In 1986 Xinxing was described as producing 5000 commodities—including meat, towels, clothing and manufactured goods—at 400 factories and farms, and by the following year it was said to run 10 joint-venture tourist hotels and to have more than 21 branches around the country, including some in Hong Kong. The business volume in 1986 reached $80 million, and the aim was to reach $100 million in 1987; as *China Daily* put it, the corporation aimed to become 'one of China's top business giants', apparently with authorization to retain 100 per cent of its foreign exchange earnings.[51]

Contracts

Not all the reforms had to do with conversion to civilian production. In a significant move, mirroring developments in the civilian sector that change production relations, the PLA started to experiment with a contract system for weapon development work. As one commentary put it, the old 'supply-type administrative system of appropriating funds had led to the separation of responsibility, authority and profit, made it difficult to arouse the initiative of all concerned, and adversely affected the integration between scientific research and production, development and application, and military industry and manufacture of goods for civilian use'.[52] Its first use came in the building of a missile escort vessel, which, Xinhua reported in 1988 'received excellent quality rating and was delivered in half the usual time. [53] By the end of the year, *People's Daily*

[50] See, in order, FBIS for 14 Jan. 1988, pp. 18–19; 15 Mar. 1988, pp 13–14; 2 Aug. 1988, p. 50; and 20 Oct. 1989, p. 34.
[51] FBIS, 14 Apr. 1987, pp. K31–32.
[52] FBIS, 18 Dec. 1985, p. K14.
[53] FBIS, 10 Feb. 1988, pp. 19–20.

noted that contracts for 'nearly 1000 types of weapons and military equipment'—tanks, aircraft, long-range radar, anti-aircraft missiles and warships—were put out in 1988, and 80 per cent of the contracts were signed.[54]

In sum, the reforms, including the industrial responsibility system, came to the military–industrial sector just as they had come to the agricultural and civilian industrial sectors. As shown below, the reforms brought some unintended consequences with them as well. The reforms brought both openness and opportunism. In 1982, then Premier Zhao Ziyang said that 'today, smuggling, peddling smuggled goods, speculation, swindling, corruption, accepting bribes and other crimes . . . are much more serious than during the Three- and Five-Anti campaigns [two fierce anti-bourgeois campaigns of the 1950s]';[55] and there was spill-over into the military. It seems that many in the Army took the slogan 'To get rich is glorious' too much to heart and started to pay more attention to commercial opportunities than to military duties. Starting in the autumn of 1981, articles appeared in *Liberation Army Daily* which denounced 'the unhealthy practice of making under-the-table deals'.[56]

Stories of PLA profiteering because of its privileged position have appeared in both the Hong Kong and Chinese press. In 1985, the Hong Kong paper *Ming Bao* reported that a meeting of the Central Military Discipline Inspection Commission noted that army units had 'reaped fat profits buying up materials of which the state is extremely short and illegally trading in cars, rolled steel, color TV sets and other commodities in short supply'. False accounting practices and outright swindles were also listed.[57] Denunciations of 'illegal businesses' continue to appear in the Chinese press. In 1988 *Liberation Army Daily* warned that active-duty soldiers should not participate in businesses run by family and friends.[58]

Finally, in 1989 the Central Military Commission decreed that the army should not engage in 'pure business operations',[59] and in that same year the official China News Agency, Zhongguo Xinwenshe, laid down the new rules: 'The Army must not forget moral principles at the sight of profits and do things that are beneficial economically but disadvantageous politically'. Among the activities banned are: illegal hoarding and profiteering, importing banned goods, misappropriation of equipment for 'speculations, profiteering and smuggling', the lending or hiring out of military vehicles or cultivated land, the formation of companies to exploit special tariff privileges, and the operation of companies by officers and men while on active duty.[60]

[54] FBIS, 19 Dec. 1988, p. 40.
[55] FBIS, 1 Apr. 1982, pp. 1 ff.
[56] FBIS, 15 Sep. 1981, pp. K12–13.
[57] FBIS, 26 Feb. 1985, pp. W3–4.
[58] FBIS, 16 May 1988, p. 30.
[59] Xinhua news agency, in FBIS, 12 Apr. 1989, p. 27.
[60] FBIS, 2 May 1989, p. 116.

Summary

Perhaps the main reason for these mixed results and problems is a lack of strategic clarity in the effort. It is not clear whether the Chinese authorities want to *convert* defence industries—that is, use existing defence industry assets to produce civilian goods—or to allow these industries to *diversify*—that is, to make new investments (or acquisitions) to move into the civilian market. As shown above, in the economic free-for-all unleashed by the Deng reforms, it would seem that the defense industries and PLA Departments are doing both. Furthermore, regardless of the strategy, the leadership seems be hold the comfortable belief that somehow the advanced capabilities of the defence industries can be transferred or spun off easily to the civilian sector. The evidence is, in fact, that spin-off strategies are deeply flawed and are a misapprehension of technological and industrial reality. Defence technologies and manufacturing standards are, after all, highly specialized, simply may not respond to civilian market needs, and may require substantial and difficult modification before they can be applied to civilian production.[61] In any case, in China the debate continues.

In late 1991 COSTIND sponsored a conference on the conversion of military industrial technology, and many of the papers by Chinese participants were frank about the problems—bureaucratic, managerial and technical—to be overcome. One paper advocated setting up small, entrepreneurial, risk-taking enterprises for the 'exploitation and trial production of new type products', along with the establishment of enterprise groups, strengthening the responsibility and contract systems, and limiting the size of staff. Another paper criticized the military industries for wastage of energy and materials, sluggish investment, long R&D times and overall high costs.[62]

It is clear that effective transformation and reform will require greater efforts. Only by considering the political ramifications of casting China's huge military–industrial complex onto the mercies of the market can the difficulties facing China's reformers perhaps be appreciated.

Employment in the defence sector

The precise number of people employed in the Chinese defence industry is unknown, but, as with most issues involving people in China, even conservative estimates will be large. State Statistical Bureau figures show that workers in state-owned industry represent about one-third of the workers in all state enter-

[61] See Alic, J. A., *et al.*, *Beyond Spinoff* (Harvard Business School Press: Cambridge, 1992) for an incisive critique of spin-off strategies.

[62] Li Xiping and Zhang Heng, 'Analyses and suggestions on the policies for civilianizing the war industry' and Jin Zhude and Chai Benliang, 'Strategic thinking on China's conversion in the 1990's', both papers delivered at the International Conference on International Cooperation in Peaceful Use of Military Industrial Technology, Beijing, 1991. See Wulf, H. *et al.*, 'Arms production', *SIPRI Yearbook 1992: World Armaments and Disarmament* (Oxford University Press: Oxford, 1992), chapter 9, for a useful summary of the papers.

Table 14.6. Work-force statistics for China, 1990

Figures are in millions.

Type of work-force	Number employed
Total number of workers	**567.4**
In state-owned enterprises	140.6 (24.8% of total)
In industry	96.97 (17% of total)
In state-owned industry	43.64 (31% of workers in state enterprises; 45% of industrial workers)
Scientists and technicians	1.517
In state-owned organizations	1.483 (97.8% of scientists and technicians)
Scientific researchers	1.196
In state organizations	1.187 (80% of scientific researchers)

Source: *China Statistical Yearbook, 1991* (State Statistical Bureau: Beijing, 1991).

prises and just under one-half of all industrial workers. Defence industry workers are contained within that set (see table 14.6).

If we take the 1986 comment by a Shaanxi official that 20 per cent of the work-force of 'enterprises under public ownership' in that province worked in the defence sector and mechanically extend that proportion to the country as a whole, using the base of workers in state-owned industry, we arrive at an estimate of 9 million in the defence industry; if we use the base of workers in all state-owned enterprises, we arrive at an estimate of 28 million. Since these figures differ by a factor of three (or more), it is clear that they are simply uncertain estimates.

The present author suspects that even the lower figure is too high, since we would have to subtract the many non-industrial employees in industry (including teachers, nursery attendants, medical staff, gardeners, housing supervisors, etc.) and the number of personnel devoted to non-aerospace production (up to 80 per cent in some aircraft factories, according to Jane's.[63]

Even a realistic estimate would be in the range of 3–5 million. This is suggested by what has been revealed about parts of the military–industrial complex. For instance, in 1987 Li Peng is reported to have said that about 500 000 workers were employed in the Aviation Ministry;[64] as noted above, the civilian goods arm of the aerospace industry (CASCPC) employs 800 000 people.

The much smaller number of scientists (including engineers) and technicians employed in the PRC should also be noted: 1.5 million, or less than 1 per cent of the total work-force (compare this with the 9.3 million employed in government, party and social organizations, which comprises 1.6 per cent of the work-force). Richard Suttmeier estimates that (in 1988) 384 000 scientists and engineers were spread out over 5275 state-run research institutes, of whom about 90

[63] *Jane's All the World's Aircraft 1990–91* (Jane's Information Group: Coulsdon, 1990), p. 38.
[64] FBIS, 2 Feb. 1987, p. K22.

per cent were involved in R&D. Another 541 000 scientists and engineers were in universities.[65] Given the technology-intensive nature of the defence industries it is likely that much of the R&D performed in China has a national security application and that many if not most of the scientists and engineers not accounted for above—about 575 000—are directly employed in the military–industrial complex. Since many if not most of the R&D institutes are defence-related, it may be reasonable to suggest that about 750 000 scientists and engineers are involved in defence work of some kind.

In other words, the military and, to a lesser degree, the educational sectors have had first call on scientists and engineers. The argument could be made that the Chinese defence establishment has siphoned off the technical talent, both workers and engineers, needed to develop the civilian economy—a point implicit in the Deng regime's push towards civilian production and, more generally, one common in discussions of defence expenditure in the West.[66] Indeed, given the the condition of Chinese civilian industry, the unavailability of technical expertise would have a disproportionately negative impact. This factor undoubtedly contributes to the regime's insistence on civilian conversion.

Arms exports: size and customers

Involvement in the world arms trade was also part of the Chinese military–industrial complex's response to the Deng reforms of the 1980s.[67] According to a report in a Hong Kong political magazine, a PLA General Logistics Department (GLD) document indicates that in 1990 Chinese arms exports were valued at $3.2 billion, 2.6 times greater than the 1980 value of $1.2 billion.[68] (It should be noted here, however, that this figure greatly exceeds the estimates for the same year available from other sources; SIPRI, for instance, values 1990 sales at $953 million. SIPRI calculates value on deliveries of major conventional weapon systems only; the GLD figure, which may not be accurate, may include small arms and probably represents the value of contracts signed.)

[65] Suttmeier, R., 'China's high technology: programs, problems and prospects', *China's Economic Dilemmas in the 1990s* (note 30), p. 548.

[66] For views on this point see Gansler, 'Defense spending and the economy', *Affording Defense* (note 38), chapter 4, pp. 79–94.

[67] It is difficult to be precise about Chinese arms exports. The Chinese media are silent on the subject; even the seemingly authoritative *Survey of World Military Industries* omits any discussion of the subject. Most information on the topic comes either from the SIPRI data base or from US Government sources, mostly importantly ACDA. ACDA data correspond to, but do not exactly match, the SIPRI count. For instance, SIPRI estimates total Chinese sales for 1984–88 to be worth $9.75 million (as expressed in 1990 US dollars); ACDA puts the figure at $9.34 (expressed in current US dollars). The number of systems transferred also does not match. In general ACDA dollar numbers tend to be higher than SIPRI's; SIPRI's system numbers tend to be higher than ACDA's. For a general discussion of China's arms trade, see Kan, S., 'China's arms sales: overview and outlook for the 1990s', in *China's Economic Dilemmas in the 1990s* (note 30).

[68] Cheng Zihua, 'Zhonggong de Qiangpao Waijiao' [Chinese Communist's Foreign Arms Traffic], *Jiushi Niandai* [The Nineties], May 1991, p. 52. $3.2 billion would represent 5% of total Chinese exports for the year; this proportion matches ACDA calculations.

In previous years China had been an arms exporter, but on a small scale. If the arms were not grants, then they were sold at 'friendship prices'. The Chinese leadership viewed arms sales as a political tool and regularly denounced the superpowers for their leadership in the arms trade. In the 1960s and 1970s, Chinese arms went to Egypt, North Korea, Pakistan, Sub-Saharan Africa and Viet Nam. Between 1976 and 1980, according to US Arms Control and Disarmament Agency (ACDA) figures, Chinese arms exports were valued at $812 million—less than 2 per cent of Chinese exports and about 0.7 per cent of the world trade in arms. A fairly wide spectrum of weapon systems were exported (jet fighters, tanks, artillery pieces and naval vessels) but, with the exception of Pakistan and North Korea, the number of systems transferred were relatively small.[69]

In the 1980s this changed. The restructuring of the Chinese economic system and the Open Door Policy provided the policy base. The 1980–88 Iran–Iraq War provided the opportunity for China's military entrepreneurs. According to the SIPRI data base on the value of major weapon systems exported from China over the period 1982–91, roughly calculated to be $15.3 billion, China's six leading customers were Iraq, Iran, Pakistan, Saudi Arabia, Egypt and Thailand (see figure 14.1 and table 14.9 below).

In other words, by value, the two combatants absorbed almost 40 per cent of China's arms exports; adding the two neighboring Arab states, which may have helped supply Iraq and which certainly felt at risk during the war, then almost two-thirds of China's arms exports during the decade can be attributed either directly or indirectly to the Iran–Iraq War. How much leakage there was from Pakistan, with whom China has had long-standing military relations (based on mutual hostility towards India) is not known.

The major weapon systems transferred during the Iran–Iraq War, according to the SIPRI data base, are listed in table 14.7.

While the Iran–Iraq War was undoubtedly a major opportunity for China's arms traders, it would be misleading to concentrate on that single conflict for the following reasons.

1. Although China was a major supplier to the conflict, it was not the only one—the USSR and some NATO countries also were involved; indeed, the conflict served as a cover for a major US political scandal, the so-called 'Iran–

[69] Arms Control and Disarmament Agency, *World Military Expenditures and Arms Transfers* (ACDA: Washington, DC, 1980). See also the discussion in Frankenstein, J., 'People's Republic of China: defense industry, diplomacy and trade", in Katz (note 13); and the more up-to-date account in Bitzinger, R. A., *Chinese Arms Production and Sales to the Third World* (A RAND Note), (RAND Corporation: Santa Monica, Calif., 1991). R. Bates Gill, in a comprehensive review of Chinese arms exports, makes the point that even in this early period, in rank order terms China was the 'fifth largest supplier of weapons to the *developing world*' for the period 1951–79' (emphasis added). See Gill, R. B., 'Curbing Beijing's arms sales', *ORBIS*, summer 1992, pp. 379–95. Exactly how one evaluates this depends on how one segments the market (world-wide, developing world or pariah states), one's policy preferences and one's appreciation that orderings of arms sales, plotted by value, take the shape of a highly skewed distribution in which the sales of the leaders far exceed those farther down the list. For instance, if we look at SIPRI figures cited by Gill for the period 1976–79, the value of Chinese major weapon exports to the developing world totaled $1.01 billion—just 3% of Soviet sales and 3.7% of US sales to the same market.

Table 14.7. Major weapon systems transferred by China to parties in the Iran–Iraq War

Weapon system	To Iran	To Iraq
B-6 bombers	. .	4
F-6 fighters	30	50[a]
F-7 fighters	. .	80[a]
Main battle tanks (T-59, -69)	740	2 800
APCs	300	650
Major artillery	620	. .
Multiple rocket launchers	800	. .
Anti-tank missiles	6 500	. .
Shore-to-ship missiles	148	72

[a] Via Egypt.

Source: SIPRI data base.

Contra' affair, in which money for arms funnelled to Iran in contravention of US law was used to fund the activities of the anti-Sandinista forces in Nicaragua.

In a world-wide context, while China may rank among the top five arms suppliers to the parties in the conflict, the number of systems and their value are small compared to those of the top players. ACDA figures for the period 1984–88, during the Iran–Iraq War, are presented in table 14.8.

Table 14.8. Arms suppliers to parties in the Iran–Iraq War, world-wide sales, 1984–88

Figures are in current US$ billion.

Supplier	Value of arms transfers	Share of world total
USSR	101.2	*40.7*
USA	59.5	*23.9*
France	18.1	*7.3*
China	9.3	*3.7*
United Kingdom	7.3	*2.9*
FR Germany	6.7	*2.7*
Czechoslovakia	5.7	*2.3*
Poland	5.6	*2.2*
World total	**248.4**	*85.7*

Source: Arms Control and Disarmament Agency, *World Military Expenditures and Arms Transfers* (ACDA: Washington, DC, 1980).

2. Perhaps the most important reason is that, while China's total arms sales have declined—in 1987, the peak year, China sold $2.9 billion; in 1991 sales were valued at half that amount—it seems clear that China has also found new markets, closer to home, demonstrating China's role as a regional player. In other words, China is diversifying its market for arms. Table 14.9 shows the

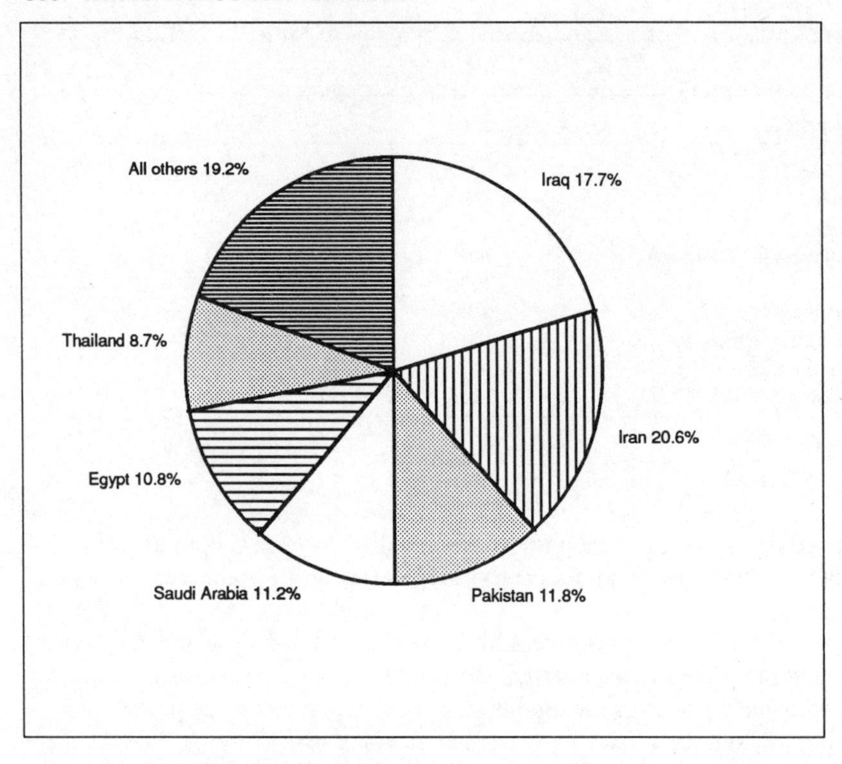

Figure 14.1. Shares of the recipients of Chinese arms, 1982–91
Source: SIPRI data base; see also table 14.9.

decline of Middle East sales (the peak for Egypt can be accounted for by arms destined for Iraq and by the sale of naval vessels, which included submarines), the relative stability of sales to Pakistan, and the rise of sales to Thailand—to sales levels close to those of Iran and Iraq earlier in the decade—and Burma.

It should be noted that Pakistan is a special case. China may have been involved with a nuclear weapon programme there; and in terms of conventional arms, Pakistan and China have developed a jet trainer, and Pakistan has licensed manufacture of the T-69 main battle bank (MBT) and has acquired one Han Class nuclear submarine. At the same time, Pakistan has received up to 150 each of the A-5 and F-7 jets, plus about 1000 T-59 MBTs. [70]

A new market appears to be building in Thailand, a front-line state in the conflict over Cambodia: sales include 59 tanks, 770 APCs, and a small number each of artillery pieces and multiple-rocket launchers. Thailand has reportedly been able to conclude some very good deals with the PRC: according to a RAND Corporation report, it has been able to purchase an advanced Chinese MBT, the T-69, for $300 000—one-fifth of the cost of the older T-59.[71]

[70] Data from Bitzinger (note 69).
[71] Bitzinger (note 69).

Table 14.9. Value of Chinese arms sales and arms recipients, 1982–91

Figures are in current US$ million. Countries are ranked by aggregate value of imports.

Recipient	1982	1983	1984	1985	1986	1987	1988	1989	1990	1991	1982–91
Iraq	249	626	550	550	474	516	186	0	0	0	**3 150**
Iran	144	156	132	368	513	649	477	60	127	77	**2 704**
Pakistan	87	197	255	181	57	197	57	231	261	281	**1 804**
Saudi Arabia	0	0	0	0	0	858	858	0	0	0	**1 715**
Egypt	292	249	721	240	143	0	0	0	0	0	**1 646**
Thailand	0	0	0	27	5	211	35	196	261	586	**1 322**
Korea, N.	135	30	21	26	175	175	231	7	0	0	**802**
Bangladesh	24	129	95	16	0	0	0	233	121	0	**619**
Zimbabwe	0	4	5	50	46	139	0	42	0	1	**288**
Burma	0	0	0	0	0	0	0	20	99	103	**222**
Sri Lanka	0	0	0	0	2	85	0	14	0	59	**161**
Tanzania	93	0	56	0	0	0	0	0	0	0	**149**
Albania	0	0	56	84	0	0	0	0	0	0	**140**
USA	0	0	0	0	0	0	64	64	0	0	**128**
Afghanistan	5	16	18	18	16	16	16	16	0	0	**120**
Sudan	0	0	0	0	0	50	0	5	0	11	**67**
Algeria	0	0	0	0	0	0	0	0	64	0	**64**
Yemen, N.	0	0	0	0	0	0	0	42	0	0	**42**
Romania	14	15	0	0	0	0	0	0	0	0	**29**
Zaire	14	0	0	0	0	10	0	0	0	0	**24**
Kampuchea	0	0	0	0	0	0	2	0	18	0	**20**
Guinea	7	5	0	0	0	0	0	0	0	0	**12**
Sierra Leone	0	0	0	0	0	10	0	0	0	0	**10**
Somalia	10	0	0	0	0	0	0	0	0	0	**10**
Peru	0	0	0	0	0	0	0	0	0	7	**7**
Oman	0	6	0	0	0	0	0	0	0	0	**6**
Laos	0	0	0	0	0	0	0	0	2	0	**2**
Guinea Bissau	0	0	2	0	0	0	0	0	0	0	**2**
Nepal	0	0	0	0	0	0	2	0	0	0	**2**
Nicaragua	0	0	0	2	0	0	0	0	0	0	**2**
Zambia	0	2	0	0	0	0	0	0	0	0	**2**
Congo	0	1	0	0	0	0	0	0	0	0	**1**
Chile	0	0	0	0	0	1	0	0	0	0	**1**
Total	**1 074**	**1 436**	**1 911**	**1 562**	**1 431**	**2 916**	**1 930**	**930**	**953**	**1 125**	**15 273**

Source: SIPRI data base.

Involvement with China, of course, also lessens Thailand's dependence on the USA. In only a few years Thailand overtook North Korea; over the period 1989–91, Thailand took 11 per cent of Chinese arms exports and North Korea took 5 per cent.

Another way to look at arms sales, however, is by product, as shown in table 14.10.

Table 14.10. Selected best-selling Chinese weapons and the major recipients, 1982–91

Weapon system	No. sold	Major recipients
Aircraft		
A-5 fighter-bombers	94	N. Korea, Bangladesh
F-6 fighters	237	N. Korea, Albania, Egypt/Iraq
F-7 fighters	178	Pakistan, Bangladesh, Egypt/Iraq
Armour		
MBTs (T-59, -69)	4 553	Iraq, Iran, Pakistan, Thailand, Burma
APCs	1 720	Iran, Thailand
Artillery and missiles		
Artillery	784	Iran, Thailand
MRL systems	1 270	Iran, Afghan Mujahideen
Silkworm missiles	316	Iran, Egypt, N. Korea, Pakistan
ATMs (Red Arrow 73)	6 570	Iran, Afghan Mujahideen

Note: MBT, main battle tank; APC, armoured personnel carrier; MRL, multiple rocket launcher; ATM, anti-tank missile.

Source: SIPRI data base.

Motives

What motivates China to move into the arms market, and why would anyone buy weapons that, by most accounts, are hardly state-of-the-art? China's rationale includes both political and economic factors. By ACDA's calculations, arms exports represent roughly 5–6 per cent of total Chinese exports in the 1980s. Since foreign exchange earnings can now be retained, these funds would presumably be funnelled back into the PLA or military industries. It also conforms to the conventional business wisdom that suggests that one seeks overseas markets to utilize excess capacity (or push unsold products) resulting from slackened domestic demand. In fact, in an interesting twist, at least one Chinese journal has suggested this, remarking that it is better to use excess military capacity to manufacture arms for export than to use that capacity for the manufacture of civilian goods for the domestic market, because the latter strategy brings the military industries into competition with the civilian sector.[72]

Involvement in the international arms trade can also be a prod to upgrade technologies. Even if the mode of technology transfer is not optimal—as, for instance, in the use of Western electronics in the F-7 Airguard export fighter—some learning and increase in technological capabilities can occur.

Politically, of course, the arms trade has often been used to curry influence. In China's highly politicalized system, politics is never very far from the surface, even if it may be temporarily overshadowed by economic considerations.

[72] Chen Hua, 'A preliminary analysis of the export of Chinese military goods', *International Trade Journal* [Guoji Maoyi Wenti], in FBIS, 31 May 1989, p. 36.

Chinese arms sales to Thailand and to Saudi Arabia would seem to fit in this category—Saudi Arabia recognized the PRC several years after the Chinese sale of missiles to the kingdom.

Why would Chinese arms be attractive on the international market? Since most are based on 1950s technology, they are, with a few exceptions, hardly competitive in terms of performance or quality with arms from the West or the former Soviet Union. China might find the post-Gulf War market drying up.

In a sense, however, these disadvantages also make Chinese arms attractive, particularly to customers in the less developed world. Chinese arms are relatively inexpensive, they are relatively simple technologically—that is, they are technologically 'appropriate'—and they are supplied with few if any strings attached.[73]

In sum, as RAND analyst Richard Bitzinger has suggested, Chinese arms, which essentially add to defensive rather than offensive capabilities, may comprise the low end of a 'high–low' defence strategy. Bitzinger cites a number of Chinese customers that appear to be following this procurement strategy: Pakistan and Egypt (both of which have F-16s and Mirages as well as J-7s), Iraq (which had Soviet T-72s), and Thailand (with its large US-supplied inventory).

In the last analysis, however, it does appear that China attaches great importance to economic considerations, and the kind of opportunistic behavior seen in domestic Chinese business practice over the past 10 years is reflected in China's overseas arms dealing. The PRC supplied both sides of the Iran–Iraq War and has moved—perhaps in response to competitive pressure on its more conventional arms sales from Brazil and other less developed suppliers—into more controversial areas. China has raised Western hackles with its manœuvring around compliance with international agreements on nuclear weapon and ballistic missile proliferation. Chinese sales of chemicals, nuclear weapon-related material and missiles to the states at the centre of potential world flashpoints—North Korea, Iran, Iraq, Libya and Syria—are seen as destabilizing, and an appropriate cause for potential trade and political sanctions from the West.[74]

The military budget

There are different estimates of the cost of defence in China. Official Chinese figures show that while the defence budget increased in an absolute amount in current yuan over time, its proportion of government expenditure and the gross national product (GNP) dipped throughout most of the 1980s. Since 1989 there have been real increases in the defence budget, even though the increases are tempered somewhat by inflation (see table 14.11).

[73] An F-7 fighter sells for about $3 million a copy; a US F-16, admittedly a much superior aircraft, has a unit cost in the range of $15–20 million.

[74] See Timmerman, K. R., 'China's comrades in arms', *Asian Wall Street Journal*, 3 Apr. 1992, p. 8.

Table 14.11. Chinese military expenditure, 1950–92

Year	Share of state expenditure	Share of the GNP	Military spending (in current b. yuan)	% change
1950	*41.1*	. .	2.8	. .
1952	*32.9*	. .	5.8	. .
1957	*18.1*	. .	5.5	. .
1965	*18.6*	. .	8.7	. .
1978	*15.1*	*4.7*	16.8	. .
1979	*17.5*	*5.6*	22.3	*32.7*
1980	*16.0*	*4.3*	19.4	*–12.9*
1981	*15.1*	*3.5*	16.8	*–13.4*
1982	*15.3*	*3.4*	17.6	*5.1*
1983	*13.7*	*3.0*	17.7	*0.4*
1984	*11.7*	*2.6*	18.1	*2.1*
1985	*10.4*	*2.2*	19.2	*6.0*
1986	*8.6*	*2.1*	20.1	*4.8*
1987	*8.6*	*1.8*	21.0	*4.4*
1988	*8.1*	*1.6*	22.0	*4.0*
1989	*8.3*	*1.6*	25.2	*15.4*
1990	*8.9*	*1.7*	29.0	*15.3*
1991[a]	*9.5*	*1.7*	32.5	*12.1*
1992[a]	*9.5*	*1.8*	37.0	*13.9*

[a] 1991 and 1992 calculations are based on estimates.

Sources: State Statistical Bureau, *China Statistical Yearbook* (various editions), (State Statistical Bureau of the People's Republic of China: Beijing, various years); 1992 budget figure from *Renmin Ribao* [People's Daily], (Beijing), 4 Apr. 1992, p. 4.

A major question yet to be answered is how these funds are allocated. A publication of the Chinese State Statistical Bureau gives a highly imprecise definition of military expenditure, although the order may give a clue as to the relative proportion taken by each activity: 'living expenses, equipment and installations, as well as the maintenance and management expenses of the military forces, construction expenses of national defence works, military building and educational expenses, expenses of military scientific research and various undertakings, combat, militia and special project expenses'.[75]

US Government analysts routinely suggest that Chinese military expenditure is at least double the public figure, with additional funds coming from other departments, arms sales and other commercial activities, including the PLA's own sideline support activities. They note a decline in operating costs—mostly pay—of about 20 per cent through the 1980s, although this allocation remains the largest single expense line in the military budget. They suggest that arms procurement has also declined somewhat—about 10 per cent—as the PLA waits for more advanced weapons to move into production. R&D funding,

[75] Cited in *China Report 1949–89* (note 13), p. 406

however, has increased by about 25 per cent; just the same, these analysts point out, R&D takes up only about 12 per cent of the budget.[76]

Given the Chinese penchant for secrecy on military matters, it seems unlikely that a more detailed picture of Chinese military spending can be derived than that presented above, although from time to time it is possible to get glimpses of the allocation process; for instance, according to a 1988 Hong Kong press interview with the Director of the PLA General Logistics Department, the PLA makes up 30 per cent of its operating expenses from local commercial activities.[77] Even if the military can get real increases in its budget—as it has over the past several years—budget growth will be remain subject to inflationary pressures, constraints imposed by Chinese Government deficits and uncertain and variable income from the commercial activities of the military industries.

Arms sales apparently do not strongly affect the defence budget. Even the high figure of $3.2 billion for arms sales in 1990 represents only roughly 6 per cent of the public number for military expenditures that year. If we accept the assumption that actual military expenditures are twice the public number, the contribution falls about 3 per cent. Furthermore, it should be recognized that the arms sales figure would simply be gross income; since the cost of goods sold is unknown, it is impossible to calculate the earnings (income or profit) realized. (Given the nature of a Stalinist economic system, with its subsidies, long depreciation schedules and 'just in case' inventory system, it may well be that no one can figure the profit, at least as it would be defined in the West.)

In addition, it is not known where the money goes. If all the funds were dedicated to R&D (and we assume that R&D takes about 12 per cent of military expenditures), the amount would represent a non-trivial fraction (25–50 per cent) of that account. However, this estimate is based on a number of conjectures, including the important one that the production of these exported goods is completely subsidized, thereby allowing this use of the money. It should also be acknowledged that the various ministries and bureaucratic subunits that make the systems exported undoubtedly have claims on the funds. The likelihood is that the money ends up being widely (and thinly) dispersed throughout the military–industrial system and probably has less of an impact than it could have if it were concentrated in a few research institutes.

Indeed, numerous commentaries, both foreign and Chinese, have noted a problem with the dispersed nature of Chinese R&D. To counter this, the leadership has instituted a 'Leading Group' for electronics (headed by Premier Li Peng) and initiated programmes that are intended to concentrate R&D efforts—the '863' programme (conceived of in March 1986)—and to accelerate the industrial application and commercialization of those efforts—the 'Torch Plan'.[78] However, funds have to be allocated across more than 5200 state

[76] See for instance, Harris, J., 'Interpreting trends in Chinese defense spending', in *China's Economic Dilemmas in the 1990s* (note 30), pp. 676–78; 'The Chinese economy in 1988 and 1989' (Directorate of Intelligence, CIA: Washington, DC, Aug. 1989), p. 17.

[77] Cited in Latham, 'China's defense industrial policy: looking toward the year 2000' (note 13), p. 87.

[78] See the following chapters in *China's Economic Dilemmas in the 1990s* (note 30): Baark, E., 'Fragmented innovation: China's science and technology Reforms in retrospect'; Suttmeier, R., 'China's

research institutes and 1000 institutions of higher learning across China; one-third of '863' funding goes to biological research, about one-quarter each to materials and communications, just over 10 per cent to energy and 5 per cent to automation. The intense vertical integration of the highly bureaucratized Chinese S&T (science and technology) and industrial system intensifies the competition for scarce resources and will only serve to impede attempts at collaboration and integration of these R&D efforts.

What is next?: 'high technology with special Chinese characteristics'

The future direction of Chinese military industrial production and R&D will be affected by several factors. As the PLA continues to professionalize, and as its leadership sheds the old 'rifles plus millet' people's war mentality, the PLA will increase its demands on the military–industrial complex for systems that can cope with 'warfare under modern conditions'. Indeed, even those latter conditions have been redefined, away from a massive central nuclear conflict and invasion to intense but local and limited conflict. China has further identified the loci of these likely conflicts as being along China's 'strategic borders': both land borders and territorial seas and islands. In order to prevail, many analysts suggest that China will have to upgrade force projection and logistics capabilities. Other forms of conflict the Chinese say they must prepare for include surprise air attacks and 'punitive counterattacks'. China's most recent experiment with 'punitive counterattacks' came, of course, in 1979 in its border war with Viet Nam—a war that revealed serious weaknesses in equipment, command and control, intelligence and logistics and which, indeed, has helped the PLA to reframe its requirements.[79]

However, in the early 1990s, problems along China's land borders do not appear to pose major dangers to Chinese national security. The Russian—or Commonwealth of Independent States (CIS)—menace along the northern border is minimal. To be sure, the spread of racialist nationalism and Islamic fundamentalism in the former Soviet republics along China's borders may pose a threat to Beijing's perception of China's territorial integrity, and questions about the Russian Federation's stability may also be a cause for concern. Indeed, there have already been disturbances along the border in Xinjiang, raising concerns that—as was once thought during the Cultural Revolution—the Uighers would rather join their Kazakh brethren on the other side of the border than remain subject to the Han of the PRC.

Just how serious a danger to Chinese national security—as opposed to internal order—these problems are seen to be is not known. There have been a few comments in the press about the desirability of a strong border defence in the

high technology: programs, problems and prospects'; Simon, D., 'China's acquisition and assimilation of foreign technology'.

[79] In *China's Economic Dilemmas in the 1990s* (note 30), see Godwin, P., 'Chinese defense policy and military strategy in the 1990s', and Skebo, R., *et al.*, 'Chinese military capabilities: problems and prospects', in *China's Economic Dilemmas in the 1990s* (note 30).

west. While in the past these discussions were covers for esoteric political discussions, revolving around whether the Self-Strengtheners—for whom read contemporary Chinese reformers—sold the country out to foreign interests, these articles appear to be more comments on the news than masked polemics. In any case, existing forces and technology should be adequate to cope with these internal border areas (unless Kazakhstan or Mongolia takes the nuclear road).

Paradoxically, it is along China's eastern borders that there is an uncertain security situation combined with great opportunities for economic development. There is the hot-and-cold status on the Korean peninsula, where the North appears to be preparing for a transition in leadership and a new, less hostile, relationship with the South. China has a Korean minority, and it seems unlikely they would trade PRC rule for that of the Kim family in Pyongyang. At the same time, the so-called Tumen River Economic Zone, which would link Vladivostok in the Russian Maritime Provinces with cities in China's industrialized north-eastern provinces and North Korea—and which would attract capital from all over Asia, including Japan and South Korea—gives everyone an incentive to avoid armed conflict.[80] The establishment of diplomatic relations between South Korea (an important trading partner and regional economic power) and the PRC in August 1992, while further isolating North Korea, should increase China's interest in regional stability.

Similarly, other potential conflict points present economic opportunities. Here we must look at 'Greater China'—the potent combination of Hong Kong, Taiwan and the southern coastal provinces of China that has produced the fastest growing economy in the world. To be sure, 'economic rationality' is a weak predictor of Chinese political behavior. The fires of irredentism and of old political rivalries burn hard and deep, particularly with respect to Taiwan. PRC military action against the island—a short and limited 'punitive attack', say, arising from growing Taiwanese independence sentiment—is not out of the question and perhaps even within China's current air and naval strike capabilities, but the international reaction would undoubtedly be even more severe than the moves taken in response to the 1989 Tianamen Square tragedy.

The most immediate area of concern, however, is considerably more complex. In legislation passed in February 1992, China formally claimed large areas of the South China Sea and the Diaoyutai Islands north of Taiwan as national territory. Both Taiwan and Japan have claims on the Diaoyutai; Brunei, Indonesia, Malaysia, the Philippines, Taiwan and Viet Nam have conflicting claims on the Spratleys and Paracels in the South China Sea. The real issue is not the islands themselves, but rather the economic potential—oil and gas—of both areas. To back up its claims, China has stationed troops in the Spratleys, placed territorial markers on some of the islands and even built an airstrip in the Paracels. There have been armed clashes with Viet Nam in the region and reports that China has fired on Japanese trawlers. Regional concerns

[80] *The Economist*, 21 Mar. 1992, p. 26.

over tensions in the area prompted the drafting of a Declaration on the South China Sea at the July 1992 ASEAN (Association of South-East Asian Nations) ministerial meeting in Manila.[81]

Thus China's national security strategies are shifting seaward, and one can expect to see continued Chinese naval development. Indeed, over the years China has increased the size and capabilities of its navy—particularly in submarines and naval missiles—and has developed it into a force that, if it hardly is a match for the US Seventh Fleet, is superior to other regional naval forces. Just the same, the requirements of operating in the South China Sea pose problems. For instance, without in-flight refuelling, China's land-based naval aviation has extremely limited 'time over target' capabilities; the Paracels airstrip probably does not provide enough additional reach. China's recent purchase of Su-27 jet fighters from Russia in part may have been stimulated by this concern.[82] The purchase of an aircraft-carrier from Ukraine, the subject of some press speculation, would fit this strategy, although it must be pointed out that China would not only have to outfit the ship and acquire the appropriate aircraft (helicopters and V/STOL [vertical/short take-off and landing] aircraft) but also have to develop a cadre of naval aviators—certainly a very long-term and expensive, budget-draining proposition. If one takes a longer view, one would have to conclude that the reversion of Hong Kong, one of the world's busiest ports, to PRC control in 1997 could only heighten China's concern with maritime security.

Reactions to the Persian Gulf War

We can get another picture of the future as seen by Beijing from Chinese reactions to the Gulf War. That conflict certainly would seem to fit Chinese conceptions of limited war. Two basic themes emerge: the overwhelming importance of high technology and China's need to modernize its weapons, and the continuing importance of the 'human factor'.

A March 1991 *Liberation Army Daily* article, 'New mechanist weapons and the war of the future',[83] describes new and emergent military technology—if it is not, in some way, an indication of what the Chinese military industry is working on, it certainly shows that the Chinese have been paying attention. The article highlights directed laser and particle beam weapons, hyper-velocity, high kinetic energy projectiles, systems using artificial intelligence and biological weapons, and their development in the USA and the USSR. These are, of course, not incompatible with the broad technological areas emphasized

[81] See 'Worries about Chinaa', *Asiaweek*, 7 Aug. 1992, pp. 2—24.

[82] See Tai Ming Cheung, 'Loaded weapons: China on arms buying spree in former Soviet Union', *Far Eastern Economic Review*, 3 Sep. 1992, p. 21. Chinese interests here reportedly include MiG-31 fighters, Tu-22 bombers, missiles and over-the-horizon radars, plus the Soviet military industrial technologies. No doubt other deals, both in counter-trade for needed consumer goods and for cash, are in the works; and, given the entrepreneurial zeal shown by the Chinese military–industrial complex, it would not be surprising to find that some of these Russian systems are resold to Chinese clients.

[83] FBIS, 20 Mar. 1991, pp. 31–33.

in the '863' and 'Torch' programmes—communications, materials, automation and electronics.

While giving lip-service to the possible civilian spill-overs, the article uses an echo from a poem by Mao to underline the urgency of advanced military R&D: 'In the early 21st century, the extensive deployment of weapons with new mechanisms . . . will naturally become an indispensable and important means to ensure a big power's status, enhance strategic deterrent power, and protect one's security. . . . The research and development of high technology weapons has a long cycle and it is necessary to make early decisions and seize the time in order not to lose the initiative.'[84]

Other post-Gulf War official commentary has covered the full range of conventional weapons. In April 1991 Beijing radio quoted the deputy commander of the PLA Air Force on increased research into 'high-power and low-energy-consumption turbojets' and 'high-reliability microelectronics'.[85] Several articles have mentioned the importance of and need for advanced armour. Conferences convened by the PLA and industrial ministries have also covered such subjects, plus improved naval artillery and anti-ship guided missile defence systems.[86] The many defence technologies mentioned by the press to be under development or recently put into operation include robotics for munitions manufacture, radar/video anti-aircraft fire control systems, training simulators of all kinds, navigation aids, communications and computer security devices. Such a review gives the impression of a large amount of R&D activity covering the full spectrum of defence requirements. The real issue, however, must be the capacity and the resources to move the hardware from the test bench to production and deployment.

At the same time, it is clear that Chinese military planners understand that there is more to advanced military technology than hardware. A February 1990 article in *Liberation Army Daily* noted that the 'competition of military power . . . [is] essentially the competition of science and technology' and that the 'domain of science and technology for national defense' includes 'not only the two major objective factors, "persons and things," but also the intermediate objective factor, i. e. . . . management (including organization, education, policy decision, advice and so on).'[87]

Other commentary highlights the urgent importance of the 'human factor' and often uses the old Maoist phrase that 'man is the decisive factor'. Rather than praising ideological awareness (although certainly not denigrating it), however, it is technical expertise that is key. In April 1991 *Liberation Army Daily* ran a series of articles entitled 'High tech war and military strategy' that stressed the 'higher demands' put upon 'commanders in formulating measures and drawing up tactics' by technology.[88] The key to the future will be 'to train talented

[84] See note 83.
[85] FBIS, 8 Apr. 1991, p. 7.
[86] *China Ship Engineering*, 7 Apr. 1991.
[87] FBIS, 2 Mar. 1990, p. 26.
[88] FBIS, 29 May 1991, pp. 47 ff.

people so that they can cultivate the correct ideas for combat, acquire good scientific and cultural knowledge, and attain high qualifications in tactics and skills. This is . . . a task of top priority.' Indeed, another *Liberation Army Daily* article, published in July 1991, stated that 'a commander without modern science and cultural knowledge is doomed to failure'.[89] This is a far cry from finding the solution to every problem in the thought of a political sage.

Following the conclusion of the war, the Central Military Commission issued instructions to all PLA departments, academies and institutes to study the war, its strategy and technology. According to Hong Kong press reports, a Chinese National Defence University study of the war concluded that the PLA had to improve troop quality, give priority to conventional weapon R&D and form quick-response units. Other areas needing emphasis included night combat, force mechanization, electronic warfare and long-distance logistics.[90] Indeed, the old argument about developmental priorities appears to have re-emerged. *Da Gong Bao*, a PRC mouth-piece in Hong Kong, reported in April 1991 high-level meetings at which 'the Chinese leaders' high regard for the pressing need to develop high technology' for defence was expressed. The newspaper quoted PLA Chief of Staff Chi Haotian to the effect that the Gulf War demonstrated 'the importance and urgency of stepping up national defense building and raising the level of defense modernization simultaneously with the development of China's national economy.' [91]

Other reports add to the sense of unease that the Gulf War may have engendered among China's military leaders. The Hong Kong magazine *Cheng Ming*, a journal with excellent sources in the PRC, ran an article in May 1991 entitled 'Old generals complain about Deng delaying improvement of weapons, equipment' which cited the 'military's sense of danger'.[92] The article alleged that military leaders like former Defence Minister Zhang Aiping had criticized Deng Xiaoping's restrictions on the military and quoted a letter from former PLA Chief of Staff Yang Dezhi to Party Chairman Jiang Zemin: 'Compared to [the West], we lag behind at least 30 years!'. The piece also quoted an air force officer who said that if China were subjected to the same kind of electronic warfare deployed by the United States in the Gulf, 'China . . . would become a second Iraq!'.[93]

The article also reported that at these meetings General Nie Li, COSTIND Vice Chairman and daughter of old Marshal Nie Rongzhen, noted the ineffectiveness of Iraq's anti-aircraft missiles against US forces. *Cheng Ming* commented, 'What she did not say . . . was: Iraq's air defense weapons and systems are basically supplied by China. Nie Li's remarks indirectly mean that China's air defense capability cannot cope with the high-level capacity of an air force like that of the United States.'[94]

[89] FBIS, 9 Aug. 1991, pp. 32–33.
[90] *South China Morning Post*, 13 Mar. 1991, p. 7; *Wen Wei Bao*, 12 Mar. 1991.
[91] In FBIS, 21 May 1991, p. 44.
[92] In FBIS, 7 May 1991, pp. 51 ff.
[93] See note 92.
[94] See note 92.

While such 'insider' information is, of course, impossible to verify, it seems clear that the events in the Gulf accelerated and gave greater impetus to Chinese defence efforts. 'Heated discussions' had already been held among defence planners about the need to upgrade and expand the technical capability and high-tech production of the defence industries, and the war could only have further stimulated these discussions.[95] The establishment of the China Electronics Industry Group—'Chinatron'—in June 1991, which essentially breaks out the old Electronics Industry ministry from the MMBEI, may have been stimulated by the demonstration of the gap between Chinese military electronics and those of the West. The Xinhua news agency, reporting on the corporation in August 1991, said that its aim was to 'equip the nation's military with modern electronic products in the next five years'.[96]

If the high-technology component of Chinese military strength is to increase, however, funds have to be found. One way to increase these funds is by increased economic activity by military entities, by moving up-market in the production of civilian goods for export and by increasing military sales. Thus we have seen considerable diversification and what one might call 'bureaucratic entrepreneurialism', with military units investing in hotels, retail stores and even high-priced Hong Kong real estate.[97] Just the same, as a PLA researcher remarked to the present author in 1992, it would be 'naive' to think that all the funds so generated went to cover budget shortfalls.

Another, if less-favoured, way for the PLA to improve its technology is to buy it. One can thus expect to see the continued, if limited, involvement of foreign military suppliers in China, particularly in naval and aircraft systems.

Some recent China press comment mentions an industry-rationalizing 'second development' in the drive to use the defence industry for civilian production, the gist of which appears to be to concentrate on, if not give priority to, civilian products for which the defence industry has particular advantages, such as motor vehicles.[98] This would move the industry away from its previously haphazard approach, which may have produced profits, but which also used resources inappropriately and brought it into competition with already established civilian enterprises. Along with this are calls for more economic efficiency and changing the defence industry from a 'closed, "internally-oriented" system' to an 'open "externally-oriented, development-type"' industrial system.[99] Other developments along these lines, which mirror developments in the

[95] *Jiefang Jun Bao* in FBIS, 7 Sep. 1990, pp. 41 ff.

[96] FBIS, 15 Aug. 1991, p. 33.

[97] See 'The long march downtown', *Asiaweek*, 31 July 1992, p. 27, on PLA interest in real estate in central Hong Kong. Not all such PLA activity is as spectacular; for instance, a PLA motor repair vehicle unit seen in Beijing in June 1992 has gone into the commercial garage trade and advertises itself with a charmingly primitive billboard of a Mercedes sports car and the promise that it will fix any kind of imported automobile.

[98] See, for instance, the commentary 'Second development of China's weapons industry', *Liaowang*, July 1991, in FBIS, 16 July 1991, pp. 44 ff. However, counter-arguments are floated which put the emphasis back on military production. Here, see 'Military civilian integration is a major development of defense economics', *Liberation Army Daily*, 30 Jan. 1991, in FBIS, 25 Feb. 1991, pp. 34 ff.

[99] A long example is 'Modes of development for the military industry' from *Jingji Guanli* [Economic Management], in FBIS, 21 Aug. 1989, pp. 39 ff.

civilian sector, include the creation of corporations that cross bureaucratic boundaries: in 1989 the first of these, the Zhongshan Group Defence Equipment Company, was set up in Nanjing—it combines 19 units from all the military industrial ministries.[100]

China can also be expected to continue to sell arms. News reports suggest that, as in the past, China's military entrepreneurs will continue to move into appropriate technology niches—as with the joint Sino-Pakistan jet trainer—and to continue to supply regimes at the margins of international respectability—such as the military rulers of Myanmar. While it is not the intention to apologize for Chinese behaviour, it is evident that the military–industrial bureaucracies, like other parts of the Chinese system under the reforms, have attained a certain degree of independence that the centre might not be able to rein in; indeed, it is not only the Chinese Government that has problems with free-lancing agencies. Although China has pledged to honor international agreements on nuclear weapon and ballistic missile proliferation, shipments through third countries (such as North Korea) are always possible. In any case, the agreements themselves have loopholes. For instance, the 1987 Missile Technology Control Regime (MTCR) applies only to missiles with ranges longer than 300 km and a payload above 500 kg; China could ship 'downgraded' missiles which could be 'upgraded' once delivered. The issue for China is whether the political flak from the West is a bearable cost for such activities—history would suggest that it is.

Finally, the matter of China's nuclear capabilities should be addressed. To begin with, China has both fission and fusion weapons. It has a small number of land-based strategic missiles, some undoubtedly with MIRV (multiple independently targetable re-entry vehicle) capabilities, which have the range to reach European Russia and the western United States, as well as 'tactical' nuclear weapons. China has also test-launched missiles from the Xia Class ballistic-missile submarine. The latter system would require solid fuels and sophisticated navigation capabilities. It appears that China's nuclear programme as a whole is domestic, even if early model IRBMs were derived from Soviet originals.

While it would appear that China has opted for a strategy of minimum deterrence, China's accomplishments in nuclear weapon development is not inconsiderable. Even one deployed SSBN would certainly be of major concern to countries along the Pacific Rim. Although the kinds of conflict the PLA sees in the future do not apparently involve or would be appropriate for nuclear weapons, Beijing's planners cannot have overlooked the possibility of conflicts along China's periphery 'going nuclear'—say, the India–Pakistan conflict—an event which would be of profound strategic concern to China (and not only the PRC). Thus a roll-back in the Chinese nuclear weapon programme seems unlikely.

[100] Reported by *Liberation Army Daily*, in FBIS, 17 Nov. 1989, p. 35.

Feng Gueifen's vision?

It seems clear that if China's leaders are divided on strategies for economic development, they are united in the need for having a China that is militarily self-sufficient. History teaches China that the world is a dangerous place and that a major seat in the concert of nations is available only to those who are strong enough to maintain their place, even if global competition is turning from a political and military confrontation to an economic contest. The world may applaud China's integration into the world system, and we realistically cannot expect China to forgo playing the power game.

At the same time, there are those in the Chinese military leadership who see China playing a weak hand in that power game. There are difficult systemic problems to be overcome. A 1990 *Liberation Army Daily* editorial, reflecting the views of people like retired Defence Minister General Zhang Aiping and undoubtedly others, noted several 'contradictions' in technology, investment, management and talent that would confront China 'for a fairly long time':

1. '[T]he high demands of modern warfare on the advanced performance of weaponry, and [China's] relatively low level of science and technology for national defense';
2. The need for investment for national defence science and technology, and the lack of funds;
3. The need for 'improving investment returns', and 'relatively low management level and the imperfect management structure';
4. The need for highly qualified scientific and technical talent, and the 'lack of incoming competent personnel'.[101]

It should not be overlooked that the Chinese military industrial system is embedded in an industrial structure that is, if not in crisis, certainly putting a heavy drain on the state. As Vice Minister Zhu Rongji, the former mayor of Shanghai and now the Vice Minister charged with revitalizing the state sector, pointed out in the autumn of 1991, 'the number of money-losing state enterprises accounts for 40 per cent, of which a considerable portion is military enterprises'.[102]

Despite these serious issues, we can expect the large Chinese defence industry and R&D effort, even if reduced in immediate military output, to continue. The discussion over management reform and the appropriate use of these considerable economic assets will persist, just as the deeper, historically and culturally conditioned arguments over the value of expertise and the propriety of foreign input will feature in policy debates. In the short term, Feng Gueifen's vision may not be realized. In the longer term—after, as the current phrase has it, '100 years of reform'—it might be.

[101] FBIS, 2 Mar. 1990, pp. 26–27.
[102] FBIS, 3 Oct. 1991, p. 29.

15. Japan: a latent but large supplier of dual-use technology

Masako Ikegami-Andersson

I. The historical background

Although Japan is often cited as an example of a country with a restrained and small arms industry, the Japanese arms industry has a long and tortuous history, starting with a prominent attempt at militarization in the second half of the 19th century. A full comprehension of the Japanese arms industry would be insufficient without an account of pre- and post-World War II Japanese history.

The pre-World War II period

In the beginning of the Meiji era (1867–1911), under the threat of Western imperial powers, the Government conducted an extensive programme of introducing modern Western industrial technologies through major Japanese government-run firms. The government arsenals not only produced arms but also provided civilian industries with the most advanced machinery available in Japan. Compared with government industries, private companies such as Mitsubishi, Kawasaki and Ishikawajima were still small and their technological capacity was underdeveloped.[1] In the early 20th century, at the end of the Meiji era, the government revised its policy in order to promote indigenous industrialization on the basis of private enterprises, and some major government-run factories were transferred to private ownership.

Japan's arms industry was considerably expanded during the Sino-Japanese War (1894–95) and the Russo-Japanese War (1904–05),[2] in which the two victories made Japan one of the major military powers at that time. Private heavy industries had grown up, since the government supported them as 'latent arms industries' or 'key industries', that is, heavy industries such as iron and shipbuilding industries, which were basically civilian but critical for the military infrastructure.[3] Before World War II, Japan caught up with the military

[1] Koyama, Hirotake, *Nihon Gunji-Kojo no Shiteki Bunseki* [A Historical Analysis of Japanese Arsenals], (Ochanomizu-Shobo: Tokyo, 1972).

[2] The Russo-Japanese War, in which naval battles determined the tide of the war, fuelled the international warship-building race. Japan was also pressured to develop warships indigenously, despite its financial crisis. Muroyama, Yoshimasa, *Kindai Nihon no Gunji to Zaisei* [Military and Finance of Modern Japan], (Tokyo University Press: Tokyo, 1984), pp. 368–70.

[3] In 1910 the policy for promoting indigenous production was implemented by the Ministry of Agriculture and Industry, the predecessor of MITI. Johnson, C., *MITI and the Japanese Miracle: The Growth of Industrial Policy 1925–1975* (Stanford University Press: Stanford, 1982). See also Nakaoka, T.

technology of the Western developed countries; for example, it started to produce aircraft in the 1920s. The military granted orders to some major private companies on a competitive basis, and these companies manufactured major weapons designed by the military. As a result, most of the major arms industries, such as Mitsubishi and Kawasaki, developed their technological capacity and increased their production of weapons. The previous dominance of national arsenals in the period from the 1870s to the 1920s was reversed to the relative superiority of the private sector in arms production.[4] During World War II, 80–87 per cent of total state expenditure was allocated to the military, and in this way Japan managed to produce, for example, more than 20 000 fighter aircraft per year, including the well-known Zero aircraft.[5]

The post-World War II period

Japan's defeat in World War II brought large-scale disarmament to Japan under the control of the GHQ (General Headquarters of the US occupational army); that is, all the weapons and most of the surviving military facilities were either dismantled or requisitioned by the Allies or converted into civilian industries. The GHQ policy of dismantling large financial combines (*Zaibatsu*) affected the leading major arms industries such as Mitsubishi. In the aircraft industry, research and production of aircraft were totally prohibited, and all the facilities for research and development (R&D) and manufacturing were requisitioned or abolished. This demolished the pre-war form of aircraft industry in Japan. None the less, a significant portion of the surviving industrial facilities of the private sector was maintained, because of the revision of the US policy towards Japan during the cold war: from the 'demilitarization of Japan' to 'stabilization of the Japanese economy' in order to make Japan a 'subordinate and complementary industrial mobilization basis for the US military force in the Far East'.[6] Eventually, quite a few industrial facilities, engineers and skilled workers were inherited by the post-war Japanese civilian industry. The 1950–53 Korean War was a shifting-point from the disarmament to the rearmament of Japan, and the Japanese arms industry started to supply arms for the *Tokuju* (special demand) of the US military forces stationed in Japan. These US military orders (e.g., for repairs, vehicles, bullets and rocket control systems) were given directly to Japanese industries without authorization by the Japanese Government.[7] In

et al., *Kindai Nihon no Gijutsu to Gijutsu-Seisaku* [Modern Japan's Technology and Technological Policies], (Tokyo University Press: Tokyo, 1986).

[4] Koyama (note 1), pp. 191–97.

[5] Koyama (note 1).

[6] Kubo, Shinichi, 'Reisen-Taisei no Saiken to Keizai no Gunji-ka' [Reconstruction of the cold-war regime and militarization of the economy], in ed. K. Miyamoto, *Nihon Shihon-shugi Bunseki* [An Analysis of Japanese Capitalism], (Otsuki: Tokyo, 1981). This policy was explicitly stated in Jan. 1948 by Secretary of the Army K. C. Royal in San Francisco, stating that the policy towards Japan should be revised in order to make Japan a 'bastion against Communism in the Far East' by promoting Japanese economic independence. Fujiwara, Akira, *Nihon Gunji-shi, Sengo-hen* [A history of the Japanese military: the post-war era], (Nihon-Hyoron: Tokyo, 1987).

[7] Sakai, Akio, *Nihon no Gunkaku Keizai* [The Japanese Militarizing Economy], (Aoki: Tokyo, 1988).

1952 the GHQ formally admitted the reopening of arms and aircraft production facilities in Japan.

The economic boom during the Korean War offered an opportunity for major Japanese arms industries to re-establish their production basis, and in 1953 they organized a lobbying group, the Defense Production Committee (KDPC), with about 80 member companies, of *Keidanren* (the Federation of Economic Organization, Japan's largest business organization, founded in 1951). During the economic recession after the Korean War, most of the small military-related industries were eliminated, and major arms industries secured a position as rationalized and oligopolistic arms producers.

Restrictions on arms production in Japan

Japan is 'the only advanced industrial nation to renounce unilaterally both the export of weapons and the projection of military power in international affairs'.[8] After World War II, restrictive conditions were imposed on the expansion of arms production in Japan.

Japanese pacifism and Article 9 of the Constitution

Grave war experiences and post-World War II democracy brought a strong sentiment of pacifism among the Japanese people. Article 9 of the 1947 Constitution, the 'no war clause' that renounces the use of force to settle international disputes, is still strongly supported by the public. In a 1989 opinion poll conducted by the Information Department of the Prime Minister's Office (*Sorifu, Kohoshitsu*), 64 per cent of the respondents named the 'peace Constitution' as a cause of the post-war peace in Japan, whereas only 8.8 per cent named the Self Defense Force.[9] This national consensus is an ideological basis for restrictive arms production and export policies as well as for strict prohibitions on the development and possession of nuclear, biological, and chemical weapons.

Arms export restrictions

After the Korean War, Japan exported ammunition and small arms on a small scale. However, during the Viet Nam War, when Japan was an important rear base for US military forces, Japanese sentiments of pacifism were highly intensified. With such sentiments and criticism of the Japanese Parliament in mind, then Prime Minister Eisaku Sato in 1967 enunciated three restrictions on arms exports which reinforced rules that were already contained in the Export Trade Control Order of 1949 administered by the Ministry of International

[8] US Office of Technology Assessment (OTA), *Global Arms Trade* (OTA: Washington, DC, 1991), p. 6.

[9] In the 1984 poll by the *Asahi* newspaper, 71% preferred economic co-operation and diplomatic efforts as a way of defending Japan, rather than US–Japanese allied military forces (13%). *Seron-Chosa Nenkan* [Public opinion poll yearbook], (Information Department of the Prime Minister's Office [*Sorifu, Kohoshitsu*]: Tokyo, 1985), p. 469.

Trade and Industry (MITI).[10] In 1976, then Prime Minister Takeo Miki enunciated more strict Cabinet guidelines which implied the actual banning of the export of any arms or military-applicable equipment. In addition, in 1978 the MITI announced the reinforcement of the regulation by restricting military-related technologies based on the 1949 Export Trade Control Order and the Foreign Exchange and Foreign Trade Law.

The US–Japanese security treaties

The Japanese–US security arrangement (consisting of the 1960 Treaty of Mutual Co-operation and Security and the 1954 Mutual Defence Assistance Agreement)[11] is thought initially to have two major functions: to counter the potential threat posed by the Soviet Union and other communist powers in East Asia, and to prevent Japan from remilitarizing.[12] Under this security relationship, the Japanese defence structure has heavily relied on the USA for such things as planning, equipment and technology. Even during the 1980s, after expansion of the Japanese arms industry, Japan remained an important US arms customer.[13] The Japanese Defense Agency (JDA) and arms industry's intention to indigenously develop and produce arms was often subjected to US political pressure to purchase US weapons instead.[14] In domestic politics, the JDA was not so autonomous in relation to other ministries such as the Ministries of Finance, Foreign Affairs and the MITI.[15] With such political and financial restrictions, the role of the JDA as a sponsor of the Japanese arms industry was relatively limited.

II. Military expenditure

Share of the GNP and total government spending

In the early 1950s Japan's defence-related expenditure accounted for nearly 20 per cent of total government outlays—mostly for infrastructure for the US

[10] Drifte, R., *Arms Production in Japan: The Military Application of Civilian Technology* (Westview Press: London, 1985), pp. 73–75.

[11] The Japanese–US Mutual Security Treaty stipulates, for instance, co-ordinated and joint action by Japan and the USA in the event of an armed attack against Japan (Article 5), and granting the use of facilities and areas in Japan to US military forces (Article 6); *Defence of Japan 1988 (Defense Agency: Tokyo, 1988).*

[12] Teshima, Ryuichi, *Nippon FSX wo Ute* [Attack the Japanese FS-X], (Shincho-sha: Tokyo, 1991), pp. 290–92. In the post-cold war period, the latter function seems to have become more explicit: in Okinawa, a top US Navy general, H. C. Stackpole, said: 'US troops must remain in Japan in large part to prevent Japan from remilitarizing'. *International Herald Tribune*, 28 Mar. 1990.

[13] In 1987–91 Japan imported $9.8 billion of major weapon systems from the USA, of which 98% came from the USA. See SIPRI, *SIPRI Yearbook 1992: World Armaments and Disarmament* (Oxford University Press: Oxford, 1992), p. 312.

[14] Sakai (note 7), p. 255.

[15] Johnson (note 3); and Ikegami, M., *The Military–Industrial Complex: The Cases of Sweden and Japan* (Gower: Dartmouth, UK, 1992).

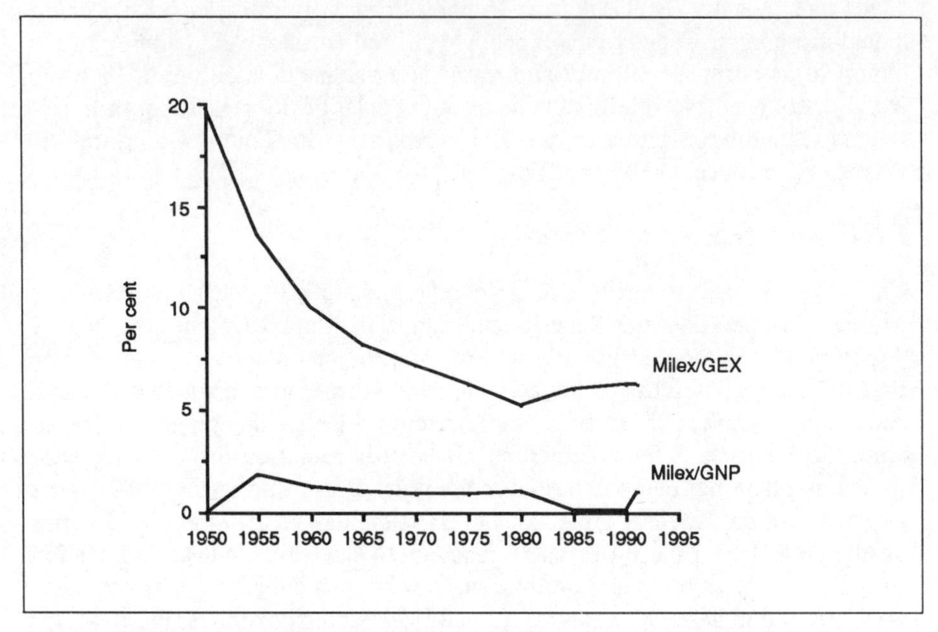

Figure 15.1. Japanese military expenditure as a share of the gross national product (GNP) and total government expenditure (GEX), FY 1950–91

Source: Boei Handbook 1991 [Defence Handbook 1991], (Asakumo Shimbun: Tokyo, 1991).

occupational army.[16] During the 1960s, in the midst of Japan's rapid economic growth, the defence budget was annually increased by over 10 per cent, although it accounted for only approximately 1.2 per cent of the gross national product (GNP).[17] In 1976 the Miki Cabinet set a ceiling of 1 per cent of the GNP as a government guideline for defence expenditure, given the economic and financial crisis in public expenditures. This ceiling was initially a decision made for economic rather than political reasons. However, it has gained such a wide national consensus that, even though the Nakasone Cabinet broke the 1 per cent ceiling guideline in 1987, Japan's later defence expenditures have been maintained at approximately 1 per cent of the GNP.

In order to strengthen its self-defence capability, in 1976 Japan formulated the National Defence Program Outline (*Boei Taiko*, which defined the country's defence goals; that is, defence forces capable of countering limited and small-scale aggression) and allowed great latitude for future development. Furthermore, in 1981 Japan announced that it would protect its territory and air/sea lines of communications (SLOCs) out to 1000 nautical miles. Based on the commitment to Japanese–US defence sharing, Japan has strengthened its own self-defence capability, especially through modernizing weapons and

[16] In the early 1950s, most of the military expenditures were for infrastructures for the US occupational army. They are therefore called 'defence-related' expenditures. Sakai (note 7), pp. 60–62.

[17] Kubo (note 6); *Boei Handbook 1991* [Defence Handbook 1991], (Asakumo Shimbun: Tokyo, 1991).

Table 15.1. Japanese military expenditure (original budget), FY 1950–91

Figures are based on budgeted expenses, in hundred million yen, at current prices.

Fiscal year	Military expenditure	Fiscal year	Military expenditure
1950	1 310	1971	6 709
1951	1 199	1972	8 002
1952	1 771	1973	9 355
1953	1 257	1974	10 930
1954	1 396	1975	13 273
1955	1 349	1976	15 124
1956	1 429	1977	16 906
1957	1 435	1978	19 010
1958	1 485	1979	20 945
1959	1 560	1980	22 302
1960	1 569	1981	24 000
1961	1 803	1982	25 861
1962	2 085	1983	27 542
1963	2 412	1984	29 346
1964	2 751	1985	31 371
1965	3 014	1986	33 435
1966	3 407	1987	35 174
1967	3 809	1988	37 003
1968	4 221	1989	39 198
1969	4 838	1990	41 593
1970	5 695	1991	43 860

Source: *Boei Handbook 1991* [Defence Handbook 1991], (Asakumo Shimbun: Tokyo, 1991).

defence systems. In the 1980s the Japanese military budget increased in nominal terms by an average of 6.5 per cent annually.[18]

Reflecting the new international environment after the end of the cold war, the military budget will be restricted in the New Medium-term Five-Year Defence Build-Up Plan (*Shin Chuki-bo,* for fiscal years (FY) 1991–95). In the previous Medium-term Plan (*Chuki-Bo*), the defence budget increased annually by 5.4 per cent on average, calculated in constant 1985 prices, whereas in the new plan, *Shin Chuki-Bo,* the average annual increase ratio is 3.0 per cent, calculated in constant 1990 prices.[19] However, the JDA will continue qualitative improvement of the Self Defense Forces to prepare for 'unpredictable crises situations in the future'.[20]

[18] *Boei Handbook 1991* [Defence Handbook 1991], (Asakumo Shimbun: Tokyo, 1991).

[19] *Boei Hakusho 1991* [Defence of Japan 1991] (Japan Defense Agency: Tokyo, 1991), p. 130. In the FY 1991 defence budget, 4 386 000 million yen is approved—an increase of 5.45% over 1990. The increase ratio is higher because of settlement of contracts from previous years. The FY 1991 defence budget was cut by 100 billion yen to help finance Japan's $9 billion contribution to the Allied forces in the Gulf War.

[20] *Jane's Defence Weekly*, 17 Aug. 1991.

Military equipment acquisition

In Japan, the share of military procurement and particularly of R&D in total military expenditure was exceptionally low among the Group of Seven (G7) countries.[21] The KDPC has been lobbying to increase the share of procurement and R&D to the level of the major Western states. In the 1980s the shares of procurement and R&D grew significantly: the ratio of capital expenditure (the sum of procurement, R&D expenses and facilities/installations improvement expenses) to the total defence budget rose from about 20 per cent in the 1970s to over 30 per cent in the late 1980s.[22] However as the basic defence capabilities outlined in the 1976 *Boei Taiko* plan were achieved in the previous *Chuki-Bo* plan (1986–90), the new *Shin Chuki-Bo* plan (1991–95) will focus on improving the defence infrastructure and on personnel needs rather than new front-line equipment. Particularly front-line procurement expenditures will decrease annually by 2.3 per cent on average, calculated in constant 1990 prices, during the *Shin Chuki-Bo* period; during the *Chuki-Bo* period the annual increase ratio was 7.7 per cent on average, calculated in 1985 constant prices.[23] In the five-year period 1991–95, two major co-production programmes will end—the McDonnell Douglas/Mitsubishi Heavy Industries F-15J and the Lockheed/Kawasaki Heavy Industries P-3C programmes—which will have a significant effect on the Japanese arms industry.[24]

Research and development

R&D expenditure, that is, the budget of the JDA Technical Research and Development Institute (TRDI),[25] has prominently increased since the mid-1980s. In 1968–84 the defence and R&D budgets increased annually by an average of 12.9 per cent and 10.8 per cent in nominal terms, respectively. In 1984–90 the R&D budget increased by an average of 15.3 per cent annually, while the defence budget's average annual increase ratio was 6.0 per cent.[26] Accordingly, the R&D share in the total defence budget has risen to 2.3 per cent in FY 1991. As the *Shin Chuki-Bo* plan focuses on replacement and modernization of the defence forces, the R&D budget will continue to grow, giving priority to qualitative improvement of major equipment such as intelligence processing and communication systems and the development of new guided missiles and the FS-X fighter aircraft. Since the 1980s the increasing R&D budget has allowed more investment for basic research on advanced and

[21] Canada, France, Germany, Italy, Japan, the UK and the USA.

[22] Keidanren, Defence Production Committee, *Defence Production in Japan* (Keidanren: Tokyo, Oct. 1991).

[23] *Boei Hakusho 1991* [Defence of Japan 1991] (note 19), pp. 130–31. For instance, the procurement number of the F-15 will decrease from 63 (in the *Chuki-Bo* period) to 42 (in the *Shin Chuki-Bo* period), and the P-3C from 50 to 8, respectively.

[24] OTA (note 8), p. 111.

[25] This is the budget for research, development, testing and evaluation (RDT&E).

[26] *Boei Nenkan 1991* [Yearbook of Defence 1991], (Boei Nenkan Publishing Co.: Tokyo, 1991), p. 549.

Table 15.2. Japanese military acquisition and R&D expenditure and total military expenditure (original budget), FY 1973–92

1973	1974	1975	1976	1977	1978	1979	1980	1981	1982	1983	1984	1985	1986	1987	1988	1989	1990	1991
Military R&D as share of military expenditure (%)																		
1.3	1.1	0.9	0.9	0.9	0.9	1.0	1.0	1.0	1.1	1.1	1.2	1.6	1.7	1.9	2.0	2.1	2.2	2.3
Military acquisition as share of military expenditure (%)																		
25.4	22.9	19.0	16.4	17.4	17.1	18.7	20.7	22.5	22.4	24.9	26.3	26.2	26.9	27.5	28.1	28.0	27.4	27.7
Maintenance as share of military expenditure (%)																		
14.2	14.2	14.4	14.4	14.5	14.5	13.9	14.1	14.7	15.8	16.2	15.5	15.1	14.4	14.2	14.1	14.2	16.1	15.9
Base countermeasures (%)																		
7.5	8.8	8.6	8.2	8.0	8.7	10.2	10.4	10.5	10.4	10.0	9.7	9.5	9.0	9.4	9.2	9.5	9.8	9.7
Supplies, total value (hundred million yen)																		
5 009	5 634	6 252	6 647	7 602	8 665	10 180	11 302	12 556	13 808	15 284	16 252	17 232	18 350	19 736	21 215	23 063	24 913	26 293
Share of military expenditure (%)																		
53.5	51.5	47.1	44.0	45.0	45.6	48.6	50.7	52.3	53.4	55.5	55.4	54.9	54.9	56.1	57.3	58.8	59.9	59.9
Personnel, provisions (hundred million yen)																		
4 346	5 296	7 201	8 477	9 304	10 345	10 765	11 000	11 444	12 053	12 258	13 094	14 140	15 086	15 439	15 789	16 136	16 680	17 568
Share of military expenditure (%)																		
46.5	48.5	52.9	56	55	54.4	51.4	49.3	47.7	46.6	44.5	44.6	45.1	45.1	43.9	42.7	41.2	40.1	40.1

Note: Figures are based on budgeted expenses: hundred million yen, at current prices. The acquisition budget includes costs for purchasing ammunition, vehicles, aircraft and shipbuilding.

Sources: *Boei Handbook 1988* and *1991* (Asakumo Shimbun: Tokyo, 1988 and 1991).

sophisticated technologies which are not always developed or procured as equipment.[27]

Despite the rapid increase of Japan's defence R&D expenditure since the 1980s, it is still significantly low compared to the other G7 states. The R&D budget for FY 1991 (original) was 102.9 billion yen ($791 million)—an increase of 10.4 per cent over 1990 in nominal terms. Comparing figures for 1989, this is only 1.7 per cent of US defence R&D expenditure and 40 per cent of that of West Germany.[28] In terms of the Government's R&D budget per agency, the TRDI budget accounted for 5.7 per cent of the total Government R&D budget in FY 1991 and only 0.85 per cent of Japan's total R&D expenditure in 1989.[29] However, the TRDI budget significantly benefits from extensive Japanese investment in commercial R&D (8.2 trillion yen in FY 1989),[30] much of which is in advanced dual technologies. TRDI's strategy is to stretch its relatively modest resources by cultivating promising technologies already under development in the private sector.[31] For instance, in the case of the development of active phased-array radar by Mitsubishi Electronics Co. (MELCo) for the FS-X fighter aircraft, the JDA did not pay for the arsenide chip technology or production process development, which lowered unit costs. However, TRDI has supported radar technology R&D at MELCo on a modest level since 1973.[32] TRDI's modest but steady support for military application, enhanced by commercial R&D of the underlying technologies, is a feature of defence R&D spending in Japan.

III. The Japanese arms industry

Overview of the major arms industries

Mitsubishi Heavy Industries (MHI) is called the 'Japanese Arsenal', which indicates the prominent character of MHI among Japanese arms manufacturers: (*a*) MHI is the only comprehensive supplier of major weapons for all the divisions of the Japanese Self Defense Forces (Air, Maritime and Ground); and (*b*) it is an exclusively dominant contractor which has won between 20 per cent (1987) and 28 per cent (1990) of total contracts awarded by the JDA in the period 1987–90. If its sister company, Mitsubishi Electric (MELCo), is included, the Mitsubishi Group shared between 27.4 per cent (1987) and 34.5 per cent (1990) of total JDA contracts in FY 1987–90.[33] MHI is a dominant contractor of most of the licensed production of major weapons (e.g., F-104J,

[27] *Boei Nenkan* (note 26), p. 561.

[28] Keidanren (note 22).

[29] *Indicators of Science and Technology 1991* (Japan Science & Technology Agency: Tokyo, 1991), p. 125.

[30] *Kagaku Gijutsu Hakusho* [White Papers on Science and Technology], (Japan Science and Technology Agency: Tokyo, 1991), p. 354.

[31] OTA (note 8), p. 116; *Boei Hakusho 1990* [Defence of Japan 1990], (Japan Defense Agency: Tokyo, 1990).

[32] OTA (note 8), p. 117.

[33] Keidanren (note 22).

F-4EJ, and F-15J aircraft, and Nike J and Patriot guided missiles) and has also indigenously developed many sophisticated weapons (e.g., the T-2 supersonic trainer aircraft, the F-1 support fighter, the ASM-1 missile and the Type 74 tank). Together with the TRDI, MHI developed the Control Configured Vehicle (CCV) in 1983 and is now a prime contractor of the FS-X fighter project, in which CCV technology is to be applied. MHI military sales accounted for 18.9 per cent of its total sales in 1990.[34] During the 1980s, sales of the Aircraft & Special Vehicles Division, where most of the defence production is carried out, have expanded 2.8-fold.[35] Major systems such as the F-15J and Patriot missile kept the production line busy in the 1980s. Even under the restrictive *Shin Chuki-Bo* defence plan, the division's sales are expected to be high for the continuous production of, for example, the F-15 and SH-60J anti-submarine helicopter[36] until FS-X fighter production starts in the late 1990s.[37] However, in 1990 MHI announced plans to de-emphasize military sales in favour of commercial products, anticipating a decline in military sales from a high of 25 per cent of total sales to 15–17 per cent over a period of two to three years.[38] MHI has played a key role in Japan's space research project; for example, it was responsible for the Japanese H-I project as system integrator and major producer.[39]

Kawasaki Heavy Industries (KHI) is a prime contractor for licensed production of the P-3C anti-submarine patrol aircraft[40] and the V-107 helicopter and is also the developer of the C-1 transport aircraft, the T-4 intermediate training aircraft as well as some anti-ship/anti-tank missiles. KHI manufactures a wide variety of rotary-wing aircraft, among which are the CH-47J and BK-117 (being developed with MBB of Germany).[41] In the vessel division, KHI is one of the two producers of submarines (the other is MHI) and the only producer of gas-turbine engines (licensed by Rolls Royce) in Japan. KHI won a share of between 9.3 per cent (1990) and 13.5 per cent (1987) of total contracts with the JDA in the period 1987–90.[42] Reportedly, KHI's aerospace and engine divisions take in about 80 per cent of its revenue from the production, overhaul and repair of military aircraft, missiles and engines.[43] KHI's dependence on military sales has increased since the demand for commercial ships was

[34] Keidanren (note 22).

[35] *Asahi Shimbun, Rupo: Gunju Sangyo* [Report: Arms Industry], (*Asahi Shimbun:* Tokyo, 1991), p. 125.

[36] SH-60J is the Japanese licensed version of the US Navy SH-60B Seahawk, featuring Japanese-developed mission avionics and a lightweight dipping sonar. *Jane's Defence Weekly*, 20 Oct. 1990.

[37] *Asahi Shimbun* (note 35), pp. 128–29.

[38] OTA (note 8), p. 113; *Aviation Week & Space Technology*, 29 July 1991.

[39] 'Japan's aero industry', *Asian Defence Journal*, vol. 20, no. 10 (Oct. 1990), pp. 82–90.

[40] Before the P-3C production, KHI had already indigenously developed the next-generation PXL anti-submarine aircraft in the 1970s. However, in spite of the R&D cost paid by the JDA, it decided to purchase the P-3C from Lockeed because of political pressure from the USA, which was suspected of being concerned with the 'Lockeed Scandal' in 1972. Kamata, Satoshi, *Nihon no Heiki-Kojo* [Arms factories in Japan], (Kodan-sha: Tokyo, 1983).

[41] *Asian Defence Journal* (note 39), p. 88.

[42] Keidanren, *Defence Production in Japan* (various editions).

[43] *Aviation Week & Space Technology*, 29 July 1991, p. 49.

stagnant in the 1970s, and sales by the Aerospace Division increased fourfold in the 1980s.[44] In the space area, KHI developed and produced the experimental satellite launched by the H-I rocket and is currently a participant in the joint development of Japan's H-II project and the US National Aeronautical and Space Agency (NASA) project to construct a space station.[45] KHI, which was counting on JDA orders to provide as much as 70 per cent of its total aerospace business by the year 2000, is suffering from cutbacks in planned military orders of the coming *Shin Chuki-Bo* plan: for example, reduced production of the P-3C. Possible compensation for this would be to develop helicopters such as the OH-X[46] to develop robots for automation of its production line because of the shortage of JDA personnel, and especially to expand commercial production and international collaboration projects such as joint production/development of the BAe Airbus (A321's fuselage) and the Boeing 777 plane.[47]

Ishikawajima Heavy Industries (IHI) is a prime producer of destroyers and is exclusively engaged in the development, manufacture, repair and overhaul of various types of jet engine (e.g., engines for the F-1, F-104J, F-4EJ, PS-1, P-3C and F-15, and indigenously developed engines for the T-1 and P2-J). IHI won between 4.2 per cent (1989) and 5.9 per cent (1987) of total JDA contracts in 1987–90.[48] It is also engaged in a large-scale national project to develop the FJR710 turbo-fan engine, using Japanese technology, and a national space development (developing a turbo-pump). IHI also plans to expand its commercial engine business, such as the V2500.[49]

Fuji Heavy Industries (FHI) was founded in 1953. Its entry into aircraft technology began in 1917, with the establishment of the aircraft research laboratory that later became Nakajima Aircraft Co., a predecessor of FHI. FHI was the largest military aircraft manufacturer in pre-war Japan; it developed the first jet aircraft in Japan in 1958 and has participated in such joint development projects as the C-1 jet transport, T-2 supersonic trainer and T-4 medium trainer. FHI also played an important role in sub-contracts such as for the F-1, F-15J and P-3C as well as the Patriot system, and in prime contracts such as licensed production of the AH1S anti-tank helicopters. The Aerospace Division has long experience in R&D of advanced composite airframe structures.[50] Since FHI's automobile division is now a subsidiary of Nissan Motor, they have co-operated in developing a missile/rocket launcher system.[51]

[44] *Asahi Shimbun* (note 35), p. 126.

[45] *Asian Defence Journal*, Oct. 1990, p. 88.

[46] The OH-X is designed to replace the McDonnell Douglas OH-6 scout/reconnaissance helicopter. The JDA will begin OH-X development in 1992. KHI is doing in-house studies on it, aiming at becoming a prime contractor. *Interavia*, 22 Jan. 1991; *Jane's Defence Weekly*, 17 Aug. 1991.

[47] *Asahi Shimbun* (note 35); *Aviation Week & Space Technology*, 7 Jan. 1991, p. 35. MHI, KHI and FHI are collaborating with Boeing on development of the Boeing 777 aircraft. They will develop and produce 21% of the parts.

[48] Keidanren, *Defence Production in Japan* (various editions).

[49] *Aviation Week & Space Technology*, 29 July 1991, p. 52.

[50] *Asian Defence Journal* (note 39).

[51] Toyama, K., 'Gunkakuka no Heiki-Sangyo' [Arms industry in expansion], and 'Nihon no Heiki Kigyo 8 sha' [Japanese eight major arms manufacturers], in ed. *Heiwa-Keizai Keikaku-Kaigi* Group

Table 15.3. The 20 leading defence contractors in Japan, FY 1990

Figures in columns 4 and 5 are in billion yen. Figures may not add up to totals due to rounding.

Rank	Company	Products	Value of defence contracts (A)	Value of total sales (B)	A/B (%)	Share of prime con- tracts (%)	Total employ- ment
1	Mitsubishi Heavy Ind.	Sh, MV, Ac, Mi	441	2 327	*18.9*	*28.1*	44 272
2	Kawasaki Heavy Ind.	Sh, Ac, Mi	146	892	*16.4*	*9.3*	17 049
3	Mitsubishi Electric Co.	El, Mi	100	2 589	*3.9*	*6.4*	48 616
4	Ishikawajima-Harima H.I.	Sh, Eng	79	731	*10.7*	*5.0*	15 280
5	Toshiba Corp.	El, Mi	60	3 228	*1.9*	*3.8*	71 921
6	NEC Corp.	El	54	2 961	*1.8*	*3.5*	38 487
7	Japan Steel Works Ltd	Ar	35	133	*26.1*	*2.2*	3 912
8	Komatsu Ltd	SA/O, MV	22	674	*3.3*	*1.4*	14 921
9	Fuji Heavy Industries	Ac	22	756	*2.9*	*1.4*	15 222
10	Hitachi Ltd	EL, MV, Mi	20	3 789	*0.5*	*1.3*	81 763
11	Oki Electric Industry	El	17	582	*2.9*	*1.1*	14 111
12	Daikin Indus.	SA/O	17	344	*4.8*	*1.1*	7 569
13	Fujitsu Ltd	El	15	2 338	*0.6*	*1.0*	50 768
14	Sumitomo Heavy Indus.	Ar, Sh	15	274	*5.5*	*1.0*	6 156
15	Shimazu Corp.	Ac	15	163	*9.0*	*0.9*	4 143
16	Cosmo Oil Co.	Oil	13	1 713	*0.8*	*0.8*	. .
17	Nissan Motor Co. Ltd	Ar, MV, Mi	13	4 175	*0.3*	*0.8*	56 873
18	Mitsubishi Precision Co.	El	12	25	*48.8*	*0.8*	. .
19	Nippon Oil Co. Ltd	Oil	12	2 193	*0.5*	*0.7*	. .
20	Nippon Koki Co. Ltd	SA/O	11	17	*65.3*	*0.7*	. .
Total			**1 119**			***71.3***	

Note: Sh, ships; Ac, aircraft; Ar, artillery; El, electronics; Eng, engines; Mi, missiles; MV, military vehicles; Oil, oil; SA/O, small arms/ordnance.

Sources: Keidanren, Defence Production Committee, *Defence Production in Japan 1991* (Keidanren: Tokyo, 1991); *Japan Companies Handbook 1991* (Toyo Keizai: Tokyo: 1991).

(Planning Committee of Peace Economy), *Kokumin no Dokusen-Hakusho* [Civilian White Paper on monopoly: arms industry], (Gunju-Sangy:. Tokyo: Ochanomizu-shobo, 1983).

However, reduced military orders under the *Shin Chuki-Bo* plan have hit the company and it is trying also to expand commercial production: for example, using Boeing 777 sub-contracts to expand capabilities and markets.[52]

In addition to these major manufactures, there are others: *Sumitomo Heavy Industries (SHI)*, a relative later-comer in the Japanese arms industry after its merger with Nittoku Metal Ind. (firearms), produces destroyers, jet engine blades, aircraft parts and gun systems. *Shin-Meiwa Ind.*, which emerged from a pre-war major aircraft producer, has since 1953 developed flying-boats, making use of its pre-war experience and technology, and is the sole producer of large short take-off and landing (STOL) amphibious aircraft such as the PS-1 anti-submarine flying boat and the US-1 rescue seaplane. Finally, *Shimazu Corporation*, produces hydraulic flight control equipment, aeronautical test equipment and head-up displays.

In the field of rockets and missiles, a few major manufacturers compete: the *Mitsubishi Group* (MHI and Mitsubishi Electric), the *Mitsui Group* (e.g., Toshiba and FHI) and the *Fuyo Group* (e.g., Nissan, Hitachi, FHI).[53] The Mitsubishi Group is dominant in licensed production of missiles such as the Nike J, Hawk, Sparrow, Sidewinder and Patriot. *MELCo* is a major contractor of missiles and radar systems (the phased array radar system for the FS-X fighter) and is to develop the new medium-range surface-to-air missile (SAM) which will replace the ageing Hawk missile. The group has also indigenously developed such missiles as the AAM-1, ASM-1 and SSM-1.

Toshiba has developed a unique infra-red homing missile, the Type 81 SAM, which won the JDA contract in competition with the European Roland missile in 1980 and is now engaged in improving the SAM. Toshiba won between 3.8 per cent (1990) and 5.9 per cent (1988) of total JDA contracts in 1987–90.[54]

NEC developed a homing system for the HAMAT missile for which the prime contractor was KHI. It also supplies electronic systems such as radars, flight data-processing systems, guidance and electro-optical equipment, satellites and sub-systems. These electronics firms are likely to benefit from the shift in procurement emphasis on electronics.

Nissan Motor's Aerospace Division has played a pivotal role for over 30 years in development and manufacture in the space field (e.g., of sounding rockets and satellite launching vehicles). Nissan has produced defence rocket systems, rocket launcher guidance and control systems, and test equipment, reportedly having acquired munitions manufacturing technology from the US Martin-Marietta Corporation.[55] The coming multiple-launcher rocket system

[52] *Asahi Shimbun* (note 35); *Aviation Week & Space Technology*, 29 July 1991, p. 55.

[53] In Japan, there are six groups (*Keiretsu*) of major private enterprises in which the companies are related to each other through e.g. presidential committee meetings, mutual share-holdings of member companies' stocks and networks through main banks. These major groups are Mitsubishi, Mitsui, Sumitomo (three of which derive from former *Zaibatus*), and Fuyo, Sannwa, Daiichi-kannginn (three of which are not derived from former *Zaibatsus*). For detail, see Baba, H. (ed.), *Shirizu Sekai Keizai IV: Nihon* [Series World Economy IV: Japan], (Ochanomizu: Tokyo, 1989).

[54] Keidanren, *Defence Production in Japan* (various editions).

[55] *Pacific Defence Reporter*, Sep. 1988, p. 32.

(MLRS) project in which Nissan will for the first time be a prime contractor will elevate Nissan to the group of the 10 leading defence producers of Japan.[56] Nissan's ambition for arms production is partly motivated by the fact that aerospace technology will be applicable to automobiles in the future and partly because the automobile market is saturated world-wide today.[57]

A similar trend is found in other producers since the 1980s.[58] *Toyota* has delivered armoured personnel carriers since 1973 and from 1992 is developing armoured vehicles for the JDA.

Honda started co-operative research on aerospace technology with Mississippi University in the USA in the 1980s, financed by the Honda Foundation.

Sony has acquired the US Transcom-Systems Co. in 1989, a major producer of audio-visual systems for aircraft.[59]

Fujitsu Electric and Hitachi Ltd. also set up divisions of defence technology in the 1980s, and produces radio communications equipment, radar and infrared night vision equipment (Fujitsu), flight simulators, data link receivers and tactical trainers (Hitachi).[60]

There are a number of manufacturers of small ammunition. The *Japan Steel Works,* which emerged from the subsidiary of both a Japanese mining company and the British Vickers Armstrong Co. in 1927, and a part of the Mitsui Group produce cannon, guns and rocket/missile launchers. The *Komatu Manufacturing Co.,* an active member of the Japan Ordnance Association, produces tanks, military vehicles, rocket launchers and bullets. *Howa Ind.* is a major producer of rifles such as the Type 64.

Significance for the national economy

The Japanese arms industrial base is limited, and the national economy is not dependent on arms production. Restrictive military export regulations have been implemented much more strictly than stipulated in the formal policy, and Japanese arms producers have not enjoyed the economies of scale in arms production. The value of defence production accounted for 0.54 per cent of the total value of industrial production in 1989, although this figure increased slightly in the 1980s.

The JDA approved 1379 defence-related manufacturers in 1991: 872 of them were small- to-medium-sized enterprises, while the 20 leading producers accounted for 71 per cent of total JDA defence contracts.[61] Since arms exports are prohibited except for exports to the USA, the volume is negligible: in 1989

[56] The MLRS is another major procurement item under the *Shin Chuki-Bo* programme. Nissan wants to produce under licence from LTV more than 50% of the units, but LTV is reportedly resisting. *Defence News,* 24 June 1991; OTA (note 8).

[57] Toyama (note 51).

[58] *Nihon Keizai Shimbun,* 2–25 Jan. 1990.

[59] Sony became the 83rd member of Keidanren's Defence Production Committee in 1984. Sakai (note 7). Sony's displays are reportedly widely used on warships and aircraft of the US military forces.

[60] Sakai (note 7).

[61] *Chotatsu Jishi-Honbu no Gaikyo* [Overview of the JDA's Procurement], FY 1991 (Defense Agency: Tokyo, 1991).

Table 15.4. Changes in the amount of Japan's defence production, FY 1978–89
Figures are the shares of defence production in total industrial production and are percentages.

Fiscal year	Share
1978	0.38
1979	0.39
1980	0.36
1981	0.35
1982	0.46
1983	0.50
1984	0.48
1985	0.51
1986	0.55
1987	0.58
1988	0.54
1989[a]	0.54

[a] The figure for 1989 is based on a preliminary report.

Sources: *Defence of Japan 1988* (Defense Agency: Tokyo, 1988); *Boei Hakusho 1991* [White Paper of Defense], (Defense Agency: Tokyo, 1991).

the value of arms exports was $110 million, while that of arms imports was $1400 million.[62]

One industry, however, shows significant dependence on military production: in the aircraft industry, arms production accounts for 74.7 per cent of total aerospace industrial output.[63] Since Japan's aircraft industry was dismantled in 1945–52, the industry's growth in the post-World War II period has relied on JDA's procurement, to the detriment of commercial production. However, the Japanese aerospace industry is growing rapidly: it achieved sales of 1 trillion yen ($7.35 billion) in 1989 (but is still less than 5 per cent of US sales). Total employment in the Japanese aerospace industry rose by about 1000 employees per year in the late 1980s,[64] mainly because of the expansion of civilian production, since the aircraft industry's dependence on arms production gradually decreased in the period 1986–89.

Major Japanese aircraft manufacturers are co-ordinating consortia supported by the Government to share the risks and to eliminate duplicate research, engineering and other costs. Japanese policy is particularly directed at supporting companies in the development of civil airframe and engine programmes, but only on an international co-development basis because this is the 'only way to cope with the enormous risk and technological innovation required to stay competitive'. For example, MHI, KHI and FHI are collaborating with Boeing on the development of the 777 aircraft; and IHI, KHI and MHI are working

[62] Arms Control and Disarmament Agency, *World Military Expenditures and Arms Transfers* (ACDA: Washington, DC, 1990), p. 110.

[63] The figure appears to be declining as the industry's commercial production expands: 82.6% in 1986, 79.6% in 1987, 77.9% in 1988, and 74.7% in 1989. Keidanren, 1991 (note 22).

[64] *Asian Defence Journal*, 1990 (note 39); *Financial Times*, 11 June 1991.

Table 15.5. Japanese industry's dependence on arms production, FY 1979–89
Figures represent shares and are percentages.

Fiscal year	Ships	Aircraft	Vehicles	Arms and ammunition	Electrical communications apparatus
1979	8.34	85.04	0.07	99.72	0.56
1980	4.37	81.86	0.08	99.60	0.70
1981	4.85	77.76	0.08	99.76	0.42
1982	4.06	76.54	0.08	99.67	0.73
1983	4.38	77.10	0.10	99.80	0.92
1984	5.16	81.54	0.09	99.88	0.55
1985	4.81	83.73	0.06	99.85	0.56
1986	7.06	82.64	0.06	99.65	0.58
1987	8.26	79.56	0.09	99.93	0.67
1989	7.16	74.68	0.07	99.62	0.65

Note: Figures are based on each industry's total amount of defence production as a share of industrial output.

Sources: Keiranren, Defence Production Committee, *Defence Production in Japan 1988, 1989, 1990, 1991* (Keidanren: Tokyo, various editions); *Boei Nenkan 1991*: [Yearbook of Defense 1991], (Boei Nenkan Publishing Co.: Tokyo, 1991), p. 513.

under the Japanese Aero Engines Corp. as part of an international consortium building V2500 turbo-fan engines.[65]

Employment

Although specific statistics of employment in arms production are not publicly available, it is estimated that over 45 000 employees are engaged in arms production, which accounts for 0.42 per cent of Japan's total industrial employees (figures are for 1989).[66]

Table 15.6 shows the estimated figures for the number of employees working in the arms production segment of the various sectors of Japanese industry. Several trends are notable.

The shipbuilding industry has become increasingly dependent on arms production, particularly since the mid-1980s. This increased dependence is indicated by the fact that, whereas total employment in shipbuilding was nearly halved—from 135 000 to 72 000—in the course of the 1980s, employment in arms production remained fairly stable, fluctuating between 6753 in 1981 and

[65] *Intervia Aerospace Review*, Apr. 1991, p. 44.

[66] Author's calculation: figures are based on *Kogyo Tokei 1989* [Industrial statistics 1989], (MITI: Tokyo, 1991); and Keidanren (note 22). This estimate is based on the sum of each industrial sector's total employment multiplied by the particular sector's arms production ratio in the following production areas: shipbuilding, aircraft, vehicles, arms and ammunition, and electronic communication equipment. This calculation may overestimate the number of employees because unit costs for arms production are on average higher than the unit costs for non-military production.

Table 15.6. Estimated arms production employment figures, FY 1980–89

	1980	1981	1982	1983	1984	1985	1986	1987	1988	1989
Shipbuilding	5 895	6 753	5 559	5 461	6 453	5 677	6 119	6 548	5 532	5 172
Aircraft	15 840	17 371	17 930	17 785	18 511	20 746	20 683	21 574	21 145	19 857
Vehicles	539	560	627	699	650	459	452	668	594	532
Arms and ammunition	1 509	3 120	3 198	1 588	1 983	2 381	5 089	5 009	3 371	7 479
Electronics	9 392	6 169	10 904	14 958	9 872	10 222	10 821	12 355	14 186	12 457
Total	**33 175**	**33 973**	**38 218**	**40 491**	**37 469**	**39 490**	**43 164**	**46 154**	**44 828**	**45 497**

Source: Keidanren, *Defence Production in Japan 1988, 1989, 1990, 1991; Kogyo Tokei 1985, 1989* [Industrial Statistics 1985, 1989], (MITI: Tokyo, 1987, 1991).

5172 in 1989. The share of arms production in this sector increased to over 7 per cent in 1989, from slightly above 4 per cent at the beginning of the 1980s.

Contrary to the shipbuilding industry, the aircraft industry has had an increase in total employment as well as in arms production employment. In 1980 the number of employees in the aircraft industry working with arms production was 15 840, whereas the number had increased by almost 25 per cent to 19 857 by the end of the decade. This represents approximately three-quarters of total employment in this sector. However, it should be added that, although the number of employees working in arms production is increasing, the aircraft industry's dependence on arms production decreased in the latter half of the 1980s as a result of a relatively larger expansion of commercial production.

Ams production employment in the Japanese vehicles industry was marginal through the 1980s, never involving more than 600 employees.

In the arms and ammunition industry, the number of employees fluctuated between 1509 employees in the first year of the decade and 7479 in 1989. Only 20–30 middle- or small-sized establishments existed in this sector, but it should be noted that the trend was a significant increase in the number of employees as well as establishments. Nearly 100 per cent of the employment in this sector is dependent on arms production.

The number of arms-producing employees in the rapidly growing electronic communication equipment industry also shows an increasing trend, although 1983 was the peak year of the decade, with 14 958 employees involved in arms production, representing less than 1 per cent of total employment in this sector

The general trend for the 1980s was that employment in the arms-producing segment of the Japanese industry—although very small—had a relatively larger increase than total industrial employment. In 1980 the total number of employees in arms production was 33 175, whereas it had increased by 37 per cent to involve 45 497 employees in 1989 (see table 15.6 above).

Features of the Japanese arms industry

The particular features of the Japanese arms industry can be summarized as follows.

1. *Less dependence on arms sales*. Japanese major arms producers are not heavily dependent on arms sales in terms of the arms sales ratio of total sales compared to large US counterpart producers: for example, the corresponding figures for MHI, KHI and IHI and McDonnell Douglas and General Dynamics are: 17 per cent, 14 per cent and 8 per cent; and 55 per cent and 82 per cent (1990), respectively.[67] This is partly because of the historical and structural constraints of the post-World War II defence industry. In addition, most of the major Japanese arms manufacturers are not independent defence companies but are part of much larger diverse industrial entities which produce other major equipment ranging from heavy industrial machinery to powerplants, transportation equipment and automobiles.[68] It is not coincident that, despite the disorganization immediately after World War II, most of these major companies returned to their pre-war history as *Zaibatsu*—monopolistic conglomerate enterprises which used to dominate all the major fields of Japanese industry.

2. *Competition and co-operation among oligopolistic manufacturers*. The five leading defence companies received 53 per cent of total JDA contracts by value, and the top 20 companies received 71 per cent of them in FY 1990. Competition among oligopolistic industries seems to be common in Japan. Five companies are engaged in the manufacture of aircraft, and seven or eight companies construct naval vessels. The JDA fosters competition by designating at least two manufacturers for most items of military hardware in the definition phase. Meanwhile, the manufacturers, when not attending to their speciality, break up and share other work or rotate orders.[69] The proportion of each company's annual procurement seems to be rather stabilized by these arrangements.[70] Usually, competition occurs between different industrial groups such as the Mitsubishi and the Mitsui Group, or between traditionally dominant producers and newcomer producers. At the same time, JDA's acquisition strategy appears to favour the manufacturers forming a separate consortium for major systems, for example, to build airframes and engines.[71]

3. *Initiative of industrial associations*. Japanese arms manufacturers form influential lobbying groups such as the KDPC and *Nihon Heiki Kogyo Kai* (the Japan Ordnance Association founded in 1951, with about 110 member companies). Many of these defence-related manufacturers are also associated with industrial associations such as *Nihon Koku-Uchu Gyokai* (the Japanese

[67] See the list of 'the 100 largest arms-producing companies in the OECD and Third World countries', in SIPRI, *SIPRI Yearbook 1992* (note 13), pp. 392–97.

[68] *Asian Defence Journal*, 1990 (note 39), p. 90.

[69] For instance, submarines have been built alternately by MHI and KHI since 1960. *Jane's Defence Weekly*, 7 Jan. 1989, p. 30.

[70] *Kokumin no Dokusen-Hakusho 1983* (note 51).

[71] *Intervia*, Sep. 1986; *Asian Defence Journal*, 1990 (note 39); *Financial Times*, 19 Sep. 1990.

Aircraft and Space Industry Association) and *Nihon Zosen Kogyokai* (the Japanese Shipbuilding Industry Association). Among them, the KDPC is sometimes called the 'private version of the Defence Agency' because of its role and its influence on the JDA.[72] Their main objectives are: (*a*) to promote self-sufficiency in arms production, (*b*) to increase the share of capital expenditure in total defence spending to the level of other major developed countries—at least to 30 per cent, and (*c*) to increase the budget for military-related R&D to at least 2–5 per cent of total defence budgets. They have taken the initiative, albeit informally, in many aspects of the Japanese military buildup in the post-World War II period. For instance, the development of missile technology, one of the most advanced Japanese military technologies, was initiated by the KDPC in the 1950s.[73] In defence R&D today, the initiative of major companies is still significant, as the TRDI budget is very limited. Usually, many years before the JDA officially begins a development project, major manufacturers fund in-house studies on a new weapon system.[74] These companies also send their engineers to an informal TRDI research group, which is the actual start of the manufacturers' competition for a prime contract. This system makes it possible for the TRDI to absorb the fruits of major defence manufacturers' R&D at a very low cost. However, once certain producers become involved in a TRDI development project, they are almost guaranteed to receive the contracts for production at the procurement stage.[75] In this way, major manufacturers' R&D expenses are shifted to high production unit costs. As the JDA is the only customer and the defence R&D budget is limited, such a procurement pattern prevails.

4. *High cost and inefficiency for self-sufficiency*. JDA programmes and procurement in the 1970s and 1980s illustrated a continued drive towards autonomy in defence production and, more recently, in R&D.[76] Almost 90 per cent of defence equipment procured by the JDA is manufactured by domestic producers.

However, this figure includes other supplies such as clothing, medicine and petroleum products.[77] The domestic production share of major weapon systems

[72] In 1951 KDPC investigated the production capacity of the Japanese arms industry for the special demands of the Korean War. The report also estimated a desirable size for the Japanese defence force, which influenced the government's medium-term plan of rearmament. Yoshihara, Koichiro, 'Mitsubishi: Gunju-Dokusen Bocho no Karakuri' [Mitsubishi: the mechanism of its expansion in the monopoly of arms production], *Keizai*, vol. 19 (1980), pp. 32–47.

[73] In 1953 KDPC set up the GM (guided missile) Consultation Section in which 17 major arms manufacturers and electronic companies got together. In 1954 the group pronounced their 'Opinion on the Research of Guided Missile', to encourage the JDA, and the JDA set up the Guided Missile Research Committee within the JDA. KDPC's GM Consultation Section was developed for the 'GM Council' by incorporating the Japan Rocket Association, with 41 member companies in 1957. Kihara, M., 'Sengo-Nihon niokeru Heikiseisan to sono Tokucho nitsuite [Production of arms in post-war Japan and its characteristics], *Journal Keizai* [The Economy], vol. 197 (1977), pp. 17–31.

[74] For instance, Toshiba reportedly paid 50 billion yen in developing the Type 81 surface-to-air guided missile, while 104 billion yen was subsidized by the TRDI over 15 years. *Heiki Bijinesu* [Arms Business], (Mainichi Shimbun: Tokyo, 1982).

[75] *Asahi Shimbun*, 1991 (note 35), pp. 139–42.

[76] OTA (note 8).

[77] See note 61.

Table 15.7. Changes in the share of domestic equipment procurement, FY 1977–89

Figures are shares of domestic procurement in total equipment procurement by the Japanese Defense Agency and are percentages.

Fiscal year	Share
1977	93.4
1978	85.4
1979	85.2
1980	88.5
1981	80.5
1982	88.6
1983	90.3
1984	90.7
1985	90.9
1986	90.8
1987	91.0
1988	91.3
1989	90.4

Sources: *Defence of Japan 1988* (Defense Agency: Tokyo, 1988); *Boei Hakusho 1991* [Defence of Japan 1991], (Defense Agency: Tokyo, 1991).

is much lower. According to the JDA, 60 per cent of Self Defense Force weapons are purchased off-the-shelf from the USA, and the remaining 40 per cent are produced in Japan either under licence or domestically.[78] This relationship is expected to change in favour of autonomous production since the JDA and major arms industries have encouraged more production of major equipment. Although much of the equipment is of US origin—defence systems produced under licence in Japan—import substitution programmes have been under way since the beginning of the post-World War II period. However, low production rates prevent economies of scale and discourage investment in automation, which leads to high costs and inefficiency in Japanese arms production. For example, procurement of the Japanese-produced Type 90 main battle tank, at 1.2 billion yen ($9.6 million), is said to be almost three times the cost of the equivalent US M1A1 Abrams tank.[79]

In addition, there is no government mechanism for investigating weapon prices along the lines of the US Office of Management and Budget or the British Defence Select Committee.[80] Nevertheless, 'Japan seems to be willing to support a costly yet modest defence industry that does not depend on exports for survival'.[81] Several independent R&D projects have been launched, aimed at self-sufficiency in complete systems and enhancing negotiating leverage *vis-à-vis* the USA and other potential foreign partners: for example, medium-range surface-to-air missiles to replace the US-designed Hawk and computers to

[78] *Defence News*, 3 Dec. 1990, p. 42.
[79] *Jane's Defence Weekly*, 17 Aug. 1991, p. 283.
[80] See note 79.
[81] OTA (note 8), p. 115.

replace IBM computers in the F-15 fire control system.[82] The JDA says that technological capability to develop and produce advanced weapon systems is significant for deterrence and that, when they have the capability to develop a sophisticated weapon system, they can purchase the equivalent system from foreign suppliers at a much lower cost.[83]

5. *The dual-use nature of Japanese technology*. Although the Japanese defence industrial base is limited, Japan's strength in many dual-use technologies makes the situation more complex. The Japanese Government has promoted R&D activities for advanced technologies such as aerospace, advanced materials, superconductivity and artificial intelligence. These are regarded as key technologies which promote the entire Japanese industry and provide important bargaining power for national security. For example, in 1970 the MITI designated aerospace as one of three key technologies for the 21st century; the then director general of the JDA, Nakasone, proposed in a document of 1970 that Japan's industrial base should be maintained as a key factor in national security.[84] Many of these targeted technologies are dual-use. According to the US Department of Defense, of the 20 'critical technologies', as defined by the US Pentagon, at least 15 were dual-use and Japan was a leader in five of the technologies.[85] The dual-use nature of Japanese technology and high degree of diversification of major manufacturers blur the borderline between commercial and military products. As common for the major producers, the work-force moves flexibly between commercial and military lines, sharing equipment and know-how.[86] Many Japanese companies have become suppliers of critical technologies and products applicable for producing weapons. For instance, Japanese industries are important sources of field-effect transistors (FETs) of both types (silicon and gallium) used in high-frequency radar, and Japanese electronics companies dominate the US market for silicon FETs used in 'smart bombs'.[87] The dependence of the US military on Japanese semiconductors is also widely noted. The Japanese aerospace industry supplies new aircraft materials and products to numerous overseas aircraft manufacturers.[88] Although it is unlikely that Japan will lift its ban on arms exports, Japanese manufacturers become involved indirectly through international arms co-production.

[82] OTA (note 8), p. 109. With the completion of these missile development programmes, Japan's military will be equipped with indigenously developed missiles only, except for the Patriot. The agency predicts, however, that US industry may try to involve itself in several Japanese missile development programmes. *Aviation Week & Space Technology*, 11 Feb. 1991.

[83] *Asahi Shimbun*, 1991 (note 35), pp. 93–94.

[84] OTA (note 8), p. 42.

[85] US Department of Defense, *Critical Technologies Plan—For the Committee on Armed Services, United States Congress*, 15 Mar. 1990 (US Government Printing Office: Washington, DC, 1990).

[86] *International Herald Tribune*, 27 June 1991; OTA (note 8).

[87] *Far Eastern Economic Review*, 28 Feb. 1991, p. 61.

[88] *Asian Defence Journal*, 1990 (note 45), p. 88.

IV. US–Japanese collaboration in military technology

From one- to two-way collaboration in military R&D

Since the end of World War II, Japan has been dependent on the USA for major weapon systems and military technologies; for instance, it has obtained the greatest number of licences for US major conventional weapon systems, and the Japanese defence industry developed its military–technological capability through this licensed production.[89] As the Japanese technological base grew, however, this one-way flow of technology transfer shifted in the 1980s. Responding to requests from the USA, the Japanese and US Governments signed the Exchange of Technology Agreement in 1983 and the Detailed Arrangement for the Transfer of Military Technologies in 1985. Also in 1987, a Japan–US government-to-government agreement concerning Japan's participation in the SDI (Strategic Defense Initiative) research project was concluded. These arrangements aimed at US–Japanese co-operation in R&D of military technology, particularly enhancing military technology transfers from Japan to the USA. This was partially in contravention of Japan's military export ban guideline, but it was justified by giving priority to the Japanese–US 1960 Treaty of Mutual Cooperation and Security and the 1954 Mutual Defense Assistance Agreement[90] rather than to the spirit of Article 9 of the Japanese Constitution. In 1984–87 the Pentagon sent three investigation committees to Japan, resulting in a list of Japanese 'critical advanced military technology' which may be usable for the USA; it consists of 16 items of, for instance, new materials or micro-electronics. According to the agreements, Japan is, for example, to provide technologies for shipbuilding and remodeling to the USA. A joint Japanese–US working group, the Systems and Technology Forum, has been discussing five specific technologies for future co-development projects.[91]

The FS-X fighter: the first case of US–Japanese military co-development

The FS-X is a Japanese Fighter Support Experimental aircraft, to replace the indigenously developed F-1 support aircraft (deployed since 1977). The FS-X is to be equipped with some innovative technologies such as CCV functions, single-piece composite wings, phased array radar and stealth characteristics. The JDA started the development project in FY 1988, and production of 130–170 aircraft is to begin in the late 1990s. The JDA initially estimated the FS-X development cost at 165 billion yen, which was revised to 250 billion yen ($1.9 billion) in 1991.[92] Initially, the JDA and Japanese manufacturers intended to

[89] OTA (note 8), pp. 5, 108.

[90] The JDA states that 'The Japan–US security arrangements constitute a cornerstone of Japan's defence and an indispensable element for its security'. *Defence of Japan 1988* (note 11), p. 82.

[91] These include, for example, the 'ducted rocket', a rocket engine that draws much of its oxygen from the outside air, technology for demagnetizing submarines and ship hulls, ceramic engine for battle tanks. *Boei Hakusho 1991* (note 19).

[92] *Intervia Air Letter*, Jan. 1991; *Flight International*, 30 Jan. 1991.

have completely indigenous development of the FS-X fighter, but the USA thwarted their intention. US–Japanese economic and technological frictions lay behind this move: for example, Japan's purchase of major weapon systems from the USA is believed to be a relatively easy way to reduce Japan's large trade surplus with the USA. After tense negotiations, the FS-X fighter aircraft has become the first case of a Japanese–US co-development project of major military equipment: in 1987 it was decided that Japan would remodel more than 70 per cent of the present General Dynamics F-16 airframe design, incorporating advanced US and Japanese avionics and composites, which means designing an almost totally different aircraft.[93] In November 1988, an MoU (memorandum of understanding) on FS-X co-development was signed between Japan and the USA, but the contract was revised in 1989 because the US Congress objected to selling sensitive fighter and avionics technologies to Japan that would pose a future challenge to the US aerospace industry.

MHI has become a prime contractor and General Dynamics a principal sub-contractor. Additionally, FHI and KHI are also named as principal sub-contractors. General Dynamics and its sub-contractors will receive a 40 per cent work-share of development and production, and MHI will receive the rest. Japan is providing all funding for the FS-X project, in which bilateral technology transfers are expected, that is, single-piece composite wings technology and phased array radar technology from Japan to the USA, and General Dynamics skills in system engineering and integration to Japan.[94] However, the transfer of General Dynamic's flight-control software was suspended by the US Government, which delayed the project.

FS-X disputes show a new dimension in US–Japanese military collaboration, as follows.

1. In the FS-X negotiations, the Japanese–US trade issue was intertwined with a national security issue. At an early stage of the negotiations, agencies of US domestic economic policy making such as the US Department of Commerce got involved in the issue.[95] Underlying the US attitudes, there existed a sense of 'threat' and an intention to prevent Japan from 'catching up and becoming a meaningful competitor in the aerospace/aircraft manufacturing industry'.[96]

2. The main interest of the USA was in 'how much technology flows out and how much comes back'.[97] The US Department of Defense in particular expected some effective technology transfer from Japan which is applicable to the next-generation US Air Force advanced tactical fighter (ATF).

3. For Japan, the FS-X dispute was not the first attempt at indigenous development of a major weapon system, resisting US pressure to purchase or

[93] *Aviation Week & Space Technology*, 20 Mar. 1989.

[94] General Accounting Office, Testimony, 'US–Japan Codevelopment Program' (GAO: Washington, DC, May 1989).

[95] *Aviation Week & Space Technology*, 20 Feb. 1989, p. 16.

[96] GAO (note 94).

[97] GAO (note 94).

licence-produce US weapons: there had been earlier attempts, such as producing the PXL instead of importing the P-3C. This time, however, the friction was manifested, partly because of the intensified Japanese–US trade disputes and partly because of Japan's growing indigenous technological capacity.

US–Japanese friction or co-operation?

The implication of the FS-X project for future Japanese–US military collaboration is not certain. Officially, it is supposed to be a spring-board for future Japanese–US military co-development. However, the frictions have formed negative perceptions on both sides. Increasingly, the USA emphasizes the economic disadvantages of military technology transfer in co-operative military development/production programmes, and it would be less likely to share key defence technologies even with its allies. In Japan, the FS-X exprience is pushing the Government and industry towards even greater reliance on domestic capabilities. Japan rejected a US offer to participate in the production of the US Air Force Advanced Medium-Range Air-to-Air Missile (AMRAAM).[98]

Under the *Shin Chuki-Bo* plan, some development programmes are under way which are specifically directed to reducing the reliance of the Japanese industry on US sources, such as an active radar homing version of the AIM-7 Sparrow surface-to-air missile system to replace US-made Hawk missiles[99] and the AAM-3 air-to-air missile to replace the AIM-9L Sidewinder. However, most of these indigenous R&D projects are in dual-use technologies, where Japan is relatively superior. Technically and economically, it is not feasible for Japan to develop all kinds of military technology alone. Even so, Japanese manufacturers are worried that the USA may pressure the JDA to purchase US systems or to co-operate with US industry, which may introduce problems similar to those in the FS-X project.

Actually, in the post-cold war era Japanese major defence manufacturers are worried about constraints not from restrictive defence expenditures or fewer military orders but rather from US–Japanese economic disputes and US political attempts to check Japanese indigenous R&D in critical fields.[100] The recent 'techno-nationalism' of the USA may provoke nationalistic reactions among major Japanese arms industries. On the government level, arrangements of collaboration in military technology may be limited because, particularly in Japan, significant parts of basic technologies for weapon systems are owned by private companies, reflecting the modest TRDI budget and the leading role of major manufacturers in R&D activities.

[98] *Defense News*, 3 Dec. 1990.
[99] *Jane's Defence Weekly*, 13 Oct. 1990, p. 703.
[100] *Asahi Shimbun*, 1991 (note 35), p. 133.

V. Conclusions

Despite post-World War II constraints, the Japanese arms industry has been fostered by rapidly developing commercial technologies. Although the defence industrial base is still limited, its impact on international arms production is significant due to its dependence on strong, high-quality Japanese dual-use technologies. Although it is not an especially profitable business, arms development and production are attractive for major manufacturers in terms of technological development and stable procurement by the JDA.[101] Japan's defence market has been steadily growing parallel to the GNP growth rate. The JDA's modest but steady support has been the basis for the defence industry, particularly in the field of aerospace.

The Japanese defence industry's interest is not so much in the international arms market.[102] As Japan has enjoyed proportionally low military expenditures, the Japanese realize that the commercial market is much more profitable in the long term, with a potentially large market. Rather, the large domestic defence market seems to be the Japanese target: how to increase the share competing with the US arms industry. It is in the field of dual-use technologies that the Japanese defence industry may become involved in international arms production. Some Japanese dual-use technologies are competitive and significant for foreign arms industries. The Japanese Government does not hesitate to become involved in international co-operation for defence-related R&D in order to keep up with state-of-the-art technology.[103] Dual-use technologies may cause a substantial shift in Japan's post-war policies of restrictive arms production and export, regardless of the sentiment of pacifism among the Japanese people.

[101] According to the statement by a director of MHI; quoted in Kamata (note 40), pp. 68–70.

[102] In an interview, the then director of the KDPC stated '. . . as to arms export, I am against it. Because the arms market is not established; arms are political goods. Instead, we are aiming at international cooperation in military R&D. In the case of co-development, we have a certain work-share which guarantees a certain market. In this case, a reduced production cost per unit is also expected.' Kamata (note 40), pp. 138–40.

[103] See, for example, MITI, *Nihon no Sentaku* [The Choice of Japan], (MITI: Tokyo, 1988), pp. 111–16.

16. Australia: an emerging arms supplier?

Graeme Cheeseman

I. Introduction

At a time when many countries in Europe are scaling down their military industrial capabilities or seeking to convert them to civilian production, Australia is building up its defence and defence-related industries as part of a move towards a more self-reliant posture. This move was precipitated by President Richard Nixon's announcement, made in Guam in 1969, that the friends and allies of the United States in the Asia–Pacific region would be required to mount their own defence at least against prospective regional military threats. It was reinforced by the realization that Australia depended on the United States for much of its major military equipment.[1]

Australia's search for industrial self-reliance culminated in the publication of the Hawke Labor Government's White Paper *The Defence of Australia 1987*, which laid out a defence industry policy for the 1990s and beyond.[2] Under this new regime, Australia was to rely primarily on its civilian manufacturing industry to support its defence forces rather than maintain a separate, and highly subsidized, defence industry (although certain munitions and other specialized production establishments would be maintained by the Australian taxpayer). 'Our objective', said the then Minister for Defence, Kim Beazley, 'is to foster Australian prime contractors able to achieve high levels of local content without subsidies'.[3]

By the turn of the decade, Australia had begun to manufacture under licence the Swedish Type 471 Collins Class submarine and is about to build 10 German-designed MEKO 200 ANZAC frigates for use by the navies of Australia and New Zealand. As a result of these two projects alone, the government has claimed that Australia is 'back in the business of naval warship building'. The new policy has also resulted in a much greater proportion of Australia's outlays on defence equipment being spent at home. In fiscal year (FY) 1989/90, 62.7 per cent of Australia's expenditure on capital equipment was spent locally, compared with 34.6 per cent in FY 1986/87. Moreover, this

[1] The US Arms Control and Disarmament Agency's *World Military Expenditures and Arms Transfers 1986* (ACDA: Washington, DC, 1986) shows that in 1965–74 over 85 per cent (by value) of Australia's military imports came from the USA.

[2] Department of Defence, *The Defence of Australia 1987*, Presented to the Parliament by the Minister for Defence, the Honourable Kim C. Beazley, MP, March 1987 (Australian Government Publishing Service: Canberra, 1987), chapter 6. A concise summary of Australia's existing policy is contained in Department of Defence publication DPUBS 77/91, entitled *Defence Policy for Australian Industry: A Partnership with Industry*, Nov. 1991.

[3] House of Representatives, *Hansard*, 10 May 1989, p. 2345.

appears to have been achieved at relatively little extra cost. According to the government, the cost premium for building the ANZAC frigates in Australia rather than importing them amounted to only 4 per cent of the total project cost.[4]

To date, Australia's nascent defence industry has been financed with money spent by the Hawke Government on modernizing its armed forces, but Australia's defence needs are insufficient to support these industries indefinitely. In order to remain viable over the longer term, industry is also being encouraged to become more competitive and export-oriented. This chapter describes in turn Australia's attempts to develop an indigenous defence industry, the industry itself, and some of the problems and dilemmas confronting the government and its advisers as they seek to capture a larger share of the international arms trade.

II. Establishing an indigenous defence industry

The experience of offsets

During the 1970s, the government relied primarily on offsets to foster the development of a local defence industry, with only limited success. The 1984 *Report of the Committee of Review on Offsets* (the Inglis Committee) found that very few industries were benefitting from the new policy. The value of offsets orders placed with Australian firms between 1970 and 1984 (A$223 million) represented only 25 per cent of the obligations entered into over this time and just 7 per cent of the eligible contract value.[5] The committee suggested that the government encourage offsets in the form of technology transfers, research and development (R&D) opportunities or training packages. It also recommended that Australia's offsets policy be regularly updated and made available to industry and that the government take a tougher line in enforcing industry's offsets commitments (essentially by including penalty clauses in the contract or Deed of Agreement between the Commonwealth and the overseas supplier, withholding offsets clearances for future purchases, and publicising delinquent companies).[6]

Most of these recommendations were accepted by the Australian Government and formed the basis of a revised offsets policy that became effective on 1 March 1986, but the return on the government's offsets commitments remained poor. A 1987 report of the Joint Committee of Public Accounts, *Implementa-*

[4] Hon. Kim Beazley, 'Australian defence policy, technology and industry', address to the New Technology Conference, Australian National University, Canberra, 19 Nov. 1989, p. 13.

[5] *Report of the Committee of Review on Offsets December 1984* (Australian Government Publishing Service: Canberra, 1985), p. 41. The eligible contract value is the total value of all major contracts entered into over this period less the value of all Australian local content included in the contracts for other than providing approved offsets.

[6] To meet this last requirement, the Department of Defence publishes each year an 'Australian Defence Offsets Program List of Participants' which details those overseas companies which have entered into contracts with the Department, the approximate value of the contract and the companies' offsets obligation status.

Table 16.1. Summary of Australian Industry Participation (AIP)/defence offsets, FY 1981/82–1987/88

Figures are in A$ million.

Fiscal year	Contract value	AIP/offsets obligation	AIP/offsets achievement[b]	Outstanding AIP/offsets obligation[b]
1981/82[a]	1 787.802	581.676	259.077	342.411
1982/83[a]	130.048	40.130	49.159	7.039
1983/84[a]	121.990	54.633	30.082	19.509
1984/85[a]	107.281	38.665	29.689	17.492
1985/86[a]	854.766	317.602	53.044	262.842
1986/87	1 474.845	221.351	38.511	182.840
1987/88	166.066	32.001	*n.a.*	32.001
Total	**4 642.788**	**1 286.058**	**459.562**	**846.134**

[a] Figures relate to obligations/achievements under the AIP programme. The AIP programme, in force until 1986, covered both offsets and designated work carried out by the contractor(s).
[b] In a number of cases achievement exceeds obligation.

Source: House of Representatives, *Hansard*, 21 Dec. 1988, p. 3909.

tion of the Offsets Program, found the compliance rate for the period 1980–86 to be 29 per cent.[7] From its discussions with representatives from local industry, the Public Accounts Committee also found that many overseas suppliers 'endeavour, where possible, to minimize their offsets obligations and the costs of satisfying them' by artificially reducing the cost of imported items or by structuring their offsets requirements so that they could not be met by Australian firms. Others were reluctant to transfer technologies which could benefit their competitors and so would do business only with their local subsidiaries. In spite of the widespread knowledge of these practices, it seemed that little could be done to halt them. The committee was advised by the government and its legal advisers that the existing 'arrangements to ensure compliance are as rigorous as possible' and that the contractual 'documents go as far as they can in terms of enforceability'.[8]

In October 1989, the government again revised its guidelines for participants in the Australian Defence Offsets Program as part of a broader review of its defence industry policies (described below). Like the earlier version, the new guidelines made it clear that defence offsets compliance and clearance were important factors in the selection of successful tenders and that in most cases the offsets obligation would be expected to be discharged within one year of the contracted date of delivery of the goods or services. They also provided more details on what types of activity would be acceptable as defence offsets and how they would be evaluated. To be accepted as defence offsets, proposals

[7] Joint Committee of Public Accounts, *Implementation of the Offsets Program* (Australian Government Publishing Service: Canberra, 1987), p. 57.

[8] Joint Committee of Public Accounts (note 7), pp. 54, 55.

were to: (*a*) lead to commercially viable activities which were internationally competitive and capable of being sustained without government support; (*b*) not result in any price increase in the goods or services being acquired; (*c*) be of a level of technological sophistication at least equivalent to that of the goods and services being purchased; and (*d*) constitute 'new work'.[9] As a result of these successive changes in policy, the government's return on its offsets policy began to improve but not by much. Achievements for the period 1981–88 still represented only 36 per cent of obligations and 10 per cent of the eligible contract value (see table 16.1). Another, more effective approach was needed for building up Australia's indigenous defence industrial capabilities.

Fostering private companies in arms production

In parallel with the Inglis Committee inquiry, the government had initiated a comprehensive review of its existing defence policy for industry. This acknowledged that Australia had neither the means nor the expertise to maintain an industrial base capable of supporting all of its defence needs. Rather, self-reliance in the industrial context meant fostering only those industries and industrial capabilities that were crucial to Australia's defence effort and which could not be covered by stockholding or by seeking greater assurances from its overseas suppliers. This view was supported by Paul Dibb in his 1986 *Review of Australia's Defence Capabilities*.[10] Dibb argued that every effort should be made to limit Australia's dependence on overseas sources of repair and maintenance, including software support for its modern weapon systems. Where the cost penalty was not excessive, high-usage spares and ammunition items should be produced in Australia, the latter by government-owned establishments (with local industry involvement where this would result in cost savings and efficiencies). Beyond this, he considered that a defence manufacturing capacity should only be retained in those areas where Australian industry was broadly competitive, where there were unique Australian requirements, or where the manufacturing and assembly processes provided the equipment and skill base for subsequent maintenance, repair and refurbishment.

Dibb's recommendations formed the basis of the Hawke Government's defence industry policy spelt out in its 1987 White Paper. Under this new policy, Australia was to rely primarily on the private sector to support its defence forces. Existing government-owned defence production facilities would be rationalized and as far as possible, commercialized or privatized. Australia's

[9] Department of Defence, *Australian Defence Offsets Program: Guidelines for Participants*, Oct. 1989, pp. 23–24. 'New work' was defined as activities that were new to individual Australian firms or improved existing capabilities; resulted in local research, design, development, production or support activities which would not otherwise have been undertaken in Australia; or opened up new markets overseas for Australian products and services.

[10] Dibb, P., *Review of Australia's Defence Capabilities*, Report to the Minister for Defence, Mar. 1986 (Australian Government Publishing Service: Canberra, 1986), pp. 112–13. Dibb also considered that there may be some areas where tooling and design information could be maintained to provide what he described as 'a production potential with reduced lead time'.

manufacturing industry as a whole would be encouraged to increase its share of both the local and overseas arms markets.

The rationalization of the government's own defence production establishments was already well advanced. In December 1984, an Office of Defence Production (ODP) had been established within the Department of Defence and made responsible for managing the existing government-owned dockyards and defence production establishments and for making them more efficient, productive and commercially competitive. In May 1986, Defence Minister Beazley announced the government's intention to restructure operations at Williamstown Dockyard, and the Government Aircraft Factories in Melbourne and the Garden Island Dockyard in Sydney. While acknowledging the past contributions of these establishments, the Minister noted that they were overstaffed and that 'management and industrial practices have not kept pace with the government's major capital investment in all three establishments. As a result, productivity has fallen and heavy subsidies have been needed. This has been a serious strain on the Defence budget'.[11]

In the following 18 months a government-owned company, Aerospace Technologies of Australia (ASTA) took over the operations of the Government Aircraft Factories and the Williamstown Dockyards was sold to a private consortium, Australian Marine Engineering Corporation (AMEC), which subsequently became the prime contractor for the lucrative ANZAC frigate project. In March 1989 the Office of Defence Production itself, and its residual factories, were formed into an independent company called Australian Defence Industries (ADI) Pty Ltd and moved outside the ambit of the Defence Department. ADI was required to operate on a commercial basis and, with an expected annual turnover of more than A$400 million, would be Australia's largest dedicated supplier of defence equipment and stores.[12]

The various efficiency and restructuring measures that were introduced by the government (with the blessing of the Australian Council of Trade Unions) resulted in some 8500 positions being removed from the government payroll and, according to Beazley, financial savings of around A$200 million to A$250 million a year.[13] However, a range of ongoing costs arising from the transition from the public to the private sector were offsetting these savings. These costs included redundancy and personnel leave entitlements, new site costs, payments for existing and inefficient contracts and orders, and funds for new plant and equipment. In 1989/90 these extra payments amounted to a net outlay of A$40.2 million to ASTA and A$87.4 million to ADI. The following

[11] Department of Defence, *Defence Report 1985–86* (Australian Government Publishing Service: Canberra, 1986), p. 80. In his 1991 Annual Report, ADI's Managing Director, Ken Harris, stated that, in the five years before ADI took over, government defence production establishments drained some $1.3 billion from Australia's defence budget.

[12] ADI's operating revenue and operating profit (after tax) for 1991 was A$466.3 million and $16.2 million, respectively, compared with $425.8 million and $14.1 million, respectively, for the previous year. Taking into account both investment revenue and 'abnormal costs'—those inherited from the previous administration—the company suffered an overall loss of $4.1 million in 1991. It expects to break even by around the end of FY 1991/92.

[13] House of Representatives, *Hansard*, 22 Mar. 1988, pp. 1115–16.

Table 16.2. Australian defence expenditure, FY 1980/81–1990/91

Fiscal year	Defence expenditure[a] A$ million	Defence expenditure[a] % of GDP	% of defence expenditure on procurement[b]	% of defence expenditure on R&D[c]
1980/81	3 656	2.5	..	..
1981/82	4 261	2.6	..	2.7
1982/83	4 940	2.8	..	..
1983/84	5 537	2.8	16.8	..
1984/85	6 229	2.8	16.4	2.4
1985/86	6 981	2.8	16.7	..
1986/87	7 579	2.8	19.3	2.3
1987/88	7 754	2.5	22.2	..
1988/89	8 171	2.3	25.1	2.5
1989/90	8 905	2.3	25.9	..
1990/91	9 066	2.3	26.9	..

Sources:

[a] *Yearbook Australia 1991* (Australian Bureau of Statistics: Canberra, 1991), p. 54.

[b] Senate, *Hansard*, 21 June 1991, p. 5409. 'Procurement' covers capital equipment, replacement stores and equipment and repair and overhaul.

[c] *Yearbook Australia*, volumes 1980–91, chapters on 'Science and Technology'.

financial year the Commonwealth paid ASTA an extra A$6.8 million and ADI A$14.3 million. The 1991/92 budget has ASTA receiving A$1.4 million for redundancy payments and ADI A$15 million for redundancy payments, the disposal of hazardous materials and for measures to comply with state health and safety standards.[14]

Government support for Australia's manufacturing industry involved, first, the use of the considerable resources associated with the defence programme to foster selected industries and industry capabilities. As Beazley informed the South Australian Premier's business forum in Adelaide in October 1989, his Government planned to spend some A$25 billion over the coming 15 years on new capital equipment for Australia's armed services. Rather than buy its replacement submarines, missile-carrying frigates and other items 'off-the-shelf' from overseas suppliers, the government had directed that they be built under licence in Australia. 'Our decision to buy significant numbers of ships and submarines in single contracts rather than order them in ones and twos', said the Minister, 'will not only enable us to re-equip and achieve economies of scale, but consolidate those industries vital to our defence'.[15] As a result of this policy, the proportion of the defence budget that has been invested in local industry has steadily increased from 17 per cent in 1983/84 to 27 per cent in 1990/91 (table 16.2). The submarine and ANZAC frigate projects alone have

[14] *Budget Statements 1990–91* and *1991–92* (Australian Government Publishing Service: Canberra, 1990 and 1991), pp. 3.70 and 3.39–3.40, respectively.

[15] Hon. Kim C. Beazley, 'The $4 billion submarine project: a new militarism or building an industrial base for the 21st century?', address to the Premier's Business Forum, Adelaide, 27 Oct. 1989, p. 8.

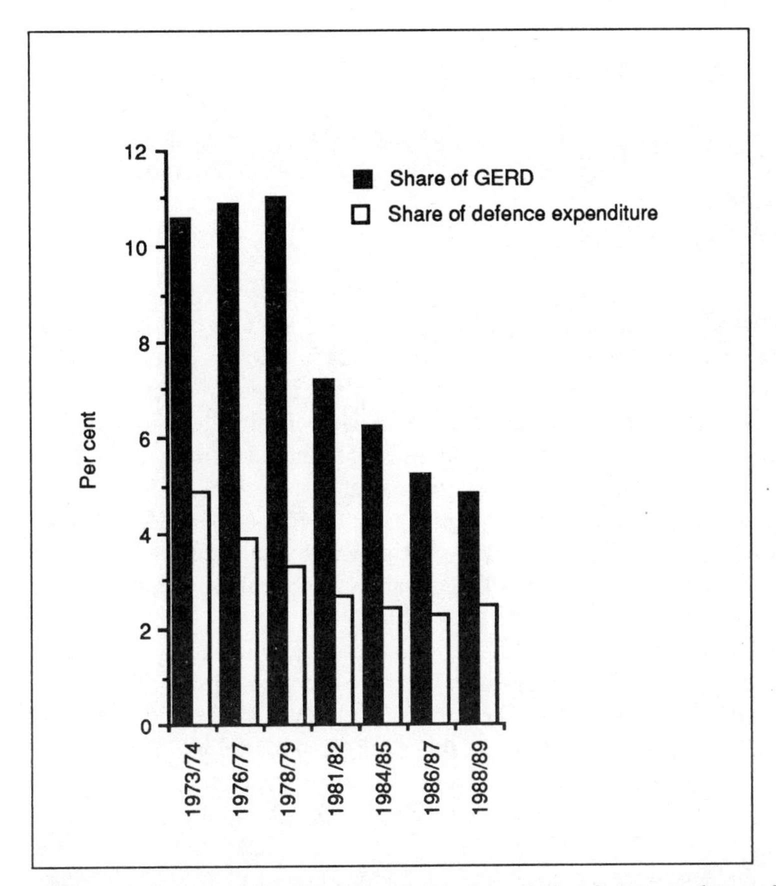

Figure 16.1. Australian defence research and development shares, FY 1973/74–1988/89

Note: GERD = gross expenditure on R&D.

Source: *Yearbook Australia*, volumes 1981 to 1991, chapters on 'Defence' and 'Science and Technology'.

provided work for more than 470 Australian and overseas companies and, in the process, created over 14 000 new jobs.

Second, the Department of Defence has been prepared to fund the 'designated work' component of particular contracts—work relating to those capabilities needed for the repair, overhaul and refurbishment of equipment used in low-level contingencies or the development of designated 'strategic' industries—where it could not be carried out competitively in Australian by the prime contractor. In FY 1985/86, Australian Industry Assistance (AIA) funds were spent on the manufacture of antennas for Air Force radar systems at Tindal and East Sale (A$6.9 million), a hydrostatic transmission system for possible use in the Army's M113 armoured personnel carriers, industry capabilities for the support and maintenance of Australia's fleet of Oberon submarines, and the development of a muzzle velocity indicator which has subsequently earned valuable export revenue.[16] In 1987/88 and 1988/89 the Defence Department spent a further A$39 million and A$43 million, respectively, on various industry assistance programmes.[17] As shown in table 16.2, the proportion of the defence budget spent on R&D has also started to increase, from 2.3 per cent in 1986/87 to 2.5 per cent in 1988/89, although it remains well short of the value for the 1970s (see figure 16.1).

Export orientation

The Hawke Government's third and most important strategy for becoming more industrially self-reliant was to encourage Australia's defence and defence-related industries to become export-oriented. Underlying this policy was the recognition that Australia's own defence needs were insufficient to sustain, without additional and costly subsidization, even the limited military industrial capabilities that were planned under the government's policy of 'defence self-reliance'. As Beazley stated in an address to Parliament in May 1989, '[d]efence, like the economy at large, is best served by an industry structure that can hold its own in the world market, capable of recognising opportunities and seizing them'.[18]

In October 1986, the Defence Minister announced a package of measures which were designed to help local defence and defence-related industries develop and export their products. These included the use of Defence Department personnel and facilities to assist in the testing and marketing of products both in Australia and overseas, the provision of Defence reports and other publications to prospective manufacturers, the provision of spares, including access to spares held by the ADF, and related support to purchasers of Australian defence products, and the evaluation of indigenous defence products.[19] In June 1988 the Hawke Cabinet approved new policy guidelines covering the export of locally produced defence *matériel*. Prepared by the Department of Defence, the guidelines affirmed the government's support for enhanced sales and relaxed the criteria for determining where such exports could go. Australia's earlier guidelines, which had been prepared in 1975, had prohibited the sale of arms to countries engaged in military conflict, into areas where military conflict could occur and to countries that were subject to United Nations or other international sanctions. They were also strongly predisposed against arms exports to coun-

[16] Department of Defence, *Defence Report 1985–86* (Australian Government Publishing Service: Canberra, 1986), pp. 61–62.

[17] Department of Defence, *Defence Report 1987–88* (Australian Government Publishing Service: Canberra, 1988), p. 32; and Department of Defence, *Defence Report 1988–89* (Australian Government Publishing Service: Canberra, 1989), p. 40.

[18] House of Representatives, *Hansard*, 10 May 1989, p. 2345.

[19] According to a departmental information booklet entitled *Doing Australian Defence Business* (Australian Government Publishing Service: Canberra, 1987), p. 20, the 'prospects for overseas sales of Australian-developed defence products is enhanced if the ADF has favourably evaluated and endorsed the equipment and, particularly, has it in service'. Many of the policy changes had been proposed by Robert J. Cooksey in his *Review of Australia's Defence Exports and Defence Industry*, Report to the Minister for Defence (Australian Government Publishing Service: Canberra, 1986).

tries where there was evidence that they could be used to suppress the local populace. Under the new guidelines, Australia would not export controlled military or related goods to countries that are subject to UN or other international arms embargoes (currently Iraq, Libya, South Africa and Yugoslavia), to countries 'with policies or interests that are inimical to the strategic interests of Australia or its friends and allies'; to governments that 'seriously violate their citizens' rights unless there is no reasonable risk that the goods might be used against those citizens'; in particular cases where specific foreign policy interests outweigh export benefits or where the export results in a reduction in any Australian military advantage; and if the export would involve the unacceptable release of classified or third country information or technology.[20]

The new guidelines also gave the Minister for Defence and his Department control over the development and implementation of export policies (this responsibility had previously been shared with the Minister for Foreign Affairs and his Department). They delegated the bulk of the responsibility for export approvals to Defence Ministry officials (the earlier guidelines provided for only very limited delegation below the level of the two ministers), required that all applications be responded to within 21 days, and placed the onus on government regulators to demonstrate why export applications should not proceed.[21] The Department of Defence was still required to consult with the Department of Foreign Affairs and Trade, but only 'if there are sensitivities regarding strategic, international or human rights issues'.[22]

The following year, the government amended the legislation covering the items that could not be exported from Australia without permission. The revised schedule of goods differed from its predecessor in two respects: (*a*) it no longer covered a range of items that could be classified as non-lethal or of minor military significance (maintenance equipment, personal appliances and equipment, certain spare parts, and, most importantly, factory or other production tools); and (*b*) it divided the original list of restricted goods into four separate categories, as follows:

Part 1A. Goods designed or adapted for military purposes and inherently lethal, incapacitating or destructive; and

Part 1B. Other military or related goods.

Part 2. Cryptographic and related electronic equipment and software (military and commercial) such as wideband radio transmitters and receivers and electronic information processing equipment).

[20] Department of Defence, *Australian Controls on the Export of Defence and Related Goods—Guidelines for Exporters* (Canberra, July 1989), p. 2.

[21] For a summary of these and subsequent changes to Australia's arms export control regime, see Raath, N., 'Australian arms exports: prospects and problems', Parliamentary Research Service Background Paper, Department of the Parliamentary Library, Parliament of Australia, 5 June 1991.

[22] Senate, *Hansard*, 30 Nov. 1989, p. 3740. The principal mechanism for considering such applications was to be a Standing Interdepartmental Committee on Defence Exports (SICDE) which met formally eight times in 1988 and only four times in 1989. See Senate, *Hansard*, 6 Mar. 1991, p. 1348.

Part 3. Non-military lethal goods such as commercial firearms and ammunition, propellant materials and explosives.[23]

Under the revised legislation, the export of items in Parts 1A, 1B and 2 of the schedule were automatically prohibited unless written permission was given by the Minister for Defence or his delegate. Those Part 1B goods which were 'determined not to be of major military significance, and do not involve the unauthorised release of classified or third country information or technology' would normally be approved.[24] Items in Part 3 can be exported under licence.

III. The defence industry and arms sales: record and dilemmas

Defence industry

Australia's defence industry presently comprises a mixture of public and private sector establishments.[25] The former include defence force workshops and bases, which carry out intermediate and some depot-level maintenance and repairs, and a small number of government-owned production establishments. As explained above, the most important of these are Aerospace Technologies of Australia Pty Ltd which is privately run and operates two factories and an airfield in Melbourne, and Australian Defence Industries Pty Ltd which is government-owned and operated and controls the existing government weapons, munitions and clothing factories as well as the Garden Island dockyard in Sydney.

As shown in table 16.3, both ASTA and ADI are heavily involved in defence production. ASTA assembled most of Australia's current fleet of F/A-18 fighter aircraft and is presently assembling under licence the Sikorsky Seahawk helicopter for the Royal Australian Navy. It continues to manufacture the locally designed Jindivick target aircraft and the Ikara anti-submarine missile system, both of which have been successfully exported. As the Government Aircraft Factories, it designed and produced the Nomad light transport aircraft and the Waimira trainer—neither of which, however, proved to be commercially successful. ADI produces a range of explosives, ammunition, munitions and weapons for the Australian Defence Force including the new Steyr 5.56-mm rifle (manufactured under licence from Steyr Mannlicher of Austria) and the Hamel 105-mm light field gun which is being exported to New Zealand as part of a A$15.5 million contract. It is also a major sub-contractor for the submarine and ANZAC frigate projects.

[23] *Australian Controls on the Export of Defence and Related Goods*, p. 1.

[24] Note 23, p. 2.

[25] See Bennett, F., 'Defence industry and research and development', in eds D. Ball and C. Downes, *Security and Defence: Pacific and global perspectives* (Allen and Unwin: Sydney, 1990), pp. 317–27; and Kettle, St J., *Australia's Arms Exports: Keeping them out of Repressive Hands*, Peace Research Centre Monograph no. 6 (Australian National University: Canberra, 1989), chapter 6.

Table 16.3. Australia's publicly owned defence industries, 1991

Company	Subsidiaries/division	Products/current involvement
Aerospace Technologies of Australia (ASTA) Pty Ltd	ASTA Aircraft Services Pty Ltd (29.1% owned by Hong Kong Aircraft Engineering Company) Pacific Aerospace Corporation Ltd	*Military* F/A–18 fighter assembly and modifications, Pilatus PC9 trainer construction, Sikorsky Sea Hawk helicopter production, Australian Army Nomad construction and modification, Jindivick target aircraft production and upgrade, Ikara guided missile production, Nulka expendable decoy *Civilian* Nomad aircraft construction, CT4 air-trainer production and upgrade, modification and maintenance of Boeing 747 aircraft, Boeing and A330/A340 airbus component production and repair, publications, training, environmental testing of component parts Sales for 1990/91 valued at A$119.5 million. Operating profit A$1.8 m. comprising (+A$11.1 m. for ASTA P/L, –A$6.7 m. for ASTAAS, –A$2.6 m. for PAC Ltd) 1758 employees
Australian Defence Industries (ADI) Pty Ltd	Ammunition and Missiles Division (1500 employees and A$155.5 m. turnover in 1991)	Small arms ammunition and components, explosives, propellants, gun ammunition up to 127 mm, aircraft bombs up to 910 kg, demolition stores, rockets, practice ammunition, pyrotechnics, commercial explosives and pyrotechnics, target systems and ranges, precision components and tools, electronic assemblies, general engineering products
	Naval Engineering Division (2500 employees and A$158.6 m. turnover in 1991)	Navel ship refits and modernization (including Oberon submarines), diesel engine production and reconditioning
	Weapons and Engineering Division (1850 employees and A$96.7 m. turnover in 1991)	Small arms, machine-guns, AFV track, explosive munitions components, man-portable and crew-served weapon components and systems, projectiles, guns, rocket launchers, gun-barrels, armour plate, propellers and propulsion shafting for ships, heavy machinery, hydro-electric equipment, turbine couplings, alternators and other specialized equipment
	Military Clothing Division (700 employees and A$44.5 m. turnover in 1991)	Military and civilian uniforms, protective clothing, military garments, camouflage combat uniforms

Sources: Department of Defence, *Directory of Australian Industry Defence Capability DRB 29* (Directorate of Departmental Publications: Canberra, 1990); ASTA and ADI Annual Reports for 1991.

In line with the Hawke Government's defence industry objectives, ADI and ASTA in particular have also diversified into the civilian market. ASTA and its subsidiaries have recently undertaken major overhauls of Boeing 747 aircraft for both Nippon and British Airways and have contributed to the development and manufacture of key components for the A330/340 series wide-bodied passenger jet. This latter work was facilitated by earlier work done on the Royal Australian Air Force's F/A-18s. ADI has been engaged in a number of civilian projects, including the Northwest Shelf oil and gas development project and the Snowy Mountains hydro-electric scheme. Over the next five years, ADI hopes to increase the proportion of its revenue earned from non-defence work from the present level of 15 per cent to around 40 per cent of total income. It is also looking to expand its activities into the areas of industrial decontamination, training (using the technology and expertise developed by the former Dart Industries) and consultant engineering.

As mentioned above, an increasing proportion of Australia's defence and defence-related work is being done by companies or groups of companies located in the private sector. These can be divided into three groups: (*a*) those firms that specialize in defence work or maintain dedicated defence production facilities within their establishments; (*b*) those enterprises which from time to time engage in defence work; and (*c*) those general suppliers of goods and services that are not normally associated with defence. Companies falling into the first category are shown in table 16.4. Again in line with government policy, Australia's more specialized defence firms are concentrated largely within the aerospace, shipbuilding, electronics and telecommunications, and software and systems engineering sectors of its manufacturing industry. Most of the defence work has been done for Australia's armed forces, although a number of overseas sales have been made, including; patrol boats (Hawker de Havilland, Australian Shipbuilding Industries and the Australian Submarine Corporation), BARRA sonobuoys (AWA), target systems (Dart Industries which was taken over by ADI in August 1990), and components and spare parts (usually but not always resulting from offsets agreements between an overseas prime and the Australian Government). The nature of Australia's defence industry makes it almost impossible to determine the total number of workers involved exclusively in defence production at any given time. Given that local defence procurement accounts for around 1 per cent of Australia's total manufacturing output, it would be reasonable to assume that the size of the defence work-force would be around 1 per cent of the number employed in the manufacturing sector as a whole, or some 12 000 people.

Table 16.4. Private-sector companies regularly involved in Australian defence, 1990

Industry sector	Company	Principal ownership	Number of employees	Assets (A$m.)	Current involvement
Aerospace	Hawker de Havilland	UK/Australia	2 805	. .	Blackhawk helicopter, Pilatus PC9 trainer aircraft production
	AWA Defence & Aerospace Grp.	Australia	297	385	Barra sonobuoy
	BHP Aerospace and Electronics	Australia	986	16	Laser Airborne Depth Sounder (LADs)
	British Aerospace (Australia)	UK	. .	. .	AUSSAT, portable all arms calculator, LADs
Shipbuilding	Carrington Slipways		488	30	HMAS Tobruk, minehunter construction
	Cockatoo Dockyard		1 180		Oberon submarine refit
	Eglo Engineering	Australia	2 355	58	FFG frigates, survey motor launches
	NQEA Australian	Australia	700	30	Patrol boats
	Australian Marine Engineering Consolidated (AMECON)	Australia	. .	. .	FFG frigates, ANZAC frigates
	Australian Shipbuilding Indus.	Australia	461	80	Pacific patrol boats
	Australian Submarine	Australia	. .	. .	Collins Class submarines
Electronics/ Telecommunications	AWA Defence Industries	Australia	598	5	RAAF ESM, Nulka decoy, tactical navigation aids
	GEC Marconi Defence Division GEC (Australia)	UK	3 200	. .	Submarine sub-contract
	NEC Australia	Japan	1 223	28	AUSSAT
	Seimens Plessey Electronic Systems Australia	Germany	. .	. .	RAVEN, DISCON, ANZAC frigates
	Rockwell Electronics (A'Asia)	USA	214	12	F–111 update, submarines
Software engineering	Computer Sciences Australia	Australia	958	12	Collins submarines, Seahawk, ANZAC frigate
	C3	Australia	80		AUSTACCS, DDG modernization
	Ferranti Computer Systems (Aus.)	UK	68	469	Submarine sub-contractor
	Logica	UK	220	3	AUSTACCS, Jindalee OTHR

Source: Department of Defence, *Directory of Australian Industry Defence Capability DRB 29*, 1990.

Arms exports: political dilemmas and industrial requirements

The value of Australia's exports of defence and defence-related products in each year for the period 1985/86 to 1990/91 is shown in table 16.5, together with the value of its military imports during that period and the value of exports made under COCOM (the Co-ordinating Committee on Multilateral Export Controls) controls.[26] At first glance, the data appear to vindicate the government's overall strategy. The value of Australia's military imports has tended to decrease over the period in question, while there has been a significant increase over the past two years in particular in the value of both its arms exports and the sale of dual-use products overseas. However, this is not quite the case. To begin with, the amounts shown for Australia's exports are not actual sales but the value of approved applications to export goods. Not all permits that were issued led to sales. Indeed, according to the Department of Defence, the value of actual exports for 1989/90 and 1990/91 was A$115.2 million and A$122.2 million, respectively.[27] The 1990/91 figure also included revenue earned from the one-off sale of the ex-ADF Mirage fighter aircraft to Pakistan (A$36 million) and the provision of Pacific Patrol Boats to Tonga, Micronesia and the Marshall Islands (at a total value of A$14.2 million but paid for by the Australian taxpayer through Australia's military aid programme).

Table 16.5. Value of Australian arms trade, FY 1985/86–1990/91

Figures are in A$ million.

		Value of permits to export[b]	
Fiscal year	Arms imports[a]	Defence products	Dual-use products
1985/86	1 528.5	40.4	0.2
1986/87	1 526.8	42.5	38.9
1987/88	981.2	14.6	37.8
1988/89	898.1	45.4	36.8
1989/90	798.6	156.8	179.3
1990/91	956.3	168.8	279.5

[a] Department of Defence expenditure overseas whether direct or through its contractors.
[b] Not all permits issued actually led to export sales, and most permits issued for dual-use products are believed to be for commercial applications.

Source: House of Representatives, *Hansard*, 16 Oct. 1991, p. 2096.

In addition, somewhere between one-third and one-half of Australia's defence and defence-related exports were in the form of offsets against Australia's own purchases of military equipments overseas. Since they are directly linked, this

[26] Australia formally joined COCOM in Apr. 1989, although it has informally adhered to COCOM guidelines for the export of so-called 'dual-use' items since the mid-1970s.
[27] Senate, *Hansard*, 21 June 1991, p. 5409; and 15 Oct. 1991, p. 2098.

component of Australia's arms sales can be expected to fall away as the level of its military imports continues to be reduced. Finally, the significant jump in the value of export applications for dual-use products could well be due to the changes in procedures affecting COCOM. Prior to joining COCOM, Australia's controls on dual-use items were applied only to the 13 COCOM-proscribed countries.[28] After May 1989, the controls covered all sensitive destinations which would have the effect of inflating the value of approved applications without their number necessarily changing.

It would appear, then, that the real value of Australia's arms exports has not increased significantly as a result of the government's new policy regime, certainly to nothing like the extent predicted by some in the mid-1980s. This is not particularly surprising since Australia has been seeking to enhance its share of the international arms trade at a time when the demand for weapons world-wide is decreasing—largely because of the increasing indebtedness of many traditional buyers, the fall in oil and other basic commodity prices, and the cessation of a number of long-standing military conflicts, in particular the Iran–Iraq War—and when the number of countries seeking to sell arms is actually increasing.[29] Within this 'buyers' market', which will last for some time yet, even well-established weapon manufacturers are having to diversify in order to survive, and prospective customers are able to demand, and are getting, package deals involving concessions on prices, the provision of extended credit, and access to production facilities or associated technologies through such mechanisms as collaborative ventures and licensing or co-production agreements. Under these circumstances, and in view of Australia's relatively small and unsophisticated defence industry, it will be very difficult, if not impossible, for Australia to capture major new markets or to significantly expand its defence exports. In recognition of this possibility, the Minister for Defence and his department have started to talk of Australia filling specialized 'niches' in the international arms market rather than making sales across-the-board. Such a strategy, however, is dependent on Australia being able to develop and market weapon systems and technologies that are not found anywhere else. As demonstrated by Australia's now aborted attempt to produce inshore mine-hunting catamarans, such an exercise can be both risky and prohibitively expensive.[30]

The failure of the Labor Government's arms export strategy has meant that it has had to continue to subsidize its own defence industries and is under pressure from industry to provide soft financing (through its Export Finance and Insurance Corporation) and other incentives to help attract overseas cus-

[28] Afghanistan, Albania, Bulgaria, China (PRC), Czechoslovakia, German Democratic Republic, Hungary, Mongolia, North Korea, Poland, Romania, USSR and Viet Nam.

[29] See Anthony, I. and Wulf, H., SIPRI, 'The trade in major conventional weapons', *SIPRI Yearbook 1990: World Armaments and Disarmament* (Oxford University Press: Oxford, 1990), pp. 219–22; and Grimmett, R. F., *Conventional Arms Transfers to the Third World, 1983–1990* (Congressional Research Service, Library of Congress: Washington, DC, 2 Aug. 1991).

[30] See Joint Committee on Foreign Affairs and Defence, *The Priorities for Australia's Mine Countermeasure Needs* (Australian Government Publishing Service: Canberra, 1989).

tomers.[31] If the existing market conditions remain in force, the level of subsidization and further support will have to increase—directly or indirectly in the form of follow-up orders for the Australian Defence Forces—as more and more of Australia's defence and related industries are developed and complete their initial Australian orders. The latter course could serve to distort Australia's military force structure while increased subsidies would sit uneasily with the government's overall policy of reducing tariffs and other forms of protection and encouraging more open competition between trading nations. It may not be possible anyway since, following overseas experience, Australia is more than likely to cut its spending on defence in the coming years, which would further reduce the already limited funds available for new or unanticipated outlays.[32]

Without these funds, the Federal Government will come under increasing pressure from its State counterparts as well as unions and business to further relax its controls over what can be exported and to whom. The dilemma here is that just as the pressures to assist defence and defence-related industries are growing, so too are the political costs of selling arms. The 1991 Persian Gulf War, coupled with revelations that Australia has either sold or was prepared to sell defence *matériel* to countries such as Burma, Indonesia, Iraq, Pakistan, Somalia and Sri Lanka, have raised the issue on the domestic political agenda. Those who oppose the government's initiatives argue that they contradict Australia's long-held position taken at international and other arms control forums, and are serving to damage Australia's relationships with a number of countries and peoples in the region. The sale, announced by Minister for Defence Senator Robert Ray in April 1990, of 50 obsolete Mirage aircraft to Pakistan provided a good case in point. For a mere A$36 million (considerably less than the planes, additional engines and associated spare parts were worth), Australia damaged its relations with a potential major trading partner, India, provoked a heated local debate over the efficacy of its new arms export policies, left itself open to the charge of not practising what it has been preaching in the United Nations, and made itself look foolish as government ministers tried to avoid taking responsibility for the decision while at the same time attempting to placate both India and Pakistan.[33]

Public concern over such episodes was instrumental in pursuading the government to agree to a public inquiry into Australia's arms transfers to be conducted by the Senate Standing Committee on Foreign Affairs, Defence and Trade from the beginning of 1992. It also prompted a government decision, announced by the Minister for Defence on 30 May 1991, to strengthen Australia's existing arms export controls. In the future, all applications to

[31] The recent contract, worth around $215 million, between the Australian Submarine Corporation and the Philippines Navy for the supply of three fast patrol boats, for example, is being funded through low-interest loans supplied by the Australian ($145 million), German and Dutch governments. See Grazebrook, A. W., 'Naval shipbuilding triumphs', *Asia–Pacific Defence Reporter*, Feb.–Mar. 1992, p. 50; and Senate *Hansard*, 4 Dec. 1991, p. 4113.

[32] See Cheeseman, G., 'Australia's defence: White Paper in the red', *Australian Journal of International Affairs*, vol. 44, no. 2 (1990), pp. 101–18.

[33] See 'Our arms export control Mirage', *Pacific Research*, vol. 3, no. 4 (Nov. 1990), pp. 28–29.

export defence and defence-related goods to 'sensitive' destinations would be referred to the Standing Interdepartmental Committee on Defence Exports (SICDE). The guidelines relating to goods of 'major military significance' would be amended to reflect not only the Australian perspective but those of the recipient region and/or country as well.[34]

The case for stricter controls is enhanced by the fact that the sale of sophisticated weapons to neighbouring countries could serve to undermine Australia's future security by encouraging the continuing militarization of the region, or embroiling us in a military conflict between the recipient and a third party. Advanced technologies obtained as part of a licensing or co-production agreement could also be used by the recipient to develop military capabilities that are eventually used to threaten Australia or its interests. The export of unique technologies, such as those associated with the Jindalee over-the-horizon radar, runs the risk of potential enemies obtaining Australia's defence secrets.[35] On the other hand, the successful implementation of the government's much-vaunted policy of 'defence self-reliance' (and the reputations of a number of key actors in the defence establishment) depends to a large extent on Australia being able to develop and maintain a reasonably comprehensive and largely self-sufficient defence and defence-related industry which depends, in turn, on selling more arms. At present, the incentives to continue with the present defence industry policy are holding sway, but this could easily change if new markets fail to materialize and the economic and political costs of attempting to join the arms trade continue to rise.

[34] These changes were reflected in a revised version of *Australian Controls on the Export of Defence and Defence-Related Goods: Guidelines for Exporters* (Canberra, Mar. 1992). Perhaps to placate industry and other interested bodies, the new guidelines also allow certain items being able to be exported without individual permits.

[35] See, for example, Cheeseman, G. and Kettle, St J., *The New Australian Militarism: Undermining Our Future Security* (Pluto Press: Sydney, 1990); Kettle (note 25); and Steketee, M., 'Arms sales, tyrants and Australia', *Sydney Morning Herald*, 29 Jan. 1991, p. 14.

17. The 'third tier' countries: production of major weapons

*Ian Anthony**

I. Introduction

For a small but important group of countries—sometimes referred to as 'third tier' arms producers[1]—political and economic/technological developments are currently pulling in opposite directions. For this group of countries, strategic and political developments are making indigenous arms production more desirable, while the chances of developing and sustaining an arms industry are receding further into the distance for economic and technological reasons.

Writing in 1989, Andrew Ross of the US Naval War College classified the international arms market in four tiers on the basis of the value of arms exports as reported by the Arms Control and Disarmament Agency (ACDA).[2] There is no source of comparative aggregate data on the value of arms production, but it is possible to make a subjective classification. The first tier would, until recently, have consisted of the United States and the Soviet Union. It is likely that in the near future the United States will form a category of one as the only country in the world with the full range of arms production capabilities under national control.[3] The 'second tier' consists of those countries which could produce equipment across the full spectrum of military technology but choose not to for either economic or political reasons. This group would include China, France, Germany, Italy, Japan, Sweden and the United Kingdom. The 'fourth tier' would consist of those countries which have some minimal arms production activities.

The 'third tier' consists of those countries which cannot produce equipment across the full spectrum of military technology but which nevertheless have significant arms industries. While the 12 countries discussed in this chapter—Argentina, Brazil, Chile, Egypt, India, Indonesia, Israel, Pakistan, Singapore,

[1] This term is used in Kwang-il Baek and Chung-in Moon, 'Technological dependence, supplier control and strategies for recipient autonomy: the case of South Korea', in eds Kwang-il Baek, McLaurin, R. D. and Chung-in Moon, *The Dilemma of Third World Defense Industries*, Pacific and World Studies no. 3 (Westview Press: Boulder, Colo., 1989), p. 153.

[2] Ross, A. L., 'Full circle: conventional proliferation, the international arms trade and Third World arms exports', in Kwang-il Baek, McLaurin and Chung-in Moon (note 1).

[3] General Colin Powell, Chairman of the US Joint Chiefs of Staff, recently suggested that it might take Russia at least 20 years to return its arms production capacity to levels recorded in the mid-1980s; *Proceedings of the US Naval Institute*, vol. 118, no. 1073 (July 1992), p. 15.

* The author would like to thank Peter Wezeman, without whose invaluable assistance this chapter could not have been prepared.

South Africa, South Korea and Taiwan—all belong to this third tier, so do others.[4] This group would also include countries such as Australia, Czechoslovakia, Greece, Poland, Spain and Turkey—all dealt with elsewhere in this book either in their own right or as part of Western Europe. There are several other countries which have made no secret of their plans to establish an arms industry. These include Iran, Iraq and North Korea, all of which already seem to have made a significant investment in developing their arms industries. In the cases of Iran, Iraq and North Korea, there is too little reliable information available to discuss them in the same manner as the other countries dealt with in this chapter.[5]

These countries constitute an interesting group in that many of them already represent the dominant political and economic force in their region and would like to play a more instrumental role in global politics. The debate about ways in which the international system could adjust to accommodate these countries is not in itself new.[6] Neither is the focus on the impact which new centres of arms production would have on this process of adjustment.[7] However, the nature of the international system has changed fundamentally since the revolution in East–West relations which took place in 1989–91.

Arms production in a new international environment

In the post-cold war international political environment, the arguments in favour of developing indigenous arms production capacities have become more compelling.

Governments still regard the preparedness of their armed forces as the most important component of national security policy, and this automatically creates a demand for military equipment. However, for arms importers, access to arms and equipment on acceptable terms and conditions can no longer be guaranteed: some suppliers are experiencing difficulties created by the extreme political and

[4] A similar country list is produced by using the classification 'Developing Defense Industrial Nations' in US Congress, Office of Technology Assessment (OTA), *Global Arms Trade*, OTA-ISC-460 (US Government Printing Office: Washington, DC, June 1991).

[5] The Iranian arms industry is discussed by Neuman, S. G., 'Arms transfers, indigenous defence production and dependency: the case of Iran', in ed. H. Amirsadeghi, *The Security of the Persian Gulf* (Croom Helm: London, 1981); the Iraqi arms industry is discussed by Yezid Sayigh in his book *Arab Military Industry: Capability, Performance and Impact* (Centre for Arab Unity Studies and Brassey's: Cairo and London, 1992); the North Korean arms industry is discussed in Jacobs, G., 'North Korea's arms industry: development and progress', *Asian Defence Journal*, Mar. 1989, pp. 28–35; Bermudez, J. S., 'North Korea's HY-2 Silkworm Programme', *Jane's Soviet Intelligence Review*, Aug. 1990, pp. 203–207; Bermudez, J. S., 'New developments in North Korean missile programme', *Jane's Soviet Intelligence Review*, Aug. 1990, pp. 343–45.

[6] See, for example, Mellor, J. W., *Regional Power in a Multipolar World* (Westview Press: Boulder, Colo., 1979).

[7] In 1980, for example, Steven E. Miller wrote that indigenous arms production by the developing countries would affect the global distribution of power by eroding the concentration of political, economic and military power in the hands of a small number of states; Miller, S. E., *Arms and the Third World: Indigenous Weapons Production*, PSIS Occasional Paper no. 3 (Graduate Institute of International Studies: Geneva, Dec. 1980), p. 29. Stephanie Neuman subsequently contested this conclusion in 'International stratification and Third World military industries', *International Organization*, vol. 38 no. 1 (winter 1984).

economic dislocation they have experienced in the recent past. The former Soviet Union and the countries which used to make up the Warsaw Treaty Organization are the clearest examples of this. Some other arms suppliers have discussed the need to avoid 'another Iraq' either by restricting exports—to prevent any country from building an arsenal considered excessive in size or destabilizing in nature—or by placing strict conditions on the application of those arms which are supplied.

As a counter-pressure, the domestic budget environment in the arms-exporting countries has increased both the need for industry to seek foreign sales and the pressure on governments to permit those sales. However, international and domestic pressure for a cut-off in supply would be greatest in circumstances where the recipients might want to use the arms which are transferred in combat. Under these circumstances political pressure for an arms embargo might override economic incentives to continue supplies.

The fear that in a moment of crisis or during war a supplier either would not or could not provide military equipment is not a new one for recipient countries. There are several cases in which countries increased their investment in arms production after being the target of an embargo imposed by an important arms supplier. This is true for Argentina (embargo of 1982), Chile (1974), India (1965), Iran (1979), Iraq (1980), Pakistan (1965), South Africa (1963), and Turkey (1974). Egypt and Israel were the countries primarily affected by the 1950 Tripartite Declaration Regarding Security in the Near East by the Governments of the United Kingdom, France and the United States, under which arms transfers were made conditional on recipient behaviour. While not embargoed, Argentina and Brazil withdrew from US Foreign Military Sales arrangements in 1977 when the US Congress made it clear that supplies would be linked to their record on human rights. After 1979, when President Jimmy Carter personally intervened to prevent the sale of F-5G fighter aircraft to Taiwan, Taiwan increased its efforts to develop and build its own fighter aircraft.

While embargoes may raise the investment in arms production capabilities, the absence of an East–West dimension in regional conflicts increases the probability that arms embargoes will be imposed.[8] Moreover, even where arms have been transferred, suppliers can no longer be trusted not to betray confidential information about weapon system performance.

During the cold war recipient governments were expected to go to considerable lengths to protect data concerning the performance of equipment bought from members of one or another major alliance. In some cases—such as India—this involved the creation of two parallel support systems for major items such as fighter aircraft and naval vessels. West European and Soviet equipment was physically separated through deployment at different sites, and access to these sites was denied to technical advisers from the 'wrong' alliance.

[8] In the 41 years between the foundation of the United Nations and Aug. 1990 the UN Security Council imposed mandatory arms embargoes on two countries—Southern Rhodesia and South Africa. In the next 24 months the UN imposed mandatory embargoes on Iraq, Libya, Somalia and Yugoslavia.

With the end of the cold war NATO countries have almost unlimited access to conventional weapons from the former Soviet Union. The United States also has significant information about Russian nuclear weapon systems. The following evaluation of the underlying causes of allied success in the 1991 Gulf War would meet with widespread acceptance:

The fundamental reason the ground campaign was so successful was that we knew where the Iraqi forces were, before and during the operation. The transparent nature of the opponent in this war stemmed from . . . Iraq's having built its military from weapons and systems purchased from other nations. And those nations, for the most part, were willing to share the specifications of hardware they had sold to the Iraqis. So Iraq was, in effect, naked. . .[9]

The growing amount of information about weapon systems available to governments has already been used in some countries as one argument for sustaining investment in arms industries. Referring to the dominance of Soviet equipment in the Indian armed forces, an Indian commentator has noted that 'with the dismantling of the Soviet Union much of the secrecy surrounding this equipment may have been jeopardised. During the Gulf War a large quantity of Soviet-built equipment was seized by the Allied forces, of which Pakistan was a part'.[10] Arguments of this type have also been used in Israel as one reason for sustaining a domestic defence industry regardless of its profitability.[11]

Twenty-one years ago a SIPRI study concluded that efforts to develop indigenous arms industries had not only proved very expensive but had largely failed. As the study noted, most of the countries under study had 'devoted large amounts of resources to the development of weapons, particularly aircraft and missiles, which have never reached the production stage.'[12] In 1992 this conclusion remains valid.

In most countries there are insurmountable financial and technical barriers to establishing an arms industry capable of meeting more than a small proportion of the needs of the armed forces. Few countries have the diversified technological and manufacturing base needed to sustain the wide range of industrial activities on which arms production depends. More usually, countries have a narrow and specialized capability, often in the relatively low-technology activity of making the hulls of ships. As most countries have a limited domestic demand for arms, the high unit cost of producing weapon systems makes them economically uncompetitive compared to foreign equipment. As the resources devoted to defence in many countries are squeezed by other funding priorities, this factor will become even more important. Furthermore, many armed forces

[9] Blaker, J. R., 'Now what, Navy?', *Proceedings of the US Naval Institute*, vol. 118, no. 1071 (May 1992), p. 60.

[10] Satish, M., 'Defence spending: why it must be increased', *The Economic Times* (New Delhi), 11 Apr. 1992.

[11] Klieman, A. and Pedatzur, R., *Rearming Israel: Defense Procurement in the 1990s*, Jaffee Centre for Strategic Studies Study no. 17 (Jerusalem Post Press: Jerusalem, 1991), especially chapter 2.

[12] SIPRI, *The Arms Trade with the Third World* (Almqvist & Wiksell: Stockholm, 1971), p. 782.

prefer to buy foreign weapons—which are often of a higher quality and bring useful contact with the armed forces of the supplier country.[13]

Strategies for the new political conditions

The question of how to satisfy the perceived requirement for military equipment without sacrificing political autonomy is by no means new. However, it is being asked in a new international environment where it is no longer possible to exploit the mutual hostility of two ideologically opposed blocs to gain access to equipment on favourable economic terms. In the future, governments exporting arms will demand a more exacting economic or political price than nominal adherence to one or another ideology.

Some recent studies have suggested several broad strategies which might be pursued by countries facing the question posed above.

The most disturbing of these strategies would be for countries to move away from traditional forms of territorial defence towards deterrence by unconventional means. This could encompass a wide range of capabilities from the acquisition of one or more of the 'NBC' (nuclear, biological and chemical) weapons to other unconventional defence postures based on terrorism or destabilizing the society of an enemy by non-military means.

A second possible strategy would be for countries to pay the political and economic price demanded by a major power to safeguard their security. As noted above, for most of the countries in the world the technological, industrial and economic barriers to arms production are insurmountable. In other cases—including many advanced industrialized countries—maintaining a major arms production capacity is considered to be a sub-optimal security policy.

One Middle East expert has suggested that in the wake of their failure to prevent Iraqi expansionism, 'the principal Arab states, Kuwait and Saudi Arabia, have been confirmed in their belief in the need to rely on outside powers, without any illusions about regional allies or arrangements'.[14] This is reminiscent of the statement made by the then Prime Minister of Pakistan in 1956 who dismissed the idea of Muslim defence industrial co-operation on the basis that 'zero plus zero equals zero'. By 1992, however, the developing countries have many more possibilities for co-operation, and a third possibility would be for several or all of the governments facing this question to co-operate with one another. Again in the context of the Middle East, one author has recently observed:

[13] The limitations of arms industries outside the major powers have been discussed by Stephanie Neuman in several publications, most recently in *Military Assistance in Recent Wars: The Dominance of the Superpowers*, Washington Papers No. 122 (Praeger: New York, 1986). These barriers have also been usefully summarized by Andrew Ross in Ross, A. L., 'The political economy of arms production and exports: the developing world', in ed. A. Pierre, *Conventional Arms Proliferation in the 1990s* (forthcoming).

[14] Chubin, S., 'Iran and regional security in the Persian Gulf', *Survival* vol. 34, no. 3 (autumn 1992), p. 73

The logic for industrial collaboration in arms development and manufacture between Egypt and other Arab states in the region is sound. Utilization of capacity would be increased through a division of labour and larger production orders. This would lead to falling unit costs. On the political side, this could generate greater consensus from the more moderate nations of the Arab world. . . . The 1990–91 Iraq–Kuwait debacle may well prove the catalyst for greater military cohesion amongst the Gulf states, and the incentive for Arab defense industrial co-operation through the revival of the AOI.[15]

Recently there have been discussions about defence co-operation between governments which would not have spoken with one another about these issues before the end of the cold war and probably would not have spoken to one another at all. Examples of these new partners in dialogue include China and Czechoslovakia; China and Poland; China and Romania; China and Russia; India and Israel; Israel and Czechoslovakia; and Russia and Turkey.

Finally, there is a small group of countries in which the perception that an arms industry is a strategic necessity is likely to be reinforced. In particular 12 countries—Argentina, Brazil, Chile, Egypt, India, Indonesia, Israel, Pakistan, Singapore, South Africa, South Korea and Taiwan—are the primary concern of this chapter.[16] These countries have made such a significant investment in developing arms production capacities that they are unlikely to withdraw from such activities except under very extreme circumstances.

In several of these countries such extreme circumstances may be at hand. In Argentina and Chile reductions in arms production and military expenditure in general may be seen by the new governments as an important element in the transition from military to civilian government. Meanwhile, in Brazil and South Africa the governments are apparently unwilling to sustain the level of subsidy which arms industries have enjoyed in the past.

Countries discussed in this chapter

This list of countries other than the major arms producers (discussed individually in other chapters in this book) that are able to sustain defence industries is relatively short, but its membership is very stable. Of the 12 countries listed above, all but Chile and Pakistan were identified in a 1984 table describing the main Third World producers of major weapon systems.[17]

[15] Matthews, R. G., 'Egyptian defense industrialization', *Defense Analysis*, vol. 8, no. 2 (Aug. 1992), p. 130.

[16] The political, strategic and economic motives for indigenous arms production have been described elsewhere and are not dealt with here. See in particular Brzoska, M., and Ohlson, T. (eds), SIPRI, *Arms Production in the Third World* (Taylor & Francis: London, 1986); the 2 volumes edited by James Everett Katz, *Arms Production in Developing Countries* (DC Heath: Lexington, Mass., 1986) and *The Implications of Third World Military Industrialization: Sowing the Serpents Teeth* (DC Heath: Lexington; Mass., 1986); Kwang-il Baek, McLaurin, R. D. and Chung-in Moon eds. *The Dilemma of Third World Defense Industries*, Pacific and World Studies no. 3 (Westview Press: Boulder, Colo., 1989); Sanders, R., *Arms Industries: New Suppliers and Regional Security* (National Defense University Press: Washington, DC, 1990); Ball, N., 'The political economy of defense industrialization in the Third World', ed. A. Ross, *The Political Economy of Defense: Issues and Perspectives* (Greenwood Press: New York, 1991).

[17] Brzoska and Ohlson (note 16), table 2.2, p. 10.

The most striking characteristic of the group of countries dealt with here is its diversity. The group includes the small island states Singapore and Taiwan and countries of sub-continental size such as Brazil and India. Some of the countries have recently experienced war and are located in unstable regions. Others have a benign regional environment where the danger of inter-state conflict is minimal. Some of the countries enjoyed rapid and sustained economic growth throughout the 1980s while others experienced a decade of stagnant or negative economic growth. All of the countries except South Africa are currently democracies—although their constitutional arrangements differ widely. However, several have only recently emerged from military or authoritarian rule while India has been under democratic government for 45 years with one brief interruption. In terms of industrial organization, some of these countries have a mix of private and government ownership in the arms industry. In other cases the government both owns and manages the arms industry.

If one classifies these countries in the two groups identified by President François Mitterrand—those which are making progress and those where human development is stagnant or slipping backwards[18]—then certainly the 12 countries dealt with in this chapter would belong in the first group. Moreover, there are few countries outside this group of 12 which could realistically be expected to develop significant defence industries in the medium term.

II. Trends in the production of major conventional weapons

Looking at the data presented in table 17.1, there are no trends which are universal across the group of countries under consideration. However, there is a general stabilization in the estimated value of major weapon systems being delivered in the late 1980s.

The data in table 17.1 are the outcome of the following process:

Line A: Step 1. Programmes involving the production of major conventional weapons whose design was not entirely external to the producer country were identified from one of the standard reference books produced by private sector information groups such as Data Forecast Associates, Jane's Information Group and the Teal Group.

Step 2. Aggregate production has been broken down into a year-by-year production series. Where possible this has been done from specific sources, either those noted above or the specialist press and periodicals dealing with military issues. For cases where no such information is available, total production has been distributed across the duration of a programme according to a bell-shaped curve.

Step 3. A price is allocated to each system.

Step 4. The number of items delivered in each year is multiplied by this price.

[18] Statement of President Francois Mitterand before the meeting of Heads of State and Government of the United Nations Security Council, United Nations doc. S/PV.3046, 31 Jan. 1992.

Line B: Step 1. Programmes involving the production of major conventional weapons whose design was entirely external to the producer country were identified from the reference books noted above.

Step 2. Aggregate production was broken down in the manner described above.

Step 3. A price is allocated to each system.

Step 4.The number of items delivered in each year is multiplied by the price and by a fraction which denotes the percentage of foreign technology and components contained in the system.

Line C: Step 1. Values of transfers of major conventional weapons between countries are estimated according to the sources and methods described in appendix 8D of *SIPRI Yearbook 1992*.

Line D: Step 1. For each year the values of A, B and C are added together

Looking at the method by which the data in table 17.1 are derived, several limitations on the use of the value estimates are immediately obvious. First, the estimates are not a proxy either for actual production costs in the case of lines A and B or for financial flows related to the arms trade in the case of line C. It is true that the SIPRI price index is based on the average production costs of those systems for which such data are available. However, few if any of the systems produced in the sample countries would fall in this category. Therefore, the prices are based on technical comparisons of weight, speed, range, year of development and year of production between these systems and those for which production costs are available (usually systems produced in the United States).

In order to derive estimates of the true production costs of these systems it would be necessary to make a detailed study of each of the countries in the sample, taking into account the unique local conditions present in each case.

The scale of production in the countries surveyed is very different. If the data for production of major conventional weapons are aggregated across the 25 years 1966–90, then two countries—India and Israel—emerge as by far the largest producers. India did not develop its industries from scratch but in 1947 inherited a significant arms production capacity from the United Kingdom which it has built on ever since.[19] Israel, by contrast, did not emerge as a major arms producer until the mid-1970s. Looking at the pattern of production, it seems that Israel has a more uneven distribution of deliveries than is the case for India.

[19] A recent discussion of arms production in India is contained in Anthony, I., *The Arms Trade and Medium Powers: Case Studies of India and Pakistan 1947–90* (Harvester Wheatsheaf: Hemel Hempstead, 1992). The issue will also be taken up in detail in Smith, C., SIPRI, *India's Ad Hoc Arsenal* (Oxford University Press: Oxford, 1993).

Table 17.1. Estimated values of the trade in, licensed production and indigenous production of major weapons, 1965–90

	1965	1966	1967	1968	1969	1970	1971	1972	1973	1974	1975	1976	1977	1978
Argentina														
A	5	0	5	24	30	24	14	15	9	11	6	42	51	248
B	–	–	1	1	1	1	–	–	–	12	12	12	12	14
C	24	96	129	157	209	134	483	401	180	487	387	322	584	490
D	29	96	135	182	240	160	497	416	189	511	405	376	648	753
Brazil														
A	3	3	3	3	10	22	42	67	177	163	183	232	310	308
B	–	–	–	–	–	–	–	39	39	39	39	20	40	20
C	69	94	82	122	182	83	41	364	1174	134	553	622	788	778
D	72	97	85	125	192	105	84	471	1391	336	775	874	1 138	1 106
Chile														
A	–	10	10	3	–	–	5	–	–	–	–	–	–	–
B	–	–	–	3	–	–	5	–	–	–	–	–	–	–
C	25	32	16	0	66	45	146	20	256	440	102	337	99	108
D	25	42	26	6	66	45	155	20	256	440	102	337	99	108
Egypt														
A	14	14	24	24	24	13	13	13	13	13	65	65	13	13
B	7	7	7	7	7	–	–	–	–	–	–	–	–	–
C	250	748	1 866	1 196	649	1 375	1 964	2 662	2 062	499	284	567	317	171
D	270	768	1 896	1 226	679	1 388	1 976	2 674	2 074	511	349	632	329	184
India														
A	133	132	359	403	473	481	493	722	588	824	540	820	935	637
B	60	77	207	222	273	278	278	386	329	435	220	344	419	333
C	580	753	311	385	599	576	671	1 160	748	384	68	871	1936	1 208
D	773	962	877	1 010	1 345	1 335	1 443	2 269	1 665	1 643	828	2 036	3 291	2 178
Indonesia														
A	0	7	–	7	–	7	–	–	–	3	3	8	7	17
B	–	–	–	–	–	–	–	–	–	–	–	1	2	9
C	168	3	0	47	10	29	85	12	65	91	73	133	35	143
D	168	11	0	54	10	36	85	12	65	94	75	142	43	170
Israel														
A	5	4	7	64	64	139	142	497	692	838	655	445	597	618
B	–	–	–	–	–	–	–	–	–	–	–	1	8	8
C	68	79	357	716	1 296	1 704	793	1 227	3 024	1 009	1 467	14 59	1 315	1 071
D	73	83	364	780	1 360	1 843	936	1 724	3 716	1 847	2 123	1 905	1 920	1 697
Pakistan														
A	–	–	–	–	–	–	–	–	–	–	–	4	4	5
B	–	–	–	–	–	–	–	–	–	–	–	–	1	2
C	70	337	209	314	180	152	478	491	136	212	198	148	520	496
D	70	337	209	314	180	152	478	491	136	212	198	152	525	503
Singapore														
A	–	–	–	3	3	–	24	–	–	105	121	–	–	33
B	–	–	–	2	2	–	18	–	–	84	85	–	–	27
C	0	0	–	–	27	90	426	196	71	42	209	268	146	15
D	0	0	–	5	32	90	468	196	71	231	416	268	146	75

1979	1980	1981	1982	1983	1984	1985	1986	1987	1988	1989	1990	Aggregate
108	44	655	216	227	220	389	286	157	26	26	157	2 997
14	11	433	136	136	94	220	199	148	–	–	148	1 609
481	132	490	666	1 377	1 159	266	224	67	74	85	17	9 118
604	187	1 578	1 019	1 740	1 473	875	708	372	100	111	321	13 724
576	545	350	413	429	457	411	340	101	181	526	478	6 334
236	236	20	20	12	5	–	240	26	0	4	25	1 060
53	264	127	48	93	108	90	206	323	316	468	123	7 306
866	1 046	497	481	534	570	500	787	450	497	998	625	14 700
–	–	12	20	25	27	32	49	43	118	38	34	425
–	5	10	23	21	14	27	29	37	25	35	31	264
105	240	300	449	148	491	15	24	421	100	121	226	4 331
105	245	322	491	195	532	74	103	500	243	194	291	5 020
14	14	21	35	176	190	226	91	108	174	145	162	1 673
0	0	0	9	107	124	129	57	71	89	69	122	810
687	1 337	911	3 358	2 783	2 689	1 398	1 953	2 779	414	189	1 091	34 198
701	1 351	933	3 402	3 066	3 003	1 753	2 101	2 958	677	403	1 374	36 682
627	1 029	1 036	609	1 089	1 092	1 400	1 103	1 137	1 445	1 177	1 198	20 483
243	435	432	281	297	317	602	863	1 060	895	1 008	1 078	11 374
688	1 228	1 662	1 697	1790	930	19 45	4 008	4 532	3 157	3 495	594	35 977
1558	2 692	3 130	2 587	3176	2 339	3 947	5 974	6 729	5 497	5 681	2 870	67 834
19	14	30	31	36	61	94	113	112	151	143	116	978
9	9	26	27	32	11	19	37	36	72	66	43	400
453	824	734	468	178	159	240	507	353	321	195	123	5 450
481	847	790	527	246	230	354	657	502	543	404	282	6 828
826	819	947	809	784	846	715	557	435	496	596	609	13 204
8	8	8	8	25	1	25	1	1	–	–	–	104
611	1 393	2 070	1 223	400	369	249	501	1 939	604	120	228	25 293
1445	2 220	3 025	2 041	1208	1 217	989	1 059	2 375	1 099	715	837	38 601
5	4	4	4	4	4	4	4	4	4	9	9	69
2	1	1	1	1	1	1	1	1	4	22	6	45
334	602	264	746	708	687	697	466	408	308	680	508	10 349
341	606	269	751	713	692	702	471	413	316	711	522	10 463
237	277	198	237	365	199	145	227	141	133	73	93	4 921
77	78	5	5	144	–	–	81	9	9	–	–	1 663
45	52	–	0	43	6	5	103	19	19	–	2	3 557
358	406	203	243	552	205	150	411	168	160	73	95	10 141

Table 17.1. *contd*

	1965	1966	1967	1968	1969	1970	1971	1972	1973	1974	1975	1976	1977	1978
South Africa														
A	–	82	114	106	93	67	123	132	278	276	248	485	342	249
B	–	25	34	32	28	20	37	39	172	164	149	311	169	76
C	252	258	350	188	52	308	87	309	374	470	115	131	35	333
D	252	365	499	326	173	395	247	480	825	911	513	928	546	657
South Korea														
A	1	–	–	–	–	–	2	48	–	–	0	36	125	104
B	1	–	–	–	–	–	–	–	–	–	0	0	62	44
C	322	282	202	267	411	117	111	490	138	270	457	272	1 002	610
D	324	282	202	267	411	117	113	539	138	270	457	308	1 189	758
Taiwan														
A	–	–	–	0	19	24	23	22	22	172	179	183	165	203
B	–	–	–	0	14	18	17	17	17	33	76	191	174	174
C	212	253	452	204	392	444	368	159	183	421	116	207	304	224
D	212	253	452	204	426	487	408	199	222	626	372	581	643	602

Notes:

0 = less than 0.5.

Data are estimated values, expressed in US$ (1990) million.

A = Indigenous production

B = Licensed production

C = Direct imports

D = Sum of A, B and C (figures may not add up to totals due to rounding)

However, conclusions drawn from changes in the year-by-year value estimates cannot be considered safe because of the limitations of the original sources of data noted above.

Brazil and South Korea make up a second sub-group. In both cases the estimated value of production of major weapon systems began to grow in the late 1970s.

South Africa and Taiwan make up a third sub-group in which the estimated value of production of major weapon systems began to grow in the mid-1970s, although production volumes have never reached the levels recorded for Brazil and South Korea.

In the case of Israel the methodology adopted here of concentrating on major conventional weapons creates particular problems in that it understates the Israeli capacity to produce important support systems such as radios and other tactical communications systems and military electronics.

1979	1980	1981	1982	1983	1984	1985	1986	1987	1988	1989	1990	Aggregate
202	152	500	191	352	811	1 235	1 348	768	645	1 266	730	8 517
91	70	34	55	119	154	187	191	71	113	67	37	1 295
904	595	246	105	202	155	339	205	680	1 101	1 045	344	10 874
1197	817	780	351	673	1 120	1 761	1 744	1 518	1 860	2 379	1 112	20 686
270	327	424	443	442	462	388	236	85	128	186	454	4 858
204	261	307	240	264	244	251	140	96	96	107	267	3 211
98	135	370	333	443	208	493	664	433	267	243	182	7 811
573	723	1 101	1 016	1149	914	1 132	1 041	614	492	536	903	15 879
77	–	1	–	1	12	52	4	47	11	–	188	682
61	–	–	–	–	–	–	–	–	–	–	113	392
270	209	291	78	85	79	254	16	187	500	72	251	3 781
408	209	291	78	86	91	305	20	233	511	72	552	4 855

Studies of emerging arms producers conducted in the early 1980s identified the arms industries in Latin America, in particular Argentina and Brazil, as a region of considerable growth. However, in the early 1990s it is an open question whether the existing capacity, much of which is idle, can be sustained and very unlikely that major new capacities will be added.

Across a range of major systems including armoured vehicles, aircraft and ships, established production lines currently face a situation in which follow-on orders have not yet materialized. For example, only 18 Argentinian IA-63 Pampa jet trainer aircraft will be built unless new orders are obtained. This programme stretches back to 1979. Another example is the Brazilian EMB-312H 'stretched' Tucano for which there are currently no announced customers.[20] Meanwhile, the Brazilian company Engesa, which produced thousands of wheeled armoured vehicles of the Cascavel, Jararaca and Urutu types in the

[20] *Jane's All the World's Aircraft, 1992–93* (Jane's Information Group: Coulsdon, 1992), pp. 2, 15.

1980s, filed for bankruptcy in 1992 after an effort to restructure the company in 1990–91 failed to generate new business.[21]

One exception to this has been the establishment of a helicopter assembly facility in Brazil in collaboration with Aérospatiale of France. The package, in which Brazil agreed to buy French helicopters, included an offset arrangement by which France bought the Embraer Tucano trainer aircraft for its air force.

In Argentina overall spending on the military has declined throughout the 1980s both in constant dollar terms and as a percentage of Argentinian gross domestic product (GDP). In Brazil spending on the military increased after 1988 but remains below 2 per cent of the GDP. In Chile total annual military spending has rarely risen above $1.5 billion and, added together, the level of military expenditure by these three countries is roughly the same as that of Spain.[22] Even allowing for significant 'off budget' resources controlled by the armed forces, this level of spending has meant that the volume of arms production in the region has depended heavily on export orders. However, products manufactured in these three countries must be sold in a shrinking global market against intense competition from North American, European and Asian suppliers.

The country in which the downward trend in the value of production is steepest is Argentina. The value of production peaked in the early 1980s when the scale of production in Argentina briefly overtook that of Brazil. However, while the Brazilian arms industry sustained its volume of production at a fairly consistent level between 1978 and 1984–85, the value of Argentinian production of major weapon systems is closely associated with three programmes: the IA-58 Pucara attack aircraft, the TAM family of armoured vehicles and light frigates of German design. Argentina developed ambitious plans for expanding its aircraft and naval industries. However, the future of the aircraft industry is heavily dependent on the success of the indigenously designed Pampa jet trainer aircraft in foreign markets. The Pampa is one of the aircraft offered to the United States to meet its requirement for a Joint Primary Aircraft Training System (JPATS)—estimated to be worth $4 billion. As well as FMA of Argentina, Embraer of Brazil and ENAER of Chile have also prepared bids to meet this requirement. Success or failure in winning this contract will go a long way to determining whether these companies have any future as producers of military aircraft. There are current discussions of the rationalization of the Latin American aircraft industry through a strategic alliance or merger between these three companies which are the largest aircraft producers in the region. This would create a single producer with a range of trainer aircraft including the small Pillan made in Chile, the larger turbo-prop Embraer Tucano and the Pampa jet trainer. This would also make it more likely that governments would

[21] *Defense News*, 29 June–5 July 1992, p. 22. A detailed discussion of Engesa is contained in Graham, C. P., 'Technology and Third World defense manufacturing: the Brazilian firm Engesa', *Defense Analysis*, vol. 6, no. 4 (Apr. 1990), pp. 367–83.

[22] *SIPRI Yearbook 1992: World Armaments and Disarmament* (Oxford University Press: Oxford, 1992), pp. 263, 268.

buy from one another rather than going outside the region for this type of aircraft, as Chile did when buying the Spanish C-101 jet trainer.

The other country among the 12 'third tier' countries in which a consistent downward trend in the value of production of major weapon systems is noticeable is South Africa. The high point of production was reached in 1976, just before South Africa became subject to a mandatory United Nations arms embargo. With the major political changes under way in South Africa, the future pattern of arms production is unclear. On the one hand, it could be argued that the domestic requirement for major systems has declined after the withdrawal of South African forces from Angola, the end of the war in Namibia and the apparent progress made in ending the war in the Mozambique. On the other hand, many aspects of the technology deployed by the South African Defence Forces remain frozen at the level of the late 1960s. If the embargoes on sales were lifted, then South Africa—a rich country—might be expected to modernize its arms production technologies. Moreover, if South Africa was fully re-admitted to the international community it would be able to expand its foreign business.[23]

The countries in which the value of the production of major weapon systems seems to be rising include India, Indonesia, South Korea and Taiwan. The value of such production in Egypt and Israel may not be growing, but it does appear to be stable and not falling. Given the uncertain availability of foreign weapon systems at affordable prices, India and Taiwan are likely to become more reliant on their own industrial base for major items of equipment. Egypt, Israel and South Korea have secure bilateral links with the United States, and Egypt has also diversified its sources of arms supply. However, none of the governments in these countries is likely to downgrade the importance of developing its defence industries.

Four of these countries—India, Israel, South Korea and Taiwan—produce a wide range of systems rather than focusing on one specialist product sector. Each of these countries produces an indigenously developed main battle tank. India and Taiwan have developed fighter aircraft, while the Israeli Lavi fighter programme was cancelled for financial rather than technical reasons. India, Israel and Taiwan also produce a range of guided missile systems. While South Korea does not produce fighter aircraft or missiles, it does produce sophisticated electronic systems such as the fire control radar for the Vulcan anti-aircraft gun. Moreover, while not yet able to develop many advanced military systems from basic scientific research, the grasp of technology in these countries is sufficient to permit some innovation through the modification of equipment or designs of foreign origin. In the past Israel has been able to sell some of these modifications back to the country where the systems originated.

[23] South Africa held its first public show of defence equipment in an exhibition in Pretoria in Nov. 1992; *Jane's Defence Weekly*, 12 Sep. 1992, p. 28.

The scale and nature of design and production activities

Leaving aside the upward and downward trends in the data, there are important differences in the scale of production in these specific weapon categories. The aggregate value of major weapons produced in the largest country in the sample (India) in the period 1965–90 is almost 300 times that of the smallest (Pakistan). Equally, there are differences in the range of major equipment types produced in the sample countries.

Table 17.2 indicates the first year in which a major item of equipment in one of the specified product categories and designed in the producer country entered full-scale production. Systems produced under foreign licences have been excluded, as have prototypes and technology demonstrators which never reached the stage of production. In the column for helicopters, micro-light helicopters built in Indonesia in 1963 and in Argentina and Brazil since the late 1960s have been excluded. Although these would be useful in developing a general understanding of rotory-wing technology, they would be of little or no value in developing either the designs or industrial processes necessary to manufacture military helicopters.

For several of the Indian programmes, the years are shown in italics. This is because the Advanced Light Helicopter (ALH), the Akash, Nag, Prithvi and Trishul guided missiles and the Arjun main battle tank are all apparently close to but not in full-scale production at the time of writing. Most of these programmes have experienced major cost and time overruns and it would not be surprising if the dates currently set as production targets are not met.

The criteria for selection is that the item produced was designed in the country of manufacture. This does not mean that the design work was carried out exclusively by nationals from that country, and in many cases the design and development of these systems have been heavily dependent on the skills of foreigners—something which is also true for many systems manufactured by the major arms producers at various times.

All of the entries in the column for jet aircraft except for the Taiwanese jet trainer first produced in 1982 are derived from foreign aircraft designs. The Argentinian Pampa was designed by Dornier of the Federal Republic of Germany; the Brazilian AMX was designed in collaboration with Aeritalia of Italy; the Indian HF-24 was designed by a German, Kurt Tank; and finally, the Israeli Nesher was based on plans for the Dassault Mirage-3 fighter obtained in Switzerland. In the helicopter column the Indian ALH and South African Rooivalk helicopters are both heavily dependent on German and French designs, respectively. In the column for main battle tanks, the South Korean K-1 tank owes a great deal to US design. The South African Oliphant is based on the design of the British Chieftain—albeit modified to the point where it can reasonably be described as a new tank.

The information presented in table 17.2 suggests that three categories of major weapon platform are particularly difficult to develop. These categories

Table 17.2. First year of indigenous production in selected weapon categories

Country	Jet aircraft	Propellor driven aircraft	Heli-copters	Guided missiles	Main battle tanks	Other armoured vehicles	Large-calibre artillery	Radar	Major surface warships	Sub-marines
Argentina	1987	1946	..	1978	..	1980	1978	..	1943	..
Brazil	1986	1944	..	1987	1985	1974	1969	..	1983	..
Chile	..	1990	..	..	..	1984	..	..	..	..
Egypt	..	..	..	1982	..	1966	1981	..	..	..
India[a]	1963	1953	*1993*	*1993*	*1995*	..	1980	1988	1978	..
Indonesia	..	1974	..	..	..	..	..	..	..	..
Israel	1971	1972	..	1970	1977	1975	1968	1977	..	..
Pakistan	..	..	..	..	..	..	1990	..	..	..
Singapore	..	..	..	..	..	..	1986	..	..	..
S. Africa	..	1961	1990	1975	1991	1973	1979	1991	..	..
S. Korea	..	..	..	..	1987	..	1976	..	1980	1983
Taiwan	1982	1976	..	1979	..	..	1976	..	..	..

[a] Years in italics indicate that some systems are not yet in full-scale production.
Source: SIPRI arms trade data base.

are submarines—where only South Korea of the sample countries has developed an indigenous design, for a mini-submarine—helicopters and radars. Only slightly easier to design and develop are main battle tanks and major surface combatants. By contrast all but two of the countries in the sample manufacture large-calibre artillery pieces to local designs.[24] Of the countries producing jet aircraft, India ended production of the HF-24 Marut in 1977 but hopes to restart it with production of the Light Combat Aircraft in the mid-1990s. Israel ended production of the Kfir fighter in 1989 having terminated the Lavi programme in 1987 before this design reached production. While Israel continues to fund two technology demonstrator programmes based on the Kfir, the production of jet aircraft is now confined to civilian business jets.

It seems likely that the group of countries included in this sample will sustain their significant involvement in arms production. There is an economic question-mark over the ability of Egypt and Israel to support arms production. Neither has a large economic base, and both are heavily dependent on foreign economic assistance to support their military programmes. The Indian arms industry has suffered short-term problems in adjusting to the economic and political collapse of one of its most important trade partners and arms suppliers—the former Soviet Union. However, the Indian Government retains its commitment to develop the defence industrial base further, and both Russia and Ukraine have powerful incentives not to disrupt relations with an important foreign customer.

[24] The definition of large-calibre artillery applied here is taken from the 1990 Treaty on Conventional Armed Forces in Europe and includes: guns, howitzers, artillery pieces combining the characteristics of guns and howitzers, mortars and multiple launch rocket systems with a calibre of 100 mm and above. Several of these countries developed mortars and guns with a calibre of less than 100 mm earlier than the date listed in table 17.2.

In Chile, Indonesia, Pakistan and Singapore the value of major weapon systems produced indigenously is much lower than is the case for the other sample countries. However, this is to some degree a function of the type of defence industrial activity in these countries. Pakistan and Singapore have devoted much of their investment in the defence industry to the repair and maintenance of major systems which were obtained as direct imports. Chile and Singapore both have export-oriented munitions industries. None of these activities is reflected in table 17.1. The table also omits another form of defence industrial activity: the modification and improvement of major weapon platforms of foreign origin. Industries in Israel, Pakistan, Singapore, South Korea and Taiwan have all learned how to upgrade old weapon platforms by fitting new equipment retrospectively. The 'retro-fitting' of new power units and frames prolongs the active life of a system. The addition of new sub-systems (often military electronics) prolongs the life of a system and increases its effectiveness. This is an area of industrial activity which is likely to become more common.

III. The industrial organization of arms production

Clearly, a significant number of developing countries have invested sizeable human and economic resources in creating arms industries. This has only been possible because governments have taken an active role not only as customers but often also as managers. Only government capital investment made it possible to import machine tools and equipment while subsidies were required for an extended period during which the production process was being learned.

More recently, many of these countries have tried to restructure the arms industry by importing management structures and techniques from the civilian private sector. This move towards a more commercial management style is apparently intended to bring greater efficiency in the use of resources and also greater flexibility in establishing external contact with industry in other countries. Whereas in the past one of the primary motives for arms production was to reduce external dependence, in the 1980s few if any countries retain the goal of full self-sufficiency in arms production. Rather, governments have increasingly sought to use their arms procurement policy as means of pursuing an international industrial policy. While South Korea and Taiwan have been pursuing this strategy for a considerable time, it is also a logical pattern for smaller arms-producing countries such as Indonesia and Singapore, which do not intend to develop a very wide range of arms production capabilities.

It is not surprising that such developments are most visible in Singapore, South Korea and Taiwan, where an aggressive private sector has been the basis for the rapid growth in the civilian manufacturing industry. However, South Africa and the countries of Latin America have also adopted this approach.

A related trend in some of the sample countries has been towards the privatization of arms production. This has occurred or is occurring in Israel, Singapore and South Africa.

In Israel the Ministry of Defence has announced a plan to privatize four government-owned companies soon, with the intention of further privatization in the future. The first companies scheduled for privatization are Ashot Ashkelon Industries, Beit Shemesh Engines, Elta Electronic Industries and Shekem.[25] In Singapore the largest arms producers—Singapore Aerospace and Singapore Technologies (formerly Chartered Industries)—are also being privatized. In South Africa Armscor is being split into 23 smaller industrial units, each organized around a specific product or narrow range of products. There will still be some linkage between these new companies—all of which will have a common holding company called Denel.[26] However, the intention is the same: to move away from a highly centralized management system within the arms production sector.

In the future these companies will have to focus on sustaining themselves without the scale of direct government subsidies they have been accustomed to having. They are likely to do this by further developing their civilian activities—such as the overhaul of airliners—as well as by continuing to seek export opportunities. However, as the Director-General of the Israeli Defence Ministry has noted, over the long term diversification seems to be more promising than dependence on exports which is a very high risk industrial strategy.[27]

Egypt and India by contrast continue to try to manage both technology development and industrial production centrally. Where external technologies is required it is sought through the negotiation of government-to-government agreements. In the 1980s India concluded such agreements with almost all the European countries and with the United States.[28]

While the nature of the products means that there will inevitably continue to be close government involvement in the arms industry in all countries, nevertheless there is a trend among many of the 12 countries discussed towards the privatization of arms production. Governments apparently prefer management decisions to be made more on the basis of commercial considerations and less on the basis of ambitous plans prepared by national scientific and military establishments. As noted above, one commercial requirement is linkage to an international market which is itself evolving.

Arms-producing companies, like their civilian counterparts, are developing global supplier networks. This is often a consequence of offset and countertrade agreements with arms suppliers. Governments in recipient countries have increasingly insisted on the establishment of close ties between local industry and companies located in the seller country as a condition of arms agreements. Consequently, even systems whose final assembly takes place in the United States or Western Europe are likely to have a growing percentage of foreign

[25] *Jerusalem Post International Edition*, 28 July 1990, p. 20.
[26] *Jane's Defence Weekly*, 7 Mar. 1992, p. 395; *Military Technology*, May 1992, p. 110; *Defence*, May 1992, p. 9.
[27] General David Ivry interviewed in Allen-Frost, P., 'Israel's road to reform', *Jane's Defence Weekly*, 10 Mar. 1990, p. 464.
[28] Anthony (note 19).

components. Moreover, in most product sectors (but especially in aerospace) there is no fixed barrier between commercial and military-related transnational linkages. As an offset for a military aircraft purchase, for example, a company in the recipient country may be awarded an offset contract to supply components or even major sub-assemblies such as wings for commercial aircraft.

A continuing expansion of contractor–sub-contractor relationships both within the industrialized world and between industrialized and industrializing countries is a logical development. This will create a trans-national network of company-to-company ties of various kinds between arms producers. This in turn stimulates the creation of new capacities for the production of the components and sub-assemblies which make up major weapon systems and will increase the movement of such items.

IV. Arms production as an element of arms procurement

It is possible to reinforce the point that self-sufficiency is rarely a realistic goal for this group of countries by looking at the figures in table 17.3. Here, production of major conventional weapon systems is expressed as a percentage alongside direct imports and the licensed production of the same categories of equipment.

Of the larger arms producers, only South Africa has virtually eliminated direct imports of major systems and this was not through choice. In the 1980s South Korea procured, for extended periods, more than 90 per cent of its major systems from local production—some of it under foreign licences. Brazil also came close to eliminating direct imports in the period 1978–90, although 20 per cent of its major weapon systems are still bought direct from foreign suppliers; Israel has achieved a similar ratio of production-to-imports.

For other major producers (including Argentina, India and Taiwan) direct deliveries of foreign equipment or kits for local assembly normally form more than 50 per cent of total major conventional weapon deliveries.

The smaller producers are rarely able to provide more than a small percentage of major equipment from local production, in the case of Pakistan perhaps as little as 4 per cent.

V. Conclusions: 'the party's over'

While definitive conclusions cannot be drawn from the data presented here, several general statements are suggested by them.

1. The data suggest that the overall aggregate volume of production of major conventional weapons outside the traditional major arms producers is growing only very slowly, if at all.

2. Overall growth has occurred because additional countries have joined the group of producers of major conventional weapons rather than because more

traditional producers are inceasing the scale of production. Such production began on a significant scale in Chile and Indonesia only in the early 1980s. Pakistan and Singapore were even later entrants as producers of major conventional weapons. The countries in which arms production has shown a fairly consistent pattern of growth are Indonesia and Taiwan. However, Indonesian growth has been from a very low base level. The value of production recorded for traditional arms producers such as Brazil, Egypt, India and Israel did not grow in the second half of the 1980s.

3. In those countries with a major design capacity there are insufficient funds to make full-scale production viable. In Israel it proved impossible to sustain programmes such as that for the Lavi fighter aircraft, while in Argentina and Brazil the future of the Pampa and AMX aircraft depends on export orders. The same barrier exists in the case of main battle tank production. These countries face the difficulty of deciding whether future systems should be funded while it is probable that there will be no customers for those products developed successfully.

4. In a significant number of the countries in the sample, the arms industry is undergoing a major restructuring. The principal features of this process are the adoption by arms producers of forms of industrial organization and management more usually associated with civilian sectors. Among the side-effects of this restructuring are the reduction in employment by many individual arms-producing companies and efforts to reduce dependence on military contracts within overall revenues.

5. The fifth finding suggested by the data presented in this chapter is that those countries which have developed their arms production capacities are not giving it up. Only two cases—Argentina and South Africa—show a sustained downward trend in the value of this type of arms production during the period surveyed. In the case of Argentina, production peaked in 1981, while in South Africa the largest volume of production is recorded for 1976.

Some private sector arms manufacturing companies located in 'third tier' countries which can move away from arms production are doing so. Others find diversification difficult because of their high dependence on arms sales and a lack of capital to finance diversification through acquisition.

As noted in the introduction, for many of the sample countries the political and military incentives for indigenous arms production have never been greater. Consequently, investment in arms production is most likely to be sustained where the arms industry is in government ownership. In other cases structural changes in the industry may be the only way of preserving an arms production capacity. However, new management techniques and new forms of industrial organization—however innovative—cannot compensate for a basic lack of demand for the products made by the arms industry.

Table 17.3. Shares of trade in, licensed production and indigenous production of major weapons, 1965–90

Figures are percentages.

		1965	1966	1967	1968	1969	1970	1971	1972	1973	1974	1975	1976	1977	1978	1979	1980	1981	1982	1983	1984	1985	1986	1987	1988	1989	1990
Argentina	A	17	–	4	13	12	15	3	4	5	2	2	11	8	33	18	24	42	21	13	15	44	40	42	26	23	49
	B	–	–	–	1	1	1	–	–	–	2	3	3	2	2	2	6	27	13	8	6	25	28	40	–	–	46
	C	83	100	96	86	87	84	97	96	95	95	95	86	90	65	80	70	31	65	79	79	30	32	18	74	77	5
Brazil	A	4	3	3	2	5	21	51	14	13	49	24	27	27	28	67	52	70	86	80	80	82	43	22	36	53	7
	B	–	–	–	–	–	–		8	3	12	5	2	3	2	27	23	4	4	2	1		31	6	–	–	4
	C	96	97	97	98	95	79	49	77	84	40	71	71	69	70	6	25	26	10	17	19	18	26	72	64	47	20
Chile	A	–	23	38	48	–	–	3	–	–	–	–	–	–	–	–	–	4	4	13	5	43	48	9	49	20	12
	B	–	–	–	45	–	–	3	–	–	–	–	–	–	–	–	2	3	5	11	3	36	28	7	10	18	11
	C	100	77	62	7	100	100	94	100	100	100	100	100	100	100	100	98	93	91	76	92	20	24	84	41	62	78
Egypt	A	5	2	1	2	3	1	1	–	1	2	19	10	4	7	2	1	2	1	6	6	13	4	4	26	36	12
	B	2	1	–	–	1	–	–	–	–	–	–	–	–	–	–	–	–	–	–	–	–	1	1	1	1	1
	C	92	97	98	98	96	99	99	100	99	98	81	90	96	93	98	99	98	99	91	90	80	93	94	61	47	79
India	A	17	14	41	40	35	36	34	32	35	50	65	40	28	29	40	38	33	24	34	47	35	18	17	26	21	42
	B	8	8	24	22	20	21	19	17	20	26	27	17	13	15	16	16	14	11	9	14	15	14	16	16	18	38
	C	75	78	35	38	45	43	47	51	45	23	8	43	59	55	44	46	53	66	56	40	49	67	67	57	62	21
Indonesia	A	–	67	–	13	–	19	–	–	–	3	3	5	15	10	4	2	4	6	14	27	27	17	22	28	35	41
	B	–	–	–	–	–	–	–	–	–	–	–	1	5	6	2	1	3	5	13	5	5	6	7	13	16	15
	C	100	33	100	87	100	81	100	100	100	97	97	94	80	84	94	97	93	89	73	69	68	77	70	59	48	44
Israel	A	7	4	2	8	5	8	15	29	19	45	31	23	31	36	57	37	31	40	65	70	72	53	18	45	83	73
	B	–	–	–	–	–	–	–	–	–	–	–	–	–	–	1	–	–	–	2	–	3	–	–			
	C	93	96	98	92	95	92	85	71	81	55	69	77	68	63	42	63	68	60	33	30	25	47	82	55	17	27
Pakistan	A	–	–	–	–	–	–	–	–	–	–	–	3	1	1	1	1	1	1	1	1	1	1	1	1	1	2
	B	–	–	–	–	–	–	–	–	–	–	–	–	–	–	1	–	–	–	–	–	–	–	–	1	3	1
	C	100	100	100	100	100	100	100	100	100	100	100	97	99	99	98	99	98	99	99	99	99	99	99	97	96	97

Singapore	A	–	–	–	55	9	–	5	–	–	45	29	–	–	44	19	–	–	–	2	13	17	21	20	2	–	34
	B	–	–	–	45	7		4	–	–	36	20	–	–	35	15	–	–	–	–	–	–	–	–	–	–	20
	C	100	100	–	–	84	100	91	100	100	18	50	100	100	20	66	100	100	100	98	87	83	79	80	98	100	45
S. Africa	A	–	22	23	32	54	17	50	27	34	30	48	52	63	38	66	68	98	98	66	97	97	55	84	83	100	98
	B	–	7	7	10	16	5	15	8	21	18	29	33	31	12	21	19	2	2	26	–	–	20	5	5	–	–
	C	100	71	70	58	30	78	35	64	45	52	23	14	6	51	12	13	–	–	8	3	3	25	11	12	–	2
S. Korea	A	–	–	–	–	–	–	2	9	–	–	–	12	10	14	17	19	64	54	52	72	70	77	51	35	53	66
	B	–	–	–	–		–	–	–	–	–	–	–	5	6	8	9	4	16	18	14	11	11	5	6	3	3
	C	100	100	100	100	100	100	98	91	100	100	100	88	84	80	76	73	32	30	30	14	19	12	45	59	44	31
Taiwan	A	–	–	–	–	5	5	6	11	10	27	48	31	26	34	47	45	39	44	38	51	34	23	14	26	35	50
	B	–	–	–	–	3	4	4	8	8	5	20	33	27	29	36	36	28	24	23	27	22	13	16	20	20	30
	C	100	100	100	100	92	91	90	80	82	67	31	36	47	37	17	19	34	33	39	23	44	64	71	54	45	20

Notes:

A = Indigenous production

B = Licensed production

C = Direct imports

Annexes

Annexe A
Yugoslavia: arms production before the war

Herbert Wulf

Dozens of cease-fires have been agreed upon since the start of the war in Yugoslavia in mid-1991, but after each agreement the increasingly bloody war has continued. Why could the war not be stopped by cutting off the delivery of weapons? Why did the warring parties not 'run out of ammunition'? Despite an arms embargo, boycott and economic sanctions, the supply of weapons to the warring parties has continued.

This annexe presents data on the arms industry in the former Yugoslavia (see tables A1 and A2); however, it is not possible to present a clear picture of the state of the industry at the end of 1992. Arms production in some of the regions and newly created states has been affected by the war (particularly in Bosnia and Herzegovina) whereas, it can be assumed, production of weapons and ammunition is continuing in other regions and states of the former Yugoslavia (especially Serbia).

Before the beginning of the war the Serbian-dominated Yugoslav Federal Armed Forces consisted of about 180 000 soldiers, equipped with nearly 2000 main battle tanks, about 1400 armoured vehicles and armed personnel carriers, over 4000 guns and artillery pieces of different calibres, 150 fighter aircraft, almost 200 helicopters, four frigates, five regular and six midget submarines, and about six dozen patrol and fast attack craft.[1] The Yugoslav budget for military spending in 1990 amounted to 4.6 per cent of the gross national product (GNP).[2]

Before the destruction of some of the arms production sites during the war, Yugoslavia had a capable arms industry that produced most of the equipment procured by the Yugoslav armed forces and exported small arms, tanks, aircraft and other military equipment.[3] About 150 different factories produced a wide range of equipment: from pistols to primary trainer aircraft, from mortars to main battle tanks, from machine-guns to missiles, from helicopters to fighting

[1] International Institute for Strategic Studies, *The Military Balance 1990–91* (Brassey's: Oxford, 1990); *Jane's Fighting Ships 1991–92* (Jane's Publishing Co.: Coulsdon, UK, 1991).

[2] Tomovic, R. and Strujic, D., *Conversion of Military Potential for Civilian Purposes,* CSS Papers no. 5 (Center for Strategic Studies [CSS]: Belgrade, 1991), pp. 7–22.

[3] The *SIPRI Yearbooks: World Armaments and Disarmament* (Oxford University Press: Oxford, annual), 'Register of the trade in and licensed production of major conventional weapons in industrialized and developing countries'. The *SIPRI Yearbook 1992: World Armaments and Disarmament* (Oxford University Press: Oxford, 1992), pp. 335 and 341, lists the export of 6 G-4 Super Galeb jet trainer aircraft to Burma and 200 T-72 main battle tanks to Kuwait.

Table A1. Small arms and ordnance production in the former Yugoslavia, 1991

Pistols	Rifles and semi-automatic rifles	Machine-guns and machine-pistols	Mortars	Cannon, anti-tank launchers, recoilless guns
7.62-mm M57	5.56-mm M80	5.56-m M82	50-mm M8	30-mm CZ
7.62-mm M70	7.62-mm M49	7.62-mm M72B1	60-mm M57	44-mm M57
7.65-mm M70	7.62-mm M56	7.62-mm M72AB1	81-mm M31	64-mm RBR-M80
9-mm M65	7.62-mm M57	7.62-mm M77B1	81-mm M68	80-mm M60
	7.62-mm M59	7.65-mm M84	120-mm UBM52	80-mm M60
	7.62-mm M66A1	7.92-mm M53	120-mm M74	90-mm M79
	7.62-mm M70B1	12.7-mm NSV	120-mm M75	
	7.92-mm M76			

Sources: SIPRI archive and data bank; *Jane's Infantry Weapons*, several issues (Jane's Publishing Co.: Coulsdon, UK); (collated by the author).

ships. Fifty-four enterprises were organized in the Association of Industry of Armament and Defense Equipment (ADE), which relied on about 1000 subcontractors.[4] In addition, 130 laboratories and 26 testing facilities, which employed about 7000 research and development (R&D) experts, worked for the armed forces.[5]

About 60 per cent of the production sites were located in Serbia, and the backbone for small arms supplies to Serbian troops that are used in this war were the Zavodi Crvena Zastava government weapon factories near Belgrade. The next most important production sites were located in Bosnia and Herzegovina in the towns of Banja Luka, Novi Travnik and Mostar. Fighting ships were built in Croatia on the basis of designs of the Marine Technology Institute in Zagreb.[6] Since the weapon factories were spread throughout the country and since central government control over them ceased to exist, the different warring parties have had access to at least some of the weapons as long as they were produced.

Arms-producing facilities and their production have of course been affected by the war. The SOKO aircraft company in Mostar, for example, had first to defer the test-flight of its new G-4M Super Galeb fighter aircraft and then to stop production totally. With the destruction of the town it is highly unlikely that the company can produce aircraft now. It is probably equally unlikely that production of military electronics can continue in Sarajevo.

[4] Tomovic and Strujic (note 2), p. 15.

[5] Mirkovic, T., *The General Aspects of Military–Technological Potentials in Yugoslavia,* CSS Papers no. 5 (note 2), pp. 23–44.

[6] *Jane's Defence Weekly,* 4 Apr. 1992, pp. 588–90; *Jane's Defence Weekly,* 18 July 1992, pp. 27–28.

Table A2. Arms production in the former Yugoslavia, 1991

Weapon description	Weapon designation	Comment	Place of production	Country/region
MBTs and armoured vehicles	M-84	MBT: modernized Soviet T-72; exported to Kuwait	Ljubljana	Slovenia
	BVP M80A	Mechanized infantry combat vehicle: in production since 1980 (c. 600 produced)	Banja Luka Novi Travnik	Bosnia and Herzegovina Bosnia and Herzegovina
	M-60P	Armoured personnel carrier: in production since 1965		
	BOV	Wheeled armoured personnel carrier: developed in the early 1980s		
Aircraft	G2-A Galeb	Light attack and trainer aircraft: exported to Libya and Zambia	Mostar	Bosnia and Herzegovina
	G-4 Super Galeb	Light attack and trainer aircraft: 136 in the Air Force in 1989; 6 exported to Burma; production of an advanced version apparently stopped in 1991		
Helicopters	SA342L Gazelle	Helicopter: French licence; armed with Soviet missiles	Mostar	Bosnia and Herzegovina
Aircraft engines	Viper 632-41 Viper 632-46	British licence	Belgrade	Serbia
Artillery (guns and multiple rocket launchers)	M46/84	155-mm gun: conversion of a Soviet model	Banja Luka	Bosnia and Herzegovina
	M84	152-mm howitzer	Novi Travnik	Bosnia and Herzegovina
	D-30J	122-mm howitzer		
	M56	105-mm howitzer		
	LRSV M-87	262-mm MRLS: exported to Iraq		
	M-77	128-mm MRLS: exported to Iraq		
	M-85, M-63 and M-88	128-mm MRLS		

Weapon description	Weapon designation	Comment	Place of production	Country/region
Fighting ships	Sava Class	2 submarines: built in 1978 and 1981	Split	Croatia
	Heroj Class	3 submarines: 1968–70	Pula and Split	Croatia
	M 100-D Class	6 midget submarines	Split	Croatia
	Split and Kotor Classes	4 frigates: 2 imported from USSR, 2 licence-produced	Split	Croatia
	Kobra Class	4 corvettes: not completed	Kraljevica	..
	Mornar Class	2 corvettes: 1959 and 1965	Kraljevica	..
	Koncav Class 240	6 fast attack craft	Kraljevica	..
	Mirna Class 140	10 patrol craft	Kraljevica	..
Military electronic and optronic equipment	..	..	Sarajevo	Bosnia and Herzegovina
			Belgrade	Serbia
Radar	IR-3	Battlefield radar	Banja Luka	Bosnia and Herzegovina
	IFF	Identification friend or foe		
Infantry weapons[a]	..	(see table A1)	Belgrade	Serbia
			Krusevac	Serbia
			Kragujevac	Serbia
Munitions	..	..	Titovo Uzice	Serbia
			Novi Travnik	Bosnia and Herzegovina
			Sarajevo	Bosnia and Herzegovina

Note: MBT = main battle tank; MRLS = multiple rocket launcher system.

[a] Infantry weapons include pistols, submachine-guns, machine-guns, rifles, automatic cannon, anti-tank launchers, rocket launchers, mortars, grenades and mines.

Sources: SIPRI archive and data bank; *Jane's All the World's Aircraft*, several issues, *Jane's Armour and Artillery*, several issues; *Jane's Fighting Ships*, several issues (Jane's Publishing Co.: Coulsdon, UK); (collated by the author).

According to Yugoslav sources, 90 per cent of the production of arms was based on domestic R&D activities,[7] while 10 per cent of the former Yugoslav equipment is based on foreign licences or requires foreign technology input. A

[7] Tomovic and Strujic (note 2), p. 15.

version of the Soviet T-72 main battle tank, with the designation M-84, has been built since 1979; the producers have modernized the tank with fire-control systems supplied by Western companies. Armoured vehicles were produced under a licence bought from Austria during the 1960s. US technical assistance went into the production of fighter aircraft; a British company supplied a licence for an aircraft engine; and the French Gazelle helicopter has also been licence-produced in Mostar. According to public information, this co-operation has stopped as a result of the United Nations arms embargo imposed on 25 September 1991.

Since the former Yugoslav arms industry was technically capable of producing most of the equipment locally, it is obvious that an arms embargo could not be successful in the short term. Furthermore, in contrast to the 1991 Persian Gulf War, the war in Yugoslavia is not so material-intensive. The attrition rate is much lower; thus, aircraft, helicopters, tanks, guns, and so on were destroyed during the war only in small numbers. There has been speculation about the extent of illegal exports of weapons in violation of the UN embargo. Reports have been published which state that Croatia 'built a web of contacts' to import weapons with false end-user certificates to circumvent the UN arm embargo.[8] From the information available it seems, however, that illegal arms transfers have had no major impact on the way in which the war is being fought. The quantity of the equipment available both in the former Yugoslav armed forces and in stores located at many sites in different parts of the country, as well as locally produced weapons, was far greater than was previously known or than was suspected to have been smuggled to the various belligerent parties.

[8] *The Independent*, 10 Oct. 1992, p. 12; *International Herald Tribune*, 30 Jan. 1992, p. 2.

Annexe B
Bibliography of works on arms production

Prepared by Espen Gullikstad

Adam, B., Zaks, A. and De Vestel, P., *Contexte et perspectives de restructuration de l'industrie de l'armement en Wallonie* (Groupe de recherche et d'information sur la paix, GRIP: Brussels, 1991)

Adams, G., *The Iron Triangle: The Politics of Defense Contracting* (Council on Economic Priorities: New York, 1981)

Adams, G., *The Role of Defense Budgets in Civil–Military Relations* (Defense Budget Project: Washington, DC, 1992)

Adelman, K. L. and Augustine, N. R., 'Defense conversion', *Foreign Affairs*, vol. 71, no. 2 (spring 1992), pp. 26–48

Adelman, K. L. and Augustine, N. R., *The Defence Revolution: Strategy for the Brave New World by an Arms Controller and an Arms Builder* (Institute for Contemporary Studies, ICS: San Francisco, Calif., 1990)

Ahlström, M., *Försvarsrelaterad offset. En studie av kompensationsaffärer och industriell samverkan i försvarsindustrin* (Försvarets forskningsanstalt, FOA, Institution 13, FIND-projektet: Sundbyberg, Sweden, 1992)

ALADIN, 'The French armaments Industry in 1989', *Defence*, vol. 20, no. 3 (1989), pp. 187–211

Albrecht, U. and Nikutta, R., *Die sovjetische Rüstungsindustrie* (Westdeutscher Verlag: Opladen, 1989)

Albrecht, U., Lock, P. and Wulf, H., *Mit Rüstung gegen Arbeitslosigkeit?* (Rowohlt Verlag: Reinbek, 1982)

Anthony, I., *The Arms Trade and Medium Powers: Case Studies of India and Pakistan 1947–90* (Harvester Wheatsheaf: Hemel Hempstead, 1992)

Anthony, I. (ed.), SIPRI, *Arms Export Regulations* (Oxford University Press: Oxford, 1991)

Anthony, I., Courades Allebeck, A. and Wulf, H., SIPRI, *West European Arms Production: Structural Changes in the New Political Environment*, A SIPRI Research Report (SIPRI: Stockholm, 1990)

Aris, Hakki, 'The Turkish defence industry: an overview', *Military Technology*, vol. 13, no. 4 (1989), pp. 95–103

Baek, K., McLaurin, R. D. and Moon, C. (eds.), *The Dilemma of Third World Defense Industries: Supplier Control or Recipient Autonomy?* (Westview Press: Boulder, Colo., and Center for International Studies, Inha University: Inchon, Republic of Korea, 1989)

Ball, N. and Leitenberg, M. (eds), *The Structure of the Defense Industry: An International Survey* (Croom Helm: London, 1983)

Bartzokas, A., 'The developing arms industries in Greece, Portugal and Turkey', eds M. Brzoska and P. Lock, SIPRI, *Restructuring of Arms Production in Western Europe* (Oxford University Press: Oxford, 1992), pp. 166–77

Beaver, P., Porteous, H., Hwang, D. J. and Lok, J. J., 'JDW special report: Korean business', *Jane's Defence Weekly*, vol. 16, no. 20 (1991), pp. 961, 965–69

Bennett, F., 'Defence industry and research and development', in eds D. Ball and C. Downes, *Security and Defence: Pacific and Global Perspectives* (Allen and Unwin: Sydney, 1990), pp. 317–27

Berger, M. *et al.*, *Produktion von Wehrgütern in der Bundesrepublik Deutschland*, IFO Studien zur Industriewirtschaft no. 42 (IFO Institut: Munich, 1992)

Bittleston, M., *Co-operation or Competition? Defence Procurement Options for the 1990s*, Adelhi Paper no. 250 (Brassey's/International Institute for Strategic Studies: London, 1990).

Bitzinger, R. A., *Chinese Arms Production and Sales to the Third World* (Rand Corporation: Santa Monica, Calif., 1991)

Blunden, M., 'Collaboration and competition in European weapons procurement: the issue of democratic accountability', *Defense Analysis*, vol. 5, no. 4 (Dec. 1989), pp. 291–304

Boniface, P. (ed.), 'L'industrie française de défense', *Relations internationales et stratégiques*, vol. 1, no. 1 (1991), pp. 7–91

Bontrup, H. J. and Zdrowomyslaw, N., *Die deutsche Rüstungsindustrie* (Distel Verlag: Heilbronn, 1988)

Bozzo, L., (ed.), *Exporting Conflict: International Transfers of Conventional Arms* (Cultura Nuova: Florence, 1991)

Brauer, J., 'Arms production in developing nations: the relation to industrial structure, industrial diversification, and human capital formation', *Defence Economics*, no. 2 (1991)

Brigagao, C., 'The Brazilian arms industry', *Journal of International Affairs*, vol. 40, no. 1 (summer 1986), pp. 101–15

Brock, L. and Jopp, M. (eds), *Sicherheitspolitische Zusammenarbeit und Kooperation der Rüstungswirtschaft in Westeuropa* (Shriftenreihe des Arbeitskreises Europäische Integration Bd 25, Nomos Verlagsgesellscaft: Baden-Baden, 1986)

Brzoska, M. and Lock, P. (eds), SIPRI, *Restructuring of Arms Production in Western Europe* (Oxford University Press: Oxford, 1992)

Brzoska, M., Lock, P. and Wulf, H., *Rüstungsproduktion in Westeuropa*, IFSH-Forschungsberichte Heft 15 (Institut für Friedensforschung und Sicherheitspolitik: Hamburg, 1979)

Brzoska, M. and Ohlson, T. (eds), SIPRI, *Arms Production in the Third World* (Taylor & Francis: London and Philadelphia, 1986)

Butterwegge, C. and Senghass-Knobloch, E. (eds), *Von der Blockkonfrontation zur Rüstungskonversion? Die Neuordnung der internationalen Beziehungen, Abrüstung und Regionalentwicklung nach dem Kalten Krieg* (Lit-Verlag: Münster, Hamburg, 1992)

Chow, B., Forest, G. and Grant, R., *Third World Missiles: Trends, Threats, Economics and Safeguards* (RAND Çorporation: Santa Monica, Calif., 1991)

Clarke, M. and Hague, R. (eds), *European Defence Co-operation: America, Britain and NATO*, Fulbright Papers, vol. 7 (Manchester University Press: Manchester, 1990)

Collet, A., *Armements: mutation, réglementation, production, commerce* (Economica: Paris, 1989)

Cooper, J., *The Soviet Defence Industry: Conversion and Reform* (Pinter: London, 1991)

Cothier, P. and Moravcsik, A., 'Defense and the single market—the outlook for collaborative ventures', *International Defense Review*, vol. 24, no. 9 (1991), pp. 949–50, 955–63

Covington, T. G., Brendley K. W. and Chenoweth, M. E., *A Review of European Arms Collaboration and Prospects for its Expansion under the Independent European Program Group*, Report no. N-2638-ACQ (Rand Corporation: Santa Monica, Calif., July 1987)

Cowen, R. H. E., *Defense Procurement in the Federal Republic of Germany* (Westview Press: Boulder, Colo., 1986)

Creasey, P. and May, S. (eds), *The European Armaments Market and Procurement Cooperation* (Macmillan: London, 1988)

Cremasco, M., 'La collaborazione tra Spagna e Italia nei settori technologici e industriali della difesa' (Instituto affari internazionali (IAI 9017): Rome, 1990)

Délégation Generale d'Armement (DGA), *L'industrie française de défense* (DGA: Paris, 1990)

Delhauteur, D., *La coopération européenne dans le domaine des équipements militaires: la relance du GEIP* (Groupe de recherche et d'information sur la paix, GRIP: Brussels, 1991)

Draper, A. G., *European Defense Equipment Collaboration. Britain's Involvement, 1957–87* (Macmillan: London, 1990)

Drifte, R., *Arms Production in Japan: The Military Applications of Civilian Technology* (Westview Press: Boulder, Colo., and London, 1986)

Drown, J. D., Drown, C. F. and Campbell, K., *A Single European Arms Industry?* (Brassey's: London, 1990)

Dussauge, P., *L'industrie française de l'armement* (Économica: Paris, 1985)

Edgar, A. D. and Haglund, D. G., *Japanese Defence Industrialization*, Centre for International Relations Occasional Paper no. 42 (Queen's University: Kingston, Ontario, Feb. 1992)

Erdem, V., 'Defence industry policy of Turkey', *Nato's Sixteen Nations*, Special Edition, 1989/1990

Erdem, V., 'History of the Turkish defence industry', *Nato's Sixteen Nations*, Special Edition, vol. 36, no. 2 (1991)

Evans, C., 'Reappraising Third World arms production', *Survival*, vol. 28, no. 2 (Mar./Apr. 1986), pp. 99–118

Faltas, S., *Warships and the World Market* (Martinus Nijhoff: The Hague, 1986)

Feddersen, H. B. and Silva, A. P., 'The Single European Market and the defence industry', *Nato's Sixteen Nations*, vol. 37, no. 2 (1992), pp. 13–16

Ferguson, A. (rapporteur), *Report drawn on Behalf of the Political Affairs Committee on Arms Procurement within a Common Industrial Policy and Arms Sales*, document no. 1-455/83, PE 78.344/fin. (European Communities, European Parliament: Strasbourg, 27 June 1983)

Filippi, A. and Perani, G., *Bilancio della difesa 1992: i programmi di acquisizione di armamenti* (Archivio disarmo: Rome, 1991)

Folta, P. H., *From Swords to Plowshares? Defense Industry Reform in the PRC* (Westview Press: Boulder, Colo., 1992)

Fouquet, D., Kohnstam, M. and Noelke, M. (eds), 'Dual-use industries in Europe', vols I, II and Executive Summary, Study carried out by Eurostrategies for the Commission of the European Communities, DG III (Brussels, 1991)

Franko-Jones, P., *The Brazilian Defense Industry* (Westview Press: Boulder, Colo., 1992).

Gansler, J. S., *The Defense Industry* (MIT Press: Cambridge, 1980).

Gething, M. J. and Preston, A., 'Country profile: Italy's defence industry: a common-sense approach (pp. 56–58); 'Italy's aerospace industry' (pp. 58–60); 'Italy's naval industry (pp. 60–62), *Defence*, vol. 19. no. 1(1988).

Gething, M. J., Paloczi-Horvath, G., Roberts, J. and Steadman, N., 'Scandinavia's defence industry', *Defence*, vol. 70, no. 11 (Nov. 1991), pp. 49–55.

Gießmann, H. J. (ed.), *Konversion im vereinten Deutschland: Ein Land—zwei Perspektiven?* (Nomos Verlagsgesellschaft: Baden-Baden, 1992).

Graham, C. P., 'Technology and Third World defense manufacturing: the Brazilian firm Engesa', *Defense Analysis*, vol. 6, no. 4 (1990), pp. 367-83

Greenwood, D., *Report on a Policy for Promoting Defence and Technical Co-operation among West European Countries, for the Commission of the European Communities*, document no. III-1499/80 (European Communities: Brussels, 1980).

Gregory, W. H., *The Defense Procurement Mess* (Lexington Books: Lexington, Mass., 1989)

Günlük-Senesen, G., 'Yerli Silah Sanayiinin Kurulmasinin Ekonomiye Olasi Etkileri' [Probable Economic Impacts of the Installation of the (Turkish) Domestic Arms Industry], *1989 Sanayi Kongresi Bildirileri*, I, MMO/134, 1989

Hagelin, B., 'International cooperation in conventional weapons acquisition: a threat to armaments control?', *Bulletin of Peace Proposals*, vol. 9, no. 2 (1978), pp. 144–55

Hagelin, B., *Military Production in the Third World*, FOA Report C 10230-M3 (National Defence Research Institute: Stockholm, June 1983)

Hagelin, B., *Neutrality and Foreign Military Sales: Military Production and Sales Restrictions in Austria, Finland, Sweden and Switzerland* (Westview Press: Boulder, Colo., 1990).

Haglund, D. G. (ed.), *The Defense Industrial Base and the West* (Routledge: London, 1989).

Hall, D., 'European cooperation in armaments research and development', *RUSI Journal*, vol. 133, no. 2 (summer 1988), pp. 53–58.

Hartley, K., *NATO Arms Cooperation: A Study in Economics and Politics* (Allen and Unwin: London, 1983)

Hartley, K., *The Economics of Defence Policy* (Brassey's: London, 1991)

Hartley, K., Hussain, F. and Smith, R., 'The UK defence industrial base', *Political Quarterly*, vol. 58, no. 1 (1987), pp. 62–71

Hartley, K. and Sandler, T. (eds), *The Economics of Defence Spending: An International Survey* (Routledge: London, 1990).

Hébert, J.-P., *Stratégie française et industrie d'armement* (Fondation pour les Etudes de Défense Nationale: Paris, 1991)

Huffschmid, J., *Für den Frieden produzieren. Alternativen zur Kriegsproduktion* (Pahl-Rugenstein: Cologne, 1981)

Huffschmid, J. and Voß, W., 'Militärische Beschaffungen–Waffenhandel–Rüstungskonversion in der EG: Ansätze koordinierter Steuerung', *PIW-Studien Nr. 7* (Progress-Institut für Wirtschaftsforschung: Bremen, 1991)

Hug, P. and Meier, R., *Rüstungskonversion: Die Umwandlung militärabhängiger Arbeitsplätze in zivile Beschäftigung* (Verlag Rüegger: Zurich/Chur, 1992)

Independent European Programme Group (IEPG), *Towards a Stronger Europe*, A Report by an Independent Study Team Established by Defense Ministers of Nations of the Independent European Programme Group to Make Proposals to Improve the Competitiveness of Europe's Defence Equipment Industry ('Vredeling Report'), (IEPG: Brussels, 1987)

India, Ministry of Defence, *Annual Report* (Government of India: New Dehli, various years)

Jane's All the World's Aircraft (Jane's Publishing Co.: Coulsdon, UK)

Jane's Armour and Artillery (Jane's Publishing Co.: Coulsdon, UK)

Jane's Fighting Ships (Jane's Publishing Co.: Coulsdon, UK)

Jane's Infantry Weapons (Jane's Publishing Co.: Coulsdon, UK)

Jane's Miitary Vehicles & Support Equipment (Jane's Publishing Co.: Coulsdon, UK)

Jeshurun, C. (ed.), *Arms and Defence in Southeast Asia* (Institute of Southeast Asian Studies: Singapore, 1989)

Kaldor, M., *The Baroque Arsenal* (Hill and Wang: New York, 1981)

Katz, J., *Arms Production in Developing Countries* (Lexington Books: Lexington, Mass., and Toronto, 1984)

Keidanren, Office of Defence Production Committee, *Defence Production in Japan* (Keidanren: Tokyo, Oct. 1991)

Klepsch, E. (rapporteur), Report Drawn on Behalf of the Political Affairs Committee on European Cooperation in Arms Procurement, document no. 83/78, PE 50944/fin. (European Communities, European Parliament: Strasbourg, 8 May

1978); also published as *Two-Way Street: USA–European Arms Procurement* (Brassey's: London, 1979; and Crane and Russak: New York, 1979)

Köllner, L. and Huck, B. J. (eds), *Abrüstung und Konversion. Politische Voraussetzungen und wirtschaftliche Folgen in der Bundesrepublik Deutschland* (Campus: Frankfurt/Main, 1990)

Kolodziej, E. A., *The Making and Marketing of Arms: The French Experience and its Implications for the International System* (Princeton University Press: Princeton, N.J., 1987)

Korea, Republic of, Ministry of National Defence, *Defense White Paper 1991–1992* (Seoul, 1992)

Latham, R., 'China's defense industrial policy: looking toward the year 2000', ed. R. H. Yang, *SCPS PLA Yearbook: 1988/89* (Sun Yatsen University: Kaohsiung, Taiwan, 1989)

Latham, R. and Slack, M., 'The European armaments market: developments in Europe's excluded industrial sector' (Programme in Strategic Studies, University of Manitoba, Department of Political Studies: Winnipeg, 1990)

Levene, P., 'Competition and collaboration: UK defence procurement policy', *RUSI Journal*, vol. 132, no. 2 (1989), pp. 3–5

Lewis, J., W., Di, H. and Litai, X., 'Beijing's defense establishment: solving the arms-export enigma', *International Security*, vol. 15, no. 4 (1991), pp. 87–109

Leysens, A. J., 'South Africa's military strategic link with Latin America: past developments and future prospects', *International Affairs Bulletin* (Braamfontein, South Africa), vol. 15, no. 3 (1991), pp. 23–47

Looney R., *Third World Military Expenditure and Arms Production* (Macmillan: London, 1988)

Looney, R., 'Have Third World arms industries reduced arms imports?', *Current Research on Peace and Violence*, vol. 12, no. 1 (1989), pp. 15–26

Louscher, D. J. and Salomone, E., *Technology Transfer and U.S. Security Assistance: The Impact of Licensed Production* (Westview Press: Boulder, Colo., and London, 1987).

Machmud, B., 'South Korea's aerospace industry', *Asian Defence*, vol. 20, no. 12 (Dec. 1990), pp. 80–84

Mandel, R., 'The transformation of the American defense industry: corporate perceptions and preferences', paper presented at the annual national meeting of the International Studies Association, Atlanta, Georgia, Apr. 1992

McMahon, P., *The Global Military–Industrial Complex*, Report for the Honorable Ruth Coleman, Senator from Western Australia, June 1987

McWilliams, J. P., *Armscor: South Africa's Arms Merchant* (Brassey's: London, 1989)

Mehrens, K., *Alternative Produktion: Arbeitnehmerinitiatien für sinnvolle Arbeit* (Bund-Verlag: Cologne, 1985)

Melmann, S., *The Permanent War Economy: American Capitalism in Decline* (Simon and Schuster: New York, 1974)

Miller, B., 'Global links in the aerospace industry', *Interavia*, vol. 46 (June 1991), pp. 12–25

Miller, S. E., *Arms and the Third "orld: Indigenous Weapons Production*, PSIS Occasional Paper no. 3 (Graduate Institute of International Studies: Geneva, December 1980)

Mintz, A., 'The military–industrial complex: the Israeli case', ed. M. Lissak, *Israeli Society and its Defense Establishment: The Social and Political Impact of a Protracted Violent Conflict* (Frank Cass: London, 1984), pp. 103–27

Molas-Gallart, J., 'Arms production and modernization in Spain', eds M. Brzoska, and P. Lock, SIPRI, *Restructuring of Arms Production in Western Europe* (Oxford University Press: Oxford, 1992), pp. 154–66

Moravcsik, A., 'The European armaments industry at the crossroads', *Survival*, vol. 32, no. 1 (1990), pp. 65–85

Nash, T., 'The new shape of the Swedish defence industry', *Military Technology*, vol. 15, no. 8 (1991), pp. 10–21

Nativi, A., 'What future for the Italian aerospace industry?', *Military Technology*, vol. 14, no. 5 (1992), pp. 44–51

Neuman, S. G., 'International stratification and Third World military industries', *International Organization*, vol. 38, no. 1 (winter 1984), pp. 167–97

Nolan, J. E., *Military Industry in Taiwan and South Korea* (Macmillan: London, 1986)

Opitz, P., 'Rüstungsproduktion und Rüstungsexport der DDR', Arbeitspapiere der Berghof-Stiftung für Konfliktforschung, no. 45 (Berghof-Stiftung für Konfliktforschung: Berlin, 1991)

Paukert, L. and Richards, P. (eds), *Defence Expenditure, Industrial Conversion and Local Employment* (International Labour Office: Geneva, 1991)

Peck, M. J. and Scherer, F. M., *The Weapons Acquisition Process: An Economic Analysis* (Boston, Graduate School of Business Administration, Harvard University: Cambridge, Mass., 1962)

Pianta, M. and Perani, G., *L'industria militare in Italia. Ascesa e declino della produzione di armamenti* (Edizione Associate: Roma, 1990)

Pollins, B. M., *Arms and Archimedes: The Newly Industrializing Countries in the Spiraling Global Arms Market* (International Institute for Comparative Social Research: Berlin, January 1982)

Raghunathan, S., 'India's move towards defense self-reliance, and the new search for defense exports', *Defense & Foreign Affairs*, vol. 18, no. 4 (Apr. 1990), pp. 29–31

Ratner, J. and Thomas, C., 'The defence industrial base and foreign supply of defence goods', *Defence Economics*, vol. 2, no. 1 (1990), pp. 57–68

Reed, J. (ed.), *The Defence Industry Yearbook 1990: UK and Europe* (Longman: London, 1989)

Reiser, S., *The Israeli Arms Industry* (Holmes & Meier: New York and London, 1989)

Renner, M., *Economic Adjustment after the Cold War: Strategies for Conversion* (United Nations Institute for Disarmament Research, UNIDIR: Aldershot, Dartmouth, 1992)

Renner, M., 'Swords into plowshares: converting to a peace economy', *Worldwatch Paper 96* (Worldwatch Institute: Washington, DC, June 1990)

Richards, P. J., 'Disarmament and employment', *Defence Economics*, vol. 2, no. 4 (1991), pp. 295–312

Ross, A. L. *Arms Production in Developing Countries: The Continuing Proliferation of Conventional Weapons* (RAND Corp.L Santa Monica, Calif., October 1981)

Ross, A. L. (ed.), *The Political Economy of Defense: Issues and Perspectives* (Greenwood Press: New York, 1991)

Rupp, R. W., 'Dual use industries', *Nato's Sixteen Nations*, vol. 37, no. 2 (1992), pp. 26–29

Sanders, R., *Arms Industries: New Suppliers and Regional Security* (National Defense University Press: Washington, DC, 1990)

Sandström, M., *Strukturförändringar inom den europeiska försvarselektronikindustrin*, FOA Rapport A10036-1.3, Sep. 1992

Sassheen, R. S., 'The Italian defence industry—a brief look', *Asian Defence Journal*, vol. 21, no. 4 (Apr. 1991), pp. 72–79

Sauerwein, B., 'Focus on the Czechoslovak defence industry', *International Defence Review*, vol. 24, no. 8 (1991), pp. 862–64

Sayigh, Y., *Arab Military Industry* (Brassey's: London, 1992)

Schomacker, K., Wilke, P. and Wulf, H., *Alternative Produktion statt Rüstung* (Bund-Verlag: Cologne, 1987)

Schomacker, K., Wilke, P. and Wulf, H., *Zivile Alternativen für die Rüstungsindustrie* (Nomos-Verlag: Baden-Baden, 1986)

Schröder, H. H., *Konversion in der UdSSR: Planungsansätze und Gesetzentwürfe* (Bundesinstitut für ostwissenschaftliche und internale Studien: Cologne, 1991)

Shambaugh, D. L., 'China's defense industries: indigenous and foreign procurement', in ed. P. Goodwin, *The Chinese Defense Establishment: Continuity and Change in the 1980s* (Westview Press: Boulder, Colo., 1983)

Singh, B., *Singapore's Defence Industries*, Canberra Papers on Strategy and Defence no. 70 (Strategic and Defence Studies Centre, Research School of Pacific Studies, Australian National University: Canberra, 1990)

Sohr, R., *La industria militar chilena* (Comision sudamericana de paz: Santiago, 1990)

Souleles, T. S., *An Industry Without Frontiers: An Economic and Political Analysis of European Defence iIndustrial Cooperation* (Hellenic Foundation for Defence and Foreign Policy: Athens, 1990)

South Africa, Republic of, Department of Defence, *White Paper on Defence and Armaments Supply 1986* (Cape Town, 1986)

Southwood, P., *Disarming Military Industries* (Macmillan Press: Houndsmill, 1991)

Southwood, P., *The UK Defence Industry at the Crossroads: An Analysis of the 'Options for Change' Study and the CFE Process* (Enterprise for Defence and Disarmament: Chippenham, 1990)

Steer, M., 'The industrial base of collaboration in defence', *Brassey's and RUSI Yearbook 1987* (Brassey's/Royal United Services Institute: London, 1987), pp. 16–79

Steinberg, J. B., *The Transformation of the European Defense Industry*, RAND Report R4141-ACQ (Rand Corp.: Santa Monica, 1992)

Stockholm International Peace Research Institute, SIPRI, *SIPRI Yearbooks 1968/69–1992:World Armaments and Disarmament* (Almqvist & Wiksell, Stockholm; Taylor & Francis, London; and Oxford University Press, Oxford, 1969–1992)

Taylor, T., *Defence, Technology and International Integration* (Frances Pinter: London, 1982)

Taylor, T., 'Defense industries in international relations', *Review of International Studies*, no. 16 (1990), pp. 59–73

Taylor, T. and Hayward, K., *The UK Defense Industrial Base: Development and Future Policy Options* (Brassey's/Royal United Services Institute: London, 1989)

Thorsson, I., *In Pursuit of Disarmament: Conversion from Military to Civil Production in Sweden* (Liber Allmänna Förlaget: Stockholm, Sweden, 1984)

Todd, D., *Defence Industries: A Global Perspective* (Routledge: London, 1988)

Tuomi, H. and Väyrynen, R., *Transnational Corporations, Armaments and Development* (Gower: Aldershot, 1982)

UK House of Commons, Defence Committee, *Anglo/French Defence Cooperation* Session 1991–92 (Her Majesty's Stationery Office: London, 1991)

UK House of Commons, Defence Committee, *Statement on the Defence Estimates 1991* (Her Majesty's Stationery Office: London, 1992)

UK National Audit Office, *Ministry of Defence: Collaborative Projects*, Report by the Comptroller and Auditor General, no. 247 (Her Majesty's Stationery Office: London, 1991)

US Congress, Office of Technology Assessment, *After the Cold War: Living with Lower Defense Spending*, OTA-ITE-524 (US Government Printing Office: Washington, DC, 1992)

US Congress, Office of Technology Assessment, *Arming our Allies: Cooperation and Competition in Defence Technology* OTA-ISC-449 (US Government Printing Office: Washington, DC, 1990)

US Congress, Office of Technology Assessment, *Redesigning Defense: Planning the Transition to the Future US Defense Industrial Base*, OTA-ISC-500 (US Government Printing Office: Washington, DC, 1991)

US General Accounting Office (GAO), *Defense Industrial Base: Industry's Investment in the Critical Technologies*, GAO/NSIAD-92-4 (GAO: Washington, DC, Jan. 1992)

US General Accounting Office (GAO), *European Initiatives: Implications for U.S. Defense Trade and Cooperation* (GAO: Washington, DC, 1991)

Voss, T., *Converting the Arms Industry: Have We the Political Will?*, Current Decisions Report no. 9 (Oxford Research Group: Oxford, Apr. 1992)

Walker, W. and Gummett, P., 'Britain and the European armaments market', *International Affairs*, vol. 65, no. 3 (1989), pp. 419–42.

Ware, R., *UK Defence Policy: Options for Change* (House of Commons Library: London, 1991)

Webb, S., *NATO and 1992: Defense Acquisition and Free Markets*, RAND Report no. R-3758-FF (Rand Corporation: Santa Monica, Calif., July 1989)

Western European Union, Assembly, *Armaments Sector of Industry in the Member Countries*, A Report Prepared by the WEU Standing Armaments Committee, document no. 1051 (WEU: Paris, 20 Jan. 1986)

Wilén, C., *Internationellt samarbete inom robotindustrin*, ROA Rapport A10035-1.3, Sep. 1992

Wolff-Casado, K., 'Spain's defence industry: in constant growth since 1975', *Defence*, vol. 21, no. 11 (Nov. 1990), pp. 721–26

Wulf, H., 'Conversion: economic adjustments in an era of arms reduction: specific issues of conversion: industries and trade', *Disarmament*, vol. 14, no. 1 (1991), pp. 95–123

Wulf, H., 'West European cooperation and competition in arms procurement: experiments, problems, prospects', *Arms Control*, vol. 7, no. 2 (1986), pp. 177–96

Zukrowska, K., *From Adjustments to Conversion of the Military Industry in East Central Europe*, Occational Papers No. 27 (Polish Institute of International Affairs: Warsaw, 1991)

About the contributors

Gordon Adams (United States) is the founder and director of the Defense Budget Project in Washington, DC. He has taught at Rutgers University, the City University of New York and Columbia University. He was a senior research associate at the Council on Economic Priorities and currently is a consultant to the Social Science Research Council. He received a Ph.D. and M.A. in political science from Columbia University and graduated from Stanford University magna cum laude and Phi Beta Kappa. Among his numerous publications are *The Iron Triangle: The Politics of Defense Contracting* (1982) and *The Role of Defense Budgets in Civil–Military Relations* (1992).

Ian Anthony (United Kingdom) is Leader of the SIPRI project on arms transfers and arms production (November 1992–). His most recent publications are *The Arms Trade and Medium Powers: Case Studies of India and Pakistan 1947–90* (1992) and *Arms Export Regulations* (editor, SIPRI, 1991). He has contributed to the *SIPRI Yearbook* since 1988.

Oldrich Cechak (Czechoslovakia) is head of the Military Economic Department of Strategic Studies in Prague. He is a graduate of the Military Academy in Bratislava, and a scientist and lector there. He is an author of several studies in economic defence, arms production and conversion, and transformation of the Czechoslovak Army.

Graeme Cheeseman (Australia) is a senior lecturer in politics at the University College, Australian Defence Force Academy. He was previously a senior research fellow in the Peace Research Centre at the Australian National University and defence adviser to the Parliamentary Joint Committee on Foreign Affairs, Defence and Trade.

Julian Cooper (United Kingdom) is the director of the Centre for Russian and East European Studies at the University of Birmingham. He is an economist and Senior Lecturer in (former) Soviet Technology in Industry. His publications include *The Soviet Defence Industry: Conversion and Reform* (1991).

Agnès Courades Allebeck (France) is a Research Assistant on the SIPRI arms transfers and arms production project. She was formerly a Research Assistant at the European Parliament, Luxembourg (1985) and has done research on the external relations of the European Community. She is a co-author of the SIPRI Research Report *West European Arms Production: Structural Changes in the New Political Environment* (1990) and has contributed to the *SIPRI Yearbook* since 1989. She is the author of several chapters in the SIPRI volume *Arms Export Regulations* (1991).

John Frankenstein (United States) is a senior lecturer at the University of Hong Kong Business School, where he teaches courses on China and international management strategies. Formerly a Chinese language officer in the US Foreign Service, he holds a Ph.D. in political science from the Massachusetts Institute of Technology. He has written on Chinese affairs for *Problems of Communism, Harvard International Review, International Studies on Organization and Management, Euro-Asia Business Review* and other journals, and has contributed to the papers of the Joint Economic Committee of the US Congress and to several books.

Espen Gullikstad (Norway) is a student scholar at the Norwegian Institute of International Affairs. His previous research is in the field of the administrative aspects of the European Community. He was research assistant for the SIPRI arms trade project in 1988–90 and the summer of 1992. He contributed to chapters in the *SIPRI Yearbooks 1989* and *1990* and is the author of several chapters in the SIPRI volume *Arms Export Regulations* (1991).

Gülay Günlük-Senesen (Turkey) is an Associate Professor in Quantitative Economics at Istanbul Technical University and the vice-president of the ITÜ Technological and Economic Research Center (TEDRC), Turkey. Apart from her works on econometrics and applied economics, she has published articles and papers on the probable economic impacts of the installation of a domestic arms industry in Turkey, on the macroeconomic and microeconomic levels, since 1989. She was a Guest Researcher at SIPRI on a Swedish Institute Grant in the autumn of 1992.

Masako Ikegami-Andersson (Japan and Sweden) is lecturer at the Shikoku-Gakuin University, Faculty of Social Sciences, Department of Sociology. Her most recent published work is *The Military–Industrial Complex: The Cases of Sweden and Japan* (1992).

Alexei Izyumov (Russia) holds M.A. degrees in economics from the Moscow State University and a Ph.D. in international economics from the Academy of Sciences of the USSR. He is a Senior Research Fellow with the Russian Academy of Sciences (on leave) and the Foreign Policy Association (Moscow). He has done research and writing on military conversion in the former USSR for the past seven years and participated in numerous national and international conferences on military conversion. In 1990 he presented a report on *Conversion in the USSR* at the United Nations Conference on Economic Adjustments in an Era of Arms Reductions, held in Moscow.

Steven M. Kosiak (United States) is senior budget analyst for the Defense Budget Project, Washington, DC. He was previously with the Center for Defense Information and the Office of the Defense Advisor at the US Mission NATO. He received a Masters degree in public affairs from the Woodrow Wilson School of Public and International Affairs and a B.A., summa cum laude, from the University of Minnesota. He is the author of the Project's annual analysis of the US defence budget submission, authorization and appropriation.

Maciej Perczynski (Poland) is professor in international economics at the Polish Institute of International Affairs (PISM). He is chairman of the Scientific Council of PISM and was formerly director of PISM and senior economic affairs officer at the United Nations Economic and Social Committee on Asia and the Pacific (ESCAP). He is a specialist on economic security affairs and is the author of numerous publications on international issues, including *Modern Capitalism* (1981), *Global Problems and Economic Security* (1989), and *International Security and Economic Cooperation* (1990).

Judith Reppy (United States) is Director of the Peace Studies Program, Cornell University. She is the co-editor, with Philip Gummett, of *The Relations Between Defence and Civil Technologies* (1988), and is the author of numerous chapters and journal articles in the field of defence economics.

Jan Selesovsky (Czechoslovakia) works at the Military Economic Department of Strategic Studies in Prague. He is a graduate of the Military Academy in Bratislava and a scientist and lector at the Military Academy in Brno. He is an author of studies and textbooks on economics and economic defence.

Elisabeth Sköns (Sweden) isa Researcher on the SIPRI arms transfers and arms production project. She was previously Researcher at the Research SErvice of the Swedish Parliament (1987–91) and Research Assistant on the SIPRI military expenditure and arms transfers projects (1978–87). She has contributed to the *SIPRI Yearbook* since 1983.

Milan Stembera (Czechoslovakia) is Senior Research Fellow at the Institute for International Relations in Prague. He is a retired army colonel, a participant in many international negotiations and conferences and is engaged in problems of disarmament, confidence- and security-building and military aspects of international relations. He is the author of articles in journals, co-author of *National Interests* (1992), editor of *Conventional Disarmament in Europe* (1991) and co-author of a special study for the UNO Secretary-General about confidence-building measures (1982).

Pawel Wieczorek (Poland) is a Research Fellow at the Polish Institute of International Affairs (PISM). His special interest focus is on the politico-military field, in particular such issues as arms production, conversion and arms exports. He is the author or co-author of seven monographs and 30 papers, mainly on economic aspects of armaments and disarmament.

Herbert Wulf (Germany) was Leader of the SIPRI project on arms transfers and arms production in 1989–92. He is an adviser to the United Nations Office of Disarmament Affairs. Prior to his work in peace research, he was Director of the German Volunteer Service in India. He is a co-author, with Ian Anthony and Agnès Courades Allebeck, of the SIPRI Research Report *West European Arms Production: Structural Changes in the New Political Environment* (1990), and the author of *Rüstungsexport aus Deutschland* (1989). He has published books and articles on security policy, arms production, arms transfers, conversion, and disarmament and development. He has contributed to the *SIPRI Yearbook* since 1985.

Index